# Quantum structure of space and time

About this book:
This book consists of selected contributions presented at the Nuffield Quantum Gravity Workshop held at Imperial College, London, in August 1981. The book is divided into three parts which correspond with the three separate themes pursued at the Workshop. Part I is concerned with the geometrical and topological aspects of quantum gravity. Part II focusses on supergravity and its application to the Grand Unified Theories of elementary particles. Part III concentrates on the early universe and cosmology. The book, therefore, covers not only supergravity, but the whole spectrum of quantum gravity research It not only presents a comprehensive picture of past successes, but spotlights all the promising new developments in this important and challenging field.

Quantum structure of space and time will be invaluable reading for researchers and postgraduates in theoretical physics, quantum gravity, particle physics, cosmology, and general relativity who wish to explore the frontier of our present knowledge of the quantum structure of space and time.

This book is published from author prepared camera ready copy. Final copy was received from the editors in early July 1982. The book was completed in September 1982.

# Quantum structure of space and time

Proceedings of the Nuffield Workshop, Imperial College London
3-21 August 1981

Edited by
M.J. DUFF and C.J. ISHAM
The Blackett Laboratory, Imperial College
University of London

Cambridge University Press
Cambridge
London New York New Rochelle
Melbourne Sydney

CAMBRIDGE UNIVERSITY PRESS
Cambridge, New York, Melbourne, Madrid, Cape Town,
Singapore, São Paulo, Delhi, Mexico City

Cambridge University Press
The Edinburgh Building, Cambridge CB2 8RU, UK

Published in the United States of America by Cambridge University Press, New York

www.cambridge.org
Information on this title: www.cambridge.org/9781107404588



First published 1982
First paperback edition 2012

*A catalogue record for this publication is available from the British Library*

*Library of Congress Catalogue Card Number: 82-9732*

ISBN 978-0-521-24732-0 Hardback
ISBN 978-1-107-40458-8 Paperback

CONTENTS

# FOREWORD

Gravity is an astonishing force. It certainly dominates our thinking on one of the frontiers of Physics - the frontier at the far reaches of the Universe at distances of order $10^{27}$ cm. Amazingly, it is likely also, to dominate the second frontier of Physics at the other end of the distance scale (of order $10^{-33}$ cm).

Gravity determines the large scale structure of the Universe firstly because of its universally attractive, long range character: secondly, and more crucially, because Einstein demonstrated the intimate link of gravity with the curvature and geometry of space and time.

However, what Einstein almost certainly did not expect, was the possibility of his theory dominating short distance scales. This may happen because Einstein's gravity, contrary to other fundamental forces, appears to have an inbuilt scale of length ((Newtonian constant)$^{\frac{1}{2}}$ in units of $\hbar \approx 10^{-33}$ cm). This length, on account of the quantum uncertainty principle, ensures the primacy of the gravitational forces at momenta of the order of $10^{19}$ GeV. Thus, combine quantum theory with Einstein's gravity which is rooted in the structure of space-time and one immediately has the motivation for the Nuffield Workshop - held at Imperial College during 3-21 August 1981 - on the quantum structure of these basic concepts.

The idea of quantising gravity is not new. However, what has given it topical urgency are the recent advances in Particle Physics Theory. One has come to recognise that the non-gravitational forces of nature are principally gauge forces. This recognition has led to a hope of unifying them, and a revival of Einstein's dream of including gravity in this unification. Added impetus for the inclusion of gravity has come from the expectation that a unified gauge theory of the non-gravitational forces may hold up to energies of order $10^{15}$ GeV or equivalently down to distance scales of order $10^{-29}$ cm.

At such distances, quantum gravity is just round the corner; its neglect in particle physics would no longer be justified.

This new impetus to quantum gravity has manifested itself in three directions, recorded in these Proceedings of the Nuffield Workshop. First is the extension to gravity of ideas developed in humbler gauge theories in flat space — $\theta$ — vacuua, topological instantons, spontaneous symmetry breaking, renormalisation, and the like.

Second are the developments associated with supersymmetry and supergravity theories and particularly extended supergravities. Here particles of spin two (gravitons) and spin one (mediating the non-gravitational fundamental forces) plus spin-zero Higgs particles (responsible for spontaneous symmetry breaking) combine with "source" matter of half-integral spin, to form one single gauge multiplet of an extended supergravity theory, thus uniting (in Einstein's phrase) the marble of gravity with the base wood of matter. With this remarkable extension of gravity theory, the geometrisation of physics which started with Einstein's association of gravity with the curvature of space-time finds its fullest flowering, within an extended higher dimensional space-time. The programme envisages two types of extensions: First are the bosonic extensions, as in (compactified) Kaluza-Klein types of theories designed to give a geometrical meaning to the hitherto mysterious internal symmetries of particle physics by associating them with the extra space-dimensions. Second are the fermionic extensions, which take the form of anti-commuting super-space coordinates. Covering as it does, an area of intense theoretical development, a large part of this volume is devoted to supergravity.

Finally, a shorter third section is concerned with the physics of the very Early Universe as it may have looked just $10^{-43}$ seconds after the Big Bang. This is where quantum gravity effects are surely expected to have manifested themselves if at all. One's appreciation for organising this timely Workshop on explosively developing subject and for its Proceedings goes to the Nuffield Foundation and to its Workshop Directors, Michael Duff and Christopher Isham.

Abdus Salam

# PREFACE

For three weeks in August 1981 about sixty theoretical physicists from various parts of the world met at Imperial College, London, to discuss what many consider to be the central quandary of theoretical physics: how to quantize the gravitational field and unite it with the other forces of Nature. However, their agreement did not extend much beyond the purpose of the meeting. For the way in which this quantization is to be brought about and the way in which quantum gravity fits into the scheme of things are matters which are still in hot dispute. Indeed, our aim in organizing this Workshop, generously financed by the Nuffield Foundation, was to bring together a variety of leading opinion in quantum gravity research. And to reduce the danger of those from the same school of thought reinforcing each other's prejudice, experts from other branches of quantum field theory, cosmology and mathematics were also invited to supply their own perspective. This diversity is reflected in these Proceedings.

A different theme was chosen for each week and this volume is divided accordingly. Part I, "Quantum Gravity, Fields and Topology", covers a multitude of different approaches to the quantum theory of gauge and matter fields in curved geometries and to the quantization of gravity itself. These twelve articles embrace canonical quantization, covariant path integrals, Regge calculus, spontaneous symmetry breaking, instantons and much more besides. The common thread, however, is the emphasis on topological and geometrical considerations, an emphasis which reflects the modern trend in quantum field theory and particle physics and which bodes well for a union with Einstein's general relativity where geometry and topology have always been to the fore.

Few developments in physics can claim as great an impact on current activity as supersymmetry, and nowhere is this Bose-Fermi symmetry more compelling than in "Supergravity", the theme of the ten contributions to Part II. All of the most urgent questions are here represented: the task of constructing all simple and extended supergravity Lagrangians together with their auxiliary fields both in components and in superspace; the progress in the understanding of ultra-violet divergences with the possibility of a completely finite quantum theory of gravity made more plausible by the finiteness of supersymmetric Yang-Mills theories; and the conceivable superunification of the strong, weak, electromagnetic and gravitational interactions together with the implications for particle physics phenomenology at current energies. Though this unification of all the forces is still speculative, the reader of Part II can be left in no doubt as to the unification of ideas brought about by supergravity, with extra dimensions, Regge Poles, and black-hole solitons each making their appearance.

The Beginning of the Universe was postponed until week three. And in Part III we find four very different views of "Cosmology and the Early Universe" which include fresh insights from quantum gravity on the old question of the cosmological constant as well as up-to-the-minute cosmology based on very recent developments in Grand Unified Theories of the elementary particles. These ideas take us right back to the first $10^{-42}$ seconds in the lifetime of the Universe and to the very Quantum Structure of Space and Time.

Michael Duff Christopher Isham

# PARTICIPANTS

B. Allen
S. Avis
M. Brown
P. Candelas
D. Capper
J. Charap
A. Chockalingham
S.M. Christensen
S. Deser
B. De Witt
S. Dowker
M. Duff
M. Fawcett
S. Ferrara
L. Ford
G. Francisco
D. Freedman
C.D. Geilker
G. Gibbons
D. Gorse
M. Green
M. Grisaru
L. Grishchuk
S. Hawking
M. Henneaux
G. Horowitz
P. Howe
C. Hull
C.J. Isham
R. Jackiw
T. Jones
A. Kakas
A. Karlhede
T. Kibble
G. Kunstatter
R. Laura
U. Lindstrom
G. Moorhouse
J. Nelson
N. Nielsen
A. Ottewill
D. Page
R. Penrose
M. Pilati
D. Pollard
C. Pope
M. Rees
M. Roček
A. Rogers
A. Salam
M. Sohnius
K. Stelle
J.G. Taylor
C. Teitelboim
D. Toms
P. Townsend
R. Tucker
P. van Nieuwenhuizen
M. Wall
N. Warner
P. West
B. Whiting
R. Williams
Ya B. Zeldovich
B. Zumino

# PART I

# QUANTUM GRAVITY, FIELDS AND TOPOLOGY

# SOME REMARKS ON GRAVITY AND QUANTUM MECHANICS

Roger Penrose,
Mathematical Institute,
Oxford

The goal of finding an appropriate union of gravitation theory with quantum mechanics has stimulated much fruitful computational activity in recent years. But I think it is also important to keep reviewing the nature of the goals themselves, so as to reduce the likelihood that too much of this activity may be aimed in the wrong direction.

Let us recall a few of the main motivations that have been often stated as reasons for wanting to quantize gravity:

(1) the unity and consistency of physics

(2) the removal of ultraviolet divergences by a gravitational cut-off at the Planck length

(3) the removal, via quantum processes, of the space-time singularities of classical general relativity.

I have no quarrel with (1) as a motivation, except that it is rather non-specific. Some unification between (essentially) the physics of the small (quantum theory) and the large (general relativity) must indeed be part of nature's design. But whether that unification takes the form of a standard quantization procedure applied to standard general relativity theory, or whether the procedure or theory must be replaced by a non-standard one, or whether we should rather be seeking some quite new type of unified scheme having ordinary quantum theory and ordinary general relativity as two distinct limits - all this seems to me to be very far from clear. Perhaps my sympathies lie most with this last suggestion.

As regards (2), it is certainly an appealing possibility that the correct way of incorporating gravity into quantum field theory might remove these divergences. However, we should bear in mind that if (2) is our goal, then it is pointless to look for a renormalizable theory of quantum gravity. What we should need would necessarily be a finite theory. My own prejudices, however, are that we should be unlikely to find such a finite theory (even with a modified general relativity such as supergravity) so long as we stick with the normal notions of space-time geometry at such small distances. The standard procedures of quantum field theory imply that we do stick with them. (The "space-time foam" calculations of the Cambridge school, for example, I am counting as non-standard.) In fact, the standard (perturbative) procedures are often essentially flat-space ones, with ordinary flat-space causality holding. I shall have more to say about that shortly.

My main quarrel, here, is with (3). I have gradually come around to the view that it is actually misguided to ask that the space-time singularities of classical relativity should disappear when standard techniques of quantum (field) theory are applied to them. Suppose, for the sake of argument, that someone is finally able to show, using some quantization procedure of a reasonably conventional kind, that "space-time singularities" do not occur - or, at least, can be avoided - in quantized gravity. In particular, suppose that the big bang has been eliminated as a singularity in the new scheme, so the universe is seen as arising from a previously collapsing phase which passes through a highly compressed but not quite singular state in order to arrive at its present state of expansion. My personal reaction to this would be one of some disappointment, since we should have lost what seems to me to be the best chance we have of explaining the mystery of the second law of thermodynamics. To explain the second law, we need, primarily, an explanation of why the universe was in such an extraordinarily "special" state in the distant past. Once we have pushed the universe non-singularly back into a previously collapsing phase we are back into the regime of "conventional physics", where there are no rules requiring that the state of the universe be anything particularly "special". A space-time singularity is, almost by definition, "a place where the known laws of physics break down". So it is not unreasonable that principles other than the ones we know and understand should come into play at a singularity. Such a principle ought to have the time-asymmetric character necessary to explain the second law of thermodynamics. Indeed, as I have pointed out elsewhere

(Penrose 1978, 1979, 1981), the "Weyl curvature hypothesis" - according to which initial singularities (TIFs) have to have vanishing Weyl curvature whereas final singularities (TIPs) are unconstrained - appears to achieve this. But if quantum gravity simply removed the singularities, then we should be unable to invoke any new principles and we should be back where we were in our attempts to understand the origin of the second law.

The big bang, whether a singularity or not, was undoubtedly a very highly constrained physical situation. Had there not been any constraining principles (such as the Weyl curvature hypothesis) the Bekenstein-Hawking formula would tell us that the probability of such a "special" geometry arising by chance is at least as small as about one part in $10^{1000B^{3/2}}$ where B is the present baryon number of the universe. If there had been a previous collapsing phase, then this figure would give a measure of how constrained that prior collapsing geometry would have had to have been.

In my own view, whatever it is that we mean by the term "quantum gravity" ought to include an explanation for the grossly time-asymmetrical constraint on whatever physical situation it is that is covered by the term "space-time singularity". In this sense, I believe that the correct theory of quantum gravity must be a time-asymmetric theory.

I should perhaps remark that I do not necessarily believe that B need be finite. (Indeed, the $k = -1$ cosmological models have a certain mathematical appeal.) An infinite B would not invalidate the "specialness" argument. In fact it would strengthen it! Another point is that even if finite, B may have a value vastly in excess of the popular figure of $B \sim 10^{80}$. It could well be that we are in a very early stage of a universe whose characteristic timescale is enormously greater than the figure of $\sim 10^{10}$ years at which we find ourselves. The arguments provided by Dicke (1962) and Carter (1974) suggesting that our temporal location in the universe should be governed by the timescales of main-sequence stars rather than by the characteristic timescale of the universe as a whole, do have some essential plausibility.

As I have explained elsewhere (Penrose 1981), there seem to be indications that this proposed quantum-gravitational time-asymmetry should, in a different guise, show up in a kind of spontaneous reductio of the state vector. My own view is that this, also, will emerge as a necessary feature of the correct theory of quantum gravity. It is highly unlikely that any too conventional approach to the quantization of gravity should be able to incorporate these features.

Just to stress my own position on these points, I should mention that the programme of twistor theory seems to be no better placed for producing these desired effects than other programmes. Perhaps there is the general point that twistor theory provides an asymmetrical picture of physics. But this asymmetry is not directly a time-asymmetry; it is rather a parity-asymmetry. Any time-asymmetry seems to have to arise more indirectly.

I have, however, been tempted towards the goal of finding a simple twistor description of the big bang. For the conformal flatness entailed by the Weyl curvature hypothesis provides that a direct description of the early geometry in twistor terms exists. The Friedmann-Robertson-Walker cosmologies give the appropriate geometry, and a neat twistor description can indeed be given (Penrose & Rindler 1983), the basic new ingredient being an expression for the space-time metric in terms of two infinity twistors.

Perhaps I ought also to state my present position with regard to the question of time-symmetry of the black-hole-evaporation and -formation processes, since this is relevant to the question of the time-symmetry of quantum gravity. Much discussion of this matter has taken place over the past few years. I think that Don Page has largely persuaded me (Page 1981) that the behaviour, with regard to black hole formation, of the contents of a box with perfectly reflecting walls is somewhat closer to being time-symmetric than I had previously thought. Nevertheless, I would still maintain that time-asymmetries ought to be present. This is not (as he seems to be suggesting) because of any belief that time-asymmetric behaviour would arise out of time-symmetric physics, but because the relevant physics in the box is, in my view, actually time-asymmetric (even in the absence of a quantum gravity theory), because the Weyl curvature hypothesis is involved in the singularity structure. This would entail that only black holes and not white holes could be present in the box. It may well be the case that (for a largish box) black hole disappearance by Hawking evaporation is essentially balanced, as Page maintains, by a production process which resembles its time-reserve. Nevertheless, manifestly classical collapses must at least occasionally take place, and their time-reverses are not physically allowed for a black hole. For a box containing a large enough mass these manifestly classical collapses would actually be the dominant collapse processes.

The strongest argument against the black hole = white hole picture has always been the incompatibility of their space-time geometries. It is worthwhile to redraw the standard conformal space-time diagram describing spherically symmetric classical collapse to a black hole, followed by its Hawking disappearance, in such a way that it looks as time-symmetric as possible (see fig. 1). There tends to be a considerable freedom in these 2-dimensional pictures since the time-scaling for the in and out light rays is arbitrary. In the familiar left-hand diagram, the "hole duration" seems much longer with respect to advanced time than it does with respect to retarded time. But this is merely an artefact of the particular shape of diagram that has been selected (the choice having been made merely to bring out its

Figure 1
Conformal space-time diagrams of spherically symmetric classical collapse to a black hole, followed by its Hawking disappearance

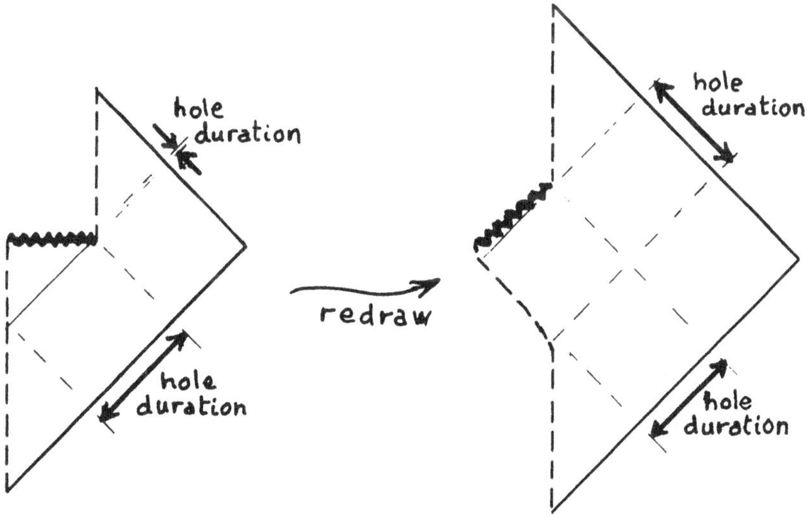

qualitative features as simply and bluntly as possible). We can keep the radial null rays at $45^{\circ}$ and readjust the in and out scales so that the hole duration seems about the same with respect to each, and we arrive at the right-hand diagram in which the space-time geometry appears to be almost time-symmetrical.

I think that from the quantitative point of view, the right-hand diagram is less misleading than the left-hand one, indicating that a time-symmetric picture of the entire physical process may well be usually quantitatively accurate (so long as the box sides are so far off that they can be ignored). The regions of space-time which violate the time-symmetry are qualitatively better illustrated by the left-hand diagram, but these normally contribute only a very small proportion of the 4-volume (say) of the entire process, the right-hand diagram illustrating this better in a quantitative way. However, for an astrophysical black hole, these 4-volumes are nevertheless rather large in absolute terms and it is not really legitimate to ignore their effects altogether.

My point is, then, that while under appropriate circumstances time-symmetry may well dominate the gross behaviour of the contents of the box, the time-asymmetric effects should still be important on an absolute rather than a relative scale (and should become relatively more important as the total mass in the box is increased). In the arguments I presented last year (Penrose 1981) at the Oxford Quantum Gravity 2 conference, I required only that there be a large number of hole-formation modes (namely manifestly classical collapses) which are not adequately balanced by evaporation modes. It is not necessary that these formation modes should be the dominant ones. Moreover, on a broad scale the classical collapse (if it proceeds at the appropriate rate) and quantum evaporation may look roughly like time-reverses of one another, as in the right-hand diagram of the figure, and the effects at large distances could be similar. Nevertheless the geometries still are different locally and a detailed examination of the implications of this fact should repay further analysis.

Finally, I should like to put in a strong plug for geometry in the investigation of quantum gravity. Even Steven Weinberg now informs me that, because of certain "no-go" theorems (Weinberg

& Witten 1980), he is no longer convinced that the anti-geometrical viewpoint is necessarily the most fruitful one! In this connection I should like to point out a result I obtained some years ago, and which is now published in Abe Taub's retirement volume (Penrose 1980). This states that even in the Schwarzschild space-time, with any matter distribution whatever covering the spatially bounded source region, it is not possible to introduce a Minkowskian background metric satisfying the following two requirements: (a) light propagation in the Schwarzschild metric does not violate the background causality, and (b) light rays for the two metrics agree at infinity. The inference to be drawn from this is that perturbation methods, using expansions about Minkowski space in which one attempts to build up a gravitational quantum field theory using the usual flat-space procedures (e.g. "field operators at spacelike-separated points commute") must necessarily break down, presumably because of divergences, even in the simplest cases. It seems to me that approaches to quantum gravity which do not fully take account of curved-space general-relativistic causality will have no chance of success - and this warning applies as much to twistor theorists as to anyone else!

## References

Carter, B. (1974) in Confrontation of Cosmological Theories with Observational Data (ed. M.S. Longair; Reidel Publ., Dordrecht).

Dicke, R.H. (1962) Nature (Lond.) **192**, 440 .

Page, D. (1981) "Black hole formation in a box", Gen. Rel. Grav. (to appear)

Penrose, R. (1978) in Theoretical Principles in Astrophysics and Relativity (eds. N.R. Lebovitz, W.H. Reid & P.O. Vandervoort; Univ. of Chicago Press).

Penrose, R. (1979) in General Relativity, an Einstein Centenary Survey (eds. S.W. Hawking & W. Israel; Cambridge Univ. Press).

Penrose, R. (1980) in Essays in General Relativity (ed. F.J. Tipler; Adacemic Press, New York).

Penrose, R. (1981) in Quantum Gravity 2, a Second Oxford Symposium (eds. C.J. Isham, R. Penrose & D.W. Sciama; Clarendon Press, Oxford).

Penrose, R. & Rindler, W. (1983) Spinors, Twistors and Space-Time Structure (to be published, Cambridge Univ. Press). For a preliminary account, see Twistor Newsletter 12 (Mathematical Institute 1981).

Weinberg, S. & Witten, E. (1980) Phys. Lett. **96B**, 59.

# AN EXPERIMENTAL TEST OF QUANTUM GRAVITY

Don N. Page
Department of Physics, The Pennsylvania State University
University Park, Pennsylvania 16802, USA

C. D. Geilker
Department of Physics, William Jewell College
Liberty, Missouri 64068, USA

The construction and elucidation of a consistent quantum theory of gravity is one of the greatest challenges of theoretical physics today (Hawking 1978). It has dominated the efforts of many scientists, several of whom have summarized their recent progress in this volume. But how do we know this challenge is a genuine physical problem, that the gravitational field really is quantized? The usual opinion is that since quantum mechanics appears to govern all non-gravitational fields (here called matter), by the unity of nature it should also apply to the gravitational field. However, until recently there had been no explicit experimental test of this. We have attempted to find indirect empirical evidence for or against quantum gravity by performing an experiment to test the simplest alternative, semiclassical general relativity (Page & Geilker 1981).

Because gravity is so weak, Feynman (1963) suggested the possibility that it need not be quantized. Also motivated by the apparent differences between gravity and other interactions, Møller (1962) and Rosenfeld (1963) proposed a specific alternative theory, semiclassical general relativity. In this theory, gravity is described by a classical field which obeys the semiclassical Einstein equations

$$G_{\mu\nu} = 8\pi\langle\psi|T_{\mu\nu}|\psi\rangle \quad . \qquad (1)$$

Here $G_{\mu\nu}$ is the Einstein tensor of the unquantized metric $g_{\alpha\beta}$, $T_{\mu\nu}$ is the stress-energy quantum operator, and $\psi$ is the wavefunction or quantum state of the matter. (One could replace $\psi$ by a density matrix or a C*-algebra state with no essential changes.) In the Heisenberg picture, which we adopt, (1) is to be supplemented by the appropriate covariant field equations and commutation relations for the quantized matter field operators in the presence of the classical metric.

The functional dependence of $g_{\alpha\beta}$ upon $\psi$ by (1) introduces a nonlinearity into the metric-dependent quantum evolution of the matter (Anderson 1962; Mielnik 1974; Kibble 1978; Kibble & Randjbar-Daemi 1980; Kibble 1981). This makes it crucial for the theory to specify what happens during a measurement. In the conventional view, the wavefunction collapses into an eigenstate of the measured variable (von Neumann 1932). It has then been argued that the semiclassical theory would permit violations of the uncertainty principle (Komar 1962), observable signals faster than light (Eppley & Hannah 1977), nonsensical jumps in the gravitational field (Kibble 1981), or other unappealing results (Belinfante 1962; Bergmann 1967). However, these predicted effects have not been ruled out experimentally, so that although they violate our present physical intuition, perhaps it is premature to use them to dismiss semiclassical gravity coupled to collapsing wavefunctions.

A more conclusive argument for the inconsistency of semiclassical general relativity and conventional measurement theory is that if $\psi$ collapses, in general the right hand side of (1) will not be conserved, whereas the left hand side is automatically conserved. That is, if $\psi = \Sigma_i c_i(x^\alpha)\psi_i$ with constant $\psi_i$'s in the Heisenberg picture but with $c_i(x^\alpha)$'s which change during a measurement, then for almost all conceivable reductions of the wavepacket,

$$8\pi\langle\psi|T^{\mu\nu}|\psi\rangle_{;\nu} = 8\pi\Sigma_{i,j}(c_i^* c_j)_{;\nu}\langle\psi_i|T^{\mu\nu}|\psi_j\rangle \neq 0 \equiv G^{\mu\nu}{}_{;\nu} \quad . \quad (2)$$

For example, if the matter wavefunction is initially a superposition of different energy states (i.e., is not stationary) and then collapses so that the expectation value of the energy changes, there will be an inconsistency in (1) with the gravitationally conserved ADM mass. One might seek to avoid the inconsistency by simply abandoning (1) during a measurement. However, one would need a replacement of (1) in order to determine the evolution of the gravitational field for any particular collapse of the wavefunction, and this would differ from the semiclassical Einstein equations. In the example in which the expectation value of the matter energy changes, one must violate the Einstein equations at arbitrarily large spatial distances to change the ADM mass appropriately if the time component of (1) is to apply both before and after the collapse of the wavefunction. Concerning the consistency of such a possible replacement of semiclassical relativity we shall have no more to say, but it seems clear that any possibility would not be a particularly simple alternative to quantum gravity.

Therefore, in order to retain (1) as the simplest semiclassical theory of gravity, we must assume the universal matter wavefunction never collapses, as in the Everett formulation of quantum mechanics (Everett 1957; Wheeler 1957; DeWitt 1968; Cooper & van Vechten 1969; DeWitt 1970, 1971 a,b; DeWitt & Graham 1973). One might think the conventional collapse view is equivalent to this, as it is in practice for linear quantum theories in which one may ignore components of the wavefunction which have negligible interference with the ones of interest. But in semiclassical gravity the metric depends upon all components of $\psi$, none of which can be ignored. Nevertheless, once the evolution of the gravitational field is determined by using the full wavefunction in (1), $\psi$ may be decomposed into linear components on that four-dimensional metric and any standard interpretation may be applied to the components. One must simply remember that the individual components do not give the full source of the gravitational field.

This formulation of quantum mechanics avoids all of the problems raised above with semiclassical relativity. The uncertainty principle for quantized matter would not be violated, because although one could use the classical gravitational field to measure certain matter expectation values arbitrarily precisely, the quantum fluctuations would not be eliminated (just as in ordinary quantum mechanics one may know the expectation values of both the position and the momentum of an electron in the hydrogen atom arbitrarily precisely with no inconsistency). Observable signals could not be sent faster than light, because one could not make the matter wavefunction collapse on a spacelike hypersurface as Eppley and Hannah (1977) assumed. With no collapse to change the right hand side of (1) suddenly, there would be no nonsensical jumps in the gravitational field. Most crucially, with $\psi$ now required to be constant in the Heisenberg picture, the conservation of the stress-energy operator implies that the right hand side of (1) is conserved and thus may be consistent with the semiclassical Einstein equations.

There does remain a problem with the short-distance behavior of semiclassical relativity in that flat spacetime seems to be unstable to fluctuations around the scale of the Planck length when one calculates the effect of the geometry on the quantum stress-energy (Horowitz & Wald 1978; Horowitz 1980). However, this is also a difficulty with the 1/N approximation to quantum gravity (Hartle & Horowitz 1981), and it is not clear how to resolve the problem in full quantum gravity. Thus we shall take the view that this interesting instability represents more a difficulty with the field-theoretic methods of determining the right hand side of (1) than a problem with semiclassical relativity itself.

Having chosen a formulation of quantum mechanics that may be meshed with (1) without creating an immediate inconsistency, we ask whether the resulting semiclassical theory may be empirically distinguished from a theory of quantum gravity in which the gravitational field is included in the full wavefunction of the universe. The key is to look for the nonlinear gravitational coupling between different components of the matter wavefunction that exists in the semiclassical theory, which would be absent in a linear quantum theory of gravity. This coupling would occur through the classical metric even if the components of $\psi$ were eigenstates of $T_{\mu\nu}$, so it is not a gravitational quantum interference effect that would be nearly impossible to detect. In order for the coupling to be observable in the semiclassical theory, the gravitational field must simply be measurably different from what it would be if each component alone (suitably normalized) were the full source of the field. This requires that $\psi$ be a superposition of components with macroscopically different stress-energy configurations, since current experimental techniques can only detect the gravitational fields of macroscopic sources.

Because of the enormous complexity of the full wavefunction of the universe, it does seem highly likely that it may have significant components in which the earth, moon, sun, and other astronomical bodies are in positions greatly different from those in our component, the relative state (Everett 1957) corresponding to our nongravitational observations. This would lead to a semiclassical gravitational field quite in conflict with gravitational observations (Bell 1981). However, it is plausible (though perhaps intrinsically unlikely) that $\psi$ has all astronomical bodies at macroscopically well-defined positions. A quick calculation then shows that the quantum-mechanical uncertainty of their positions could remain observationally negligible during their lifetimes. Hence to make a more definite test of semiclassical gravity, one needs to make certain the wavefunction does have components that would give measurably different gravitational fields.

We performed an experiment to make such a test of semiclassical gravity. A quantum-mechanical decision and amplification process was used to set the positions of certain macroscopic masses. As amplitudes for different decisions occurred, $\psi$ developed simultaneous components in which the masses were in macroscopically different configurations. The gravitational field induced by the masses was measured in our component and was found to be highly correlated with the mass distribution in our component alone. There was no indication of any gravitational coupling with other components of $\psi$. This was consistent with quantum gravity but inconsistent with the semiclassical Einstein equations.

The experiment included ten runs of a procedure which had two parts that were done at different times but were coupled by the action of the experimenter. The first part was a simultaneous 30-second measurement of $\gamma$-rays from a small cobalt-60 source by two nearby Geiger counters. The application of quantum mechanics to the radioactive decay and detection events leads to various amplitudes for all possible results of this decision and amplification process. Those in which the ratio of counts in the two counters was greater than an overall average were classified as decision $\alpha$, and those in which the ratio of counts was less than average were clasified as $\beta$. This classification was selected so that for each run the Born-Dirac square-amplitude measure on the components of $\psi$ in which $\alpha$ was registered should be roughly equal that on the components in which $\beta$ was registered.

One might object to assuming that both kinds of components existed in $\psi$, since after the measurement we in our component saw only one result and were unable to communicate with our alter egos in the other possible components who saw different results. That is, how do we know that the full wavefunction $\psi$ didn't just consist of components in which only a specific one of the two possible decisions was recorded? However, in order for the amplitudes for the other components to be zero after the measurement (assuming no collapse of the full wavefunction, in order to allow a sensible semiclassical theory as discussed above), there would have had to have been very precisely correlated amplitudes for incoming waves impinging upon the radioactive nuclei or Geiger counters to cancel the other final components of $\psi$. Although one cannot logically rule out this incredibly correlated initial state (which would be obtained by evolving our component alone backward in time and not including the interference of other components to wipe out the correlation in the past), the experimental observations supporting the second law of thermodynamics strongly suggest that such correlations do not occur in our universe. Therefore, we assumed the second law was valid during our experiment and concluded that there must have been significant amplitudes for other components of the full wavefunction in which the other decision was registered.

The second part of the experiment used a Cavendish torsion balance to measure the gravitational field induced by two macroscopic masses whose configurations were determined by the quantum decision process. The balance consisted of two small lead balls mounted 0.10 m apart (center to center) on a light horizontal rod hung in its center by a thin metallic fiber. A mirror attached to the balance reflected a light beam to a scale 12.83 m away to determine the angle of twisting which would result from a gravitational torque. The macroscopic masses were two larger lead balls, 1.497 kg each, which could be placed in stationary positions 0.0463 m in front or behind the mean equilibrium positions of the smaller balls. In position A the large balls were each on the clockwise side of the respective small ball (as seen from above); in position B they were counterclockwise. The gravitational field corresponding to the large balls in each definite position should thus exert either a clockwise or a counterclockwise torque on the torsion balance.

The procedure for each run was to generate a quantum decision with the Geiger tubes, position the macroscopic masses accordingly, and measure the gravitational field by the torsion balance. If the quantum decision was $\alpha$, we set the masses in configuration AB, meaning the four-dimensional configuration in which the large balls were placed in position A for 30 minutes of Cavendish balance measurements and then in position B for 30 minutes. If the Geiger counters gave $\beta$, we set configuration BA, meaning B first and then A. (Using a sequence of two positions rather than one increased the sensitivity and reduced the effects of slowly varying nongravitational influences on the torsion balance, but it is in principle unnecessary.)* In each experimental run, the appropriate configuration was started at a predetermined time, independent of the quantum decision.

Since the quantum process caused the wavefunction to have amplitudes of comparable weight for both decisions $\alpha$ and $\beta$, the corresponding positioning of the masses led to simultaneously occurring amplitudes for both mass configurations AB and BA. We of course assume the full wavefunction never collapses and that it includes all aspects of the positioning (including the experimenter who recorded the Geiger tube counts, calculated the ratio to classify the decision, and then placed the masses in the corresponding positions), as is necessary even to discuss the semiclassical Einstein equations consistently. We also assume the positioning was generally faithful to the quantum decision rather than being determined by some systematic effect. A refinement of the experiment might employ a completely inanimate positioning process, but this is not necessary so long as it is assumed the experimenter did not put the masses in nearly the same configuration in nearly all components of the wavefunction, disregarding the quantum measurements. With these assumptions we conclude that the wavefunction really did have a comparable measure of amplitudes for components with both mass configurations. We had no information about the complicated phase relations between these amplitudes, but that was not necessary since we were not doing an interference experiment.

Now in a quantum theory of gravity, we would predict that the quantized gravitational field would differ from component to component of the wavefunction and be highly correlated with the mass configuration. Thus we would expect the torsion balance to respond in each component according to the mass configuration in that component. But in the semiclassical theory of gravity, we would predict a definite classical four-dimensional (i.e., not necessarily static) gravitational field that would correspond to the expectation value of the stress-energy operator. Since the amplitudes for different components have rapidly varying relative phases, there would be negligible contributions from cross terms in the right hand side of (1). For our nonrelativistic configurations it would essentially be a square-amplitude weighted average over the mass distributions of the different components of the wavefunction. Because the configurations AB and BA have nearly equal weights, we would expect only a small response by the torsion balance in the semiclassical theory, and no correlations with the particular mass configuration in our component of the wavefunction.

The series of ten experimental runs gave 30-second $\gamma$-ray counts with means and standard deviations $1509.1 \pm 31.0$ and $887.6 \pm 23.0$ for the two respective Geiger counters. The fluctuations are consistent with Poisson statistics and thus were attributed to the quantum mechanics of the radioactive decays and detections. There was a negligible background count rate when the cobalt-60 source was removed. The ratios of counts in the two counters in our present component of the wavefunction gave the sequence of decisions $\alpha$, $\alpha$, $\alpha$, $\beta$, $\beta$, $\alpha$, $\beta$, $\alpha$, $\beta$, $\beta$, and the masses were set in the appropriate configurations. During each run the torsion balance responded to each repositioning of the masses and then underwent damped oscillations with a mean period of 710 seconds. By fitting the extrema of the oscillations to exponentially decaying sine waves during each half hour, the change in the equilibrium position (of the reflected light beam on the distant scale) as the large balls were moved from A to B or B to A was determined. The changes in equilibria we measured were -0.613 m, -0.639 m, -0.360 m, +0.692 m, +0.361 m, -0.488 m, +0.464 m, -0.452 m, +0.513 m, +0.596 m.

Although the sensitive torsion balance was affected by temperature changes, vibrations, and other factors not under our control, so that the data have a large scatter, they give a correlation coefficient with the quantum decisions of $r = 0.9788$. If our data came randomly from an uncorrelated population, as would be predicted by the semiclassical Einstein equations, the correlation coefficient for $N-2=8$ degrees of freedom would have a probability distribution (Bevington 1969) $P(r)dr = (35/32)(1-r^2)^3 dr$, giving a chance of only $4 \times 10^{-7}$ of being as large as ours was. Thus the correlation we observed is highly significant, as would be expected if gravity is quantized. Under that assumption the data give a gravitational constant $G = (6.1 \pm 0.4) \times 10^{-11}$ $m^3 kg^{-1} s^{-2}$, where the uncertainty represents one standard deviation of the mean. Although the accuracy is poor, this result is within 1.3 standard deviations of the accepted value and thus shows that at least most of the torque on the torsion balance can be attributed to gravity (if quantized), so there is no evidence that the strong correlation observed is likely to have arisen from nongravitational forces. Of course, the correlation is what one would intuitively expect, but it is in conflict with the predictions of the semiclassical Einstein equations (cf. Kibble 1981).

In conclusion, our theoretical arguments show that the semiclassical Einstein equations (1) are mathematically inconsistent if the matter wavefunction collapses arbitrarily during a measurement, and our experimental results show that these equations are inconsistent with nature (to a high confidence level) if the wavefunction does not collapse. (An analogous argument and experiment could easily be used to rule out the semiclassical Maxwell equations, but we already know the electromagnetic field is quantized.) Because there are presumably more complicated schemes for coupling a classical gravitational field to matter that we know is quantized, this does not prove that gravity is quantized, but it may be interpreted as indirect evidence supporting the hypothesis of quantum gravity by ruling out what is probably the simplest plausible alternative.

Since performing our experiment, we have been informed that many other people, including P. C. W. Davies (1977), D. Deutsch, T. W. B. Kibble (1981), and W. G. Unruh, have suggested such an experiment, perhaps in a modified form. S. W. Hawking has noted that previous Cavendish balance experiments may have unwittingly tested semiclassical gravity if the decisions as to where to position the active gravitational masses were subject to quantum uncertainties. However, we are not aware of any actual previous experiments that explicitly tested this semiclassical alternative to quantum gravity.

Discussions with D. Deutsch, G. N. Fleming, G. W. Gibbons, R. H. Good, S. W. Hawking, E. Kazes, J. Stachel, K. S. Thorne, W. G. Unruh, and R. M. Wald (some of whom independently voiced ideas suggested here and others of whom strongly disagreed) were helpful on the interpretation of the experiment after it was performed, particularly in suggesting references. This work was supported in part by NSF grant PHY79-18430 to The Pennsylvania State University.

## REFERENCES


Anderson, J. L. (1962). Discussion, in Møller 1962 (see below).

Belinfante, F. (1962). Discussion, in Møller 1962 (see below).

Bell, J. S. (1981). Quantum mechanics for cosmologists. In Quantum Gravity 2: A Second Oxford Symposium, eds. C. J. Isham, R. Penrose & D. W. Sciama. Oxford: Oxford University Press.

Bergmann, P. G. (1967). In The Nature of Time, ed. T. Gold, p. 242. Ithaca, N. Y.: Cornell University Press.

Bevington, P. R. (1969). Data Reduction and Error Analysis for the Physical Sciences. New York: McGraw-Hill.

Cooper, L. N. & van Vechten, D. (1969). On the interpretation of measurement within the quantum theory. Am. J. Phys. 37, 1212-20.

Davies, P. C. W. (1977). Workshop on quantum mechanics and gravity, University of British Columbia (unpublished).

DeWitt, B. S. (1968). The Everett-Wheeler interpretation of quantum mechanics. In Battelle Rencontres: 1967 Lectures in Mathematics and Physics, eds. C. M. DeWitt & J. A. Wheeler. New York: Benjamin.

DeWitt, B. S. (1970). Quantum mechanics and reality. Physics Today 23 (9), 30-5.

DeWitt, B. S. (1971 a). Quantum mechanics debate. Physics Today 24(4) 41-4.

DeWitt, B. S. (1971 b). The many-universes interpretation of quantum mechanics. In Proc. Int. School of Physics "Enrico Fermi" Course IL: Foundations of Quantum Mechanics, ed. B. d'Espagnat. New York: Academic Press.

DeWitt, B. S., & Graham, N. (1973). The Many-Worlds Interpretation of Quantum Mechanics. Princeton: Princeton University Press.

Eppley, K. & Hannah, E. (1977): The necessity of quantizing the gravitational field. Foundations of Physics 7, 51-68.

Everett, H. (1957). "Relative State" formulation of quantum mechanics. Rev. Mod. Phys. 29, 454-62.

Feynman, R. P. (1963). Lectures on Gravitation. California Institute of Technology (unpublished).

Hartle, J. B., & Horowitz, G. T. (1981). Ground-state expectation value of the metric in the 1/N or semiclassical approximation to quantum gravity. Phys. Rev. D24, 257-74.

Hawking, S. W. (1978). Quantum gravity and path integrals. Phys. Rev. D18, 1747-53.

Horowitz, G. T. (1980). Semiclassical relativity: the weak-field limit. Phys. Rev. D21, 1445-61.

Horowitz, G. T. & Wald, R. M. (1978). Dynamics of Einstein's equation modified by a higher-order derivative term. Phys. Rev. D17, 414-6.

Kibble, T. W. B. (1978). Relativistic models of nonlinear quantum mechanics. Commun. Math. Phys. 64, 73-82.

Kibble, T. W. B. (1981). Is a semiclassical theory of gravity viable? In Quantum Gravity 2: A Second Oxford Symposium, eds. C. J. Isham, R. Penrose, & D. W. Sciama. Oxford: Oxford University Press.

Kibble, T. W. B. & Randjbar-Daemi, S. (1980). Non-linear coupling of quantum theory and classical gravity. J. Phys. A13, 141-8.

Komar, A. (1962). Discussion, in Møller 1962 (see below).

Mielnik, B. (1974). Generalized quantum mechanics. Commun. Math. Phys. 37, 221-56.

Møller, C. (1962). The energy-momentum complex in general relativity and related problems. In Les Théories Relativistes de la Gravitation, eds. A. Lichnerowicz & M. A. Tonnelat, pp. 15-29. Paris: Centre National de la Recherche Scientifique.

Page, D. N., & Geilker, C. D. (1981). Indirect evidence for quantum gravity. Phys. Rev. Lett. 47, 979-82.

Rosenfeld, L. (1963). On quantization of fields. Nucl. Phys. 40, 353-6.

Von Neumann, J. (1932). Mathematische Grundlagen der Quantenmechanik. Berlin: Springer.

Wheeler, J. A. (1957). Assessment of Everett's "relative state" formulation of quantum theory. Rev. Mod. Phys. 29, 463-5.

# QUANTUM MECHANICAL ORIGIN OF THE SANDWICH THEOREM IN CLASSICAL GRAVITATION THEORY



Claudio Teitelboim
Center for Theoretical Physics, University of Texas
Austin, Texas 78712, U.S.A.

## 1 INTRODUCTION

If one adopts the point of view of path integrals (Feynman, 1948), it appears fair to say that there is no action principle in quantum mechanics.

In fact, the quantum mechanical propagation amplitude from one configuration to another is obtained by summing $(i/\hbar)$ times the action over all histories which connect the two given configurations. Thus, at the quantum level there is no privileged role for any particular history. It is only in the limiting case of typical actions much larger than Planck's constant that one approximates the path integral by the stationary phase method and the sum is dominated by histories close to that which extremizes the action.

Through this limiting process one obtains a theory (the classical theory) where the action principle plays a central role, from another one (the quantum theory) where the action principle plays, strictly speaking, no role.

We have begun this article emphasizing the above standard argument because it applies practically unchanged to a remarkable feature of classical gravitation theory, namely the fact that three-dimensional geometry carries information about time (Baierlein *et al.*, 1962).

The idea behind this statement has been lucidly discussed by Wheeler (1964) and is based on regarding four-dimensional spacetime as a stacking of three-dimensional surfaces or, in other words, as being the history of space. In this context, one gives two three-dimensional geometries and asks whether a four-dimensional spacetime can be constructed which contains two spacelike hypersurfaces whose intrinsic geometries are those initially given.

If nothing else is required there are of course infinitely

many spacetimes which will embed the two three-geometries. However, if one demands in addition that the spacetime in question must satisfy Einstein's equations then, in the generic case, the solution is unique and furthermore the location of the given three-geometries within that spacetime is completely fixed.

As a consequence, since one knows both the four-dimensional geometry and the location of the hypersurfaces within it, one may determine the pointwise distance between the two hypersurfaces along any curve, that is one may find their separation in time.

The fact that giving two three-geometries as boundary data for Einstein's equations permits one to determine the four-dimensional geometry in between them (and also outside) is known as the "sandwich conjecture" or "sandwich theorem," depending on one's level of rigor. If the three-geometries differ infinitesimally from each other, their distance in proper time is also infinitesimal and one speaks in that case of a "thin sandwich."

In order to assign a unique time separation to two three geometries one must use Einstein's equations. This necessity already shows that the sandwich theorem is inextricably linked to the action principle from which Einstein's equations follow and therefore one should not expect a quantum analog of it to exist. Rather one should expect the sandwich theorem to emerge in the classical theory "from nothing" much in the same way as the action principle itself emerges "from nothing" through the stationary phase approximation.

The idea that the sandwich theorem is only a classical notion is already present in the writings by Wheeler (1964), where an illuminating analogy with the harmonic oscillator is given. However, the precise way in which the theorem emerges from the quantum theory could not be spelled out at that time due to the lack of an appropriate expression for the quantum mechanical propagation amplitude.

It is the purpose of this article to fill in that gap in the light of some results in a proper-time approach to quantized gravitation theory (Teitelboim 1980b, 1981). Thus we shall see precisely how the semiclassical approximation to the quantum propagation amplitude from one three-geometry to another brings the sandwich theorem into play. Particular attention will be paid to the case when the two faces of the sandwich intersect each other. In that circumstance the solution of the sandwich equations is not unique. However, quantum mechanics selects--through the principle of causality--the solution to be chosen. Interestingly enough, that solution is not smooth and hence it is not the one which one would have chosen on purely classical grounds.

## 2 ACTION PRINCIPLE

The Hamiltonian action principle for Einstein's equations (Dirac, 1958;Arnowitt *et al.* 1962) is given by the statement that a functional of the form,

$$S = \int (\pi^{ij}\dot{g}_{ij} - N^{\perp}\mathcal{H}_{\perp} - N^{i}\mathcal{H}_{i})\, d^3x d\tau \equiv \int (\pi^{i}g_{ij} - N^{\mu}\mathcal{H}_{\mu}) d^3x d\tau \tag{2.1}$$

must be extremized with respect to variations of $g_{ij}$, $\pi^{ij}$ and $N^{\mu}$

The integral in (2.1) is extended over the region of spacetime included between two generic spacelike surfaces $\tau = \tau_1$, $\tau = \tau_2$. The $g_{ij}$ are the components of the metric tensor on the three-dimensional spacelike surfaces $\tau$ = constant and $\pi^{ij}$ are its canonically conjugate momenta. The functions $N^{\mu}$ are Lagrange multipliers which describe the relative position of two neighboring spacelike surfaces.

The only restriction on the variations of the functions involved in (2.1) is that the spatial metric tensor $g_{ij}$ must be fixed up to a change of spatial coordinates both at $\tau_1$ and $\tau_2$.

The constraints $\mathcal{H}_{\mu}$ are given by,

$$\mathcal{H}_{\perp} = (2\kappa)\, G_{ijk\ell}\pi^{ij}\pi^{k\ell} + (2\kappa)^{-1}g(\sigma R + 2\Lambda) \ , \tag{2.2a}$$

$$\mathcal{H}_{i} = -2\nabla_{j}\pi^{j}{}_{i} \ , \tag{2.2b}$$

with,

$$G_{ijk\ell} = \frac{1}{2}(g_{ik}g_{j\ell} + g_{i\ell}g_{jk} - g_{ij}g_{k\ell}) \ , \tag{2.3}$$

and where g, R and $\nabla$ are respectively the determinant, curvature and covariant derivative corresponding to $g_{ij}$, $\Lambda$ is the cosmological constant and $\kappa = 8\pi G/c^2$. The quantity $\sigma$ is the signature of spacetime (Teitelboim 1973, 1980a); it takes the value -1 for the hyperbolic case and +1 for the elliptic one.

As stated in passing above, the functions $N^\mu$ describe the relative position of two neighboring spacelike surfaces. More precisely, as shown in Fig. 1, $N^\perp \delta\tau$ is equal to $g^{-1/2}$ times the normal proper distance between the surfaces corresponding to values $\tau$ and $\tau + \delta\tau$ of the time coordinate and $N^i\delta\tau$ is a vector field which fixes the spatial coordinates on the second surface relative to the ones used on the first one. Therefore if a spacelike surface is given with a coordinate system defined on it and if the functions $N^\mu$ are known one can unambiguously determine the location of the $\tau + \delta\tau$ surface within the common enveloping spacetime and also the coordinate system employed on it.

The thin sandwich theorem is then formulated in terms of the functions $N^\mu$ as follows: "Given $g_{ij}$ $(x,t)$ and $g_{ij}$ $(x,\tau+\delta\tau)$ one can uniquely determine, with the help of Einstein's equations, the functions $N^\mu$."

Actually only four out of the ten Einstein's equations are utilized in the thin sandwich problem, namely the initial value equations $\mathcal{H}_\mu = 0$ obtained by extremizing the action (2.1) with respect to the $N^\mu$ themselves. When the sandwich is not thin one also needs the other six (dynamical) equations as discussed at the end of section 4 below.

Figure 1. Knowledge of $N^\mu$ fixes the position of the $\tau + \delta\tau$ surface relative to the one corresponding to the time $\tau$. The function $N^\perp$ sets the geometrical location of the new surface whereas $N^i$ lays a coordinate system on it. The normal distance between both surfaces is $\delta\tau\ g^{1/2}\ N^\perp$, so that our $N^\perp$ equals $g^{-1/2}$times the more commonly used "lapse function." This choice of weight greatly simplifies the measure in the path integral and leads to the simple formula (3.1) for the quantum propagation amplitude.

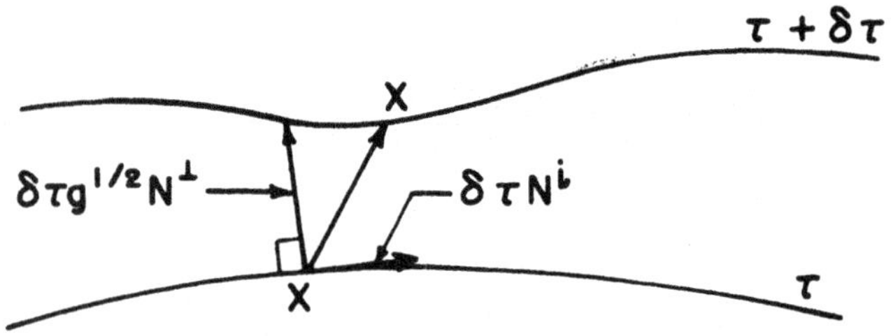

## 3 QUANTUM MECHANICAL AMPLITUDE

The action functional (2.1) is to be extremized keeping fixed the three geometries $G_1$, $G_2$ at $\tau_1$ and $\tau_2$. This means that the associated quantum mechanical amplitude will depend solely on those three-geometries and will be obtained by summing the exponential of $(i/\hbar)$ times the action evaluated for all four-dimensional geometries which have the given three-geometries for boundary at $\tau_1$ and $\tau_2$. In using the word "solely" here we have assumed that the three-geometry is compact. The other case of interest is when it is asymptotically flat. Then the end-point times $\tau_1$ and $\tau_2$ have an invariant significance--they determine the asymptotic location of the three-geometry in the Minkowski spacetime at infinity--and also appear as arguments of the quantum amplitude. The details of that case will be treated elsewhere (Teitelboim, 1982b), but the necessary changes do not affect the discussion in this article.

A simple expression for this propagation amplitude has been recently derived (Teitelboim 1980b, 1981). It reads

$$K\,[G_2,\,G_1] = \int DT^{\mu} \langle 2 \mid \exp\left[(-i/\hbar)\int d^3x\; T^{\mu}\,\mathcal{H}_{\mu}^{\,eff}\right] \mid 1\rangle\,. \quad (3.1)$$

Here the $\mathcal{H}_{\mu}^{\,eff}$ differ from the classical generators (2.2) by the addition of a quantum-mechanical "ghost" correction which involves two anticommuting scalar fields $C$, $\bar{C}$ and their conjugate momenta $\bar{P}$, $P$ given by

$$\mathcal{H}_{i}^{\,ghost} = i\left[(2\kappa)\bar{P}\,P + (2\kappa)^{-1}\sigma\, g\, g^{ij}\,\bar{C}_{,i}\,C_{,j}\right], \quad (3.2a)$$

$$\mathcal{H}_{i}^{\,ghost} = i\left[\bar{P}\,C_{,i} + P\,\bar{C}_{,i}\right] \quad (3.2b)$$

The matrix element (3.1) is understood to be taken between states with $C(1) = C(2) = \bar{C}(1) = \bar{C}(2) = 0$ while for $g_{ij}(1)$ and $g_{ij}(2)$ one may choose any set of metric coefficients corresponding to the three-geometries 1 and 2 respectively.

The measure $DT_{\mu}$ appearing in (3.1) is the product of

$$DT^{\perp} = \prod_{x} d\,\log T\,(x) \quad (3.3)$$

and of an appropriate invariant measure over the three-dimensional diffeomorphism group. The details of that measure need not concern us here and will be discussed elsewhere (Teitelboim 1982a). We just notice that the $T^i(x)$ may be considered formally as "canonical coordinates" for the diffeomorphism group and are associated to the diffeomorphism $x \rightarrow f(x)$ through the solution of the differential equation

$$\frac{\partial f^k(t,x)}{\partial t} = T^k\left(f(t,x)\right) \tag{3.4}$$

with the initial condition

$$f^k(o,x) = x^k \tag{3.5}$$

The diffeomorphism f corresponding to a given $T^k$ is then given by

$$f(x) = f(1,x). \tag{3.6}$$

The integral over $T^\mu$ in (3.1) is extended over all $T^i(x)$ (or, more precisely, over all the diffeomorphism group) and over all <u>positive</u> <u>$T^\perp(x)$</u>.

The amplitude (3.1) may be described in the following terms: The matrix element

$$\langle 2 \mid \exp\left[-(i/\hbar)\int d^3x\; T^\mu \mathcal{H}_\mu\right] \mid 1\rangle\ . \tag{3.7}$$

may be thought of as describing the propagation from three-geometry 1 to three-geometry 2 through a convenient sequence of intermediate configurations determined by the "proper time gauge conditions"

$$\dot{N} = 0\ ,\quad \dot{N}^i = 0\ , \tag{3.8}$$

and for a fixed choice of both, the total pointwise normal distance $T^\perp(x) = N^\perp(x)(\tau_2 - \tau_1)$ between $G_2$ and $G_1$ and the "position" f of coordinate system on $G_2$ relative to the one on $G_1$ given by the diffeomorphism (3.6) .

However, to obtain the physically important amplitude--the sum over histories of the exponential of (2.1)--one must add the amplitudes (3.7) for all positive values of $T^\perp(x)$. This means that all proper time separations enter the quantum amplitude in the same footing and no privileged role is played by any particular $N^{\perp}(x)$.

Similarly the fact that one must average over all possible coordinate systems on the final geometry, means that all possible choices for the coordinate system on the final geometry relative to that on the initial one are considered on the same footing and hence no privileged role is given to any particular diffeomorphism--or on account of (3.4-6) to any particular shift $N^{i}(x)$.

Incidentally, it should be mentioned here that the invariant averaging over all possible coordinate systems on $G_2$ makes the amplitude (3.1) invariant under independent changes of coordinates on both $g_{ij}(2)$ and $g_{ij}(1)$ and hence renders it a functional of $G_2$ and $G_1$. This is so because the integral of (3.7) over $T^{\perp}$ --prior to the integration over $T^{i}$--is already invariant under the same change of coordinates on both $g_{ij}(2)$ and $g_{ij}(1)$.

It is therefore apparent that there is nothing like a sandwich theorem at the quantum level. Instead of being determined, the time separation and coordinate system of $G_2$ relative to $G_1$ are absolutely "fuzzy": the quantum amplitude contains a superposition of all possible choices for them.

## 4 CLASSICAL LIMIT

One may pass to the classical limit by studying the semiclassical approximation of the quantum amplitude (3.1)

This may be done in two steps. First one may examine the behavior of (3.7) for $\hbar \to 0$. That amplitude is the result of summing over histories the exponential of $(i/\hbar)$ times an effective action which is obtained from (2.1) as follows: (*i*) a ghost kinetic term $i(\bar{P}\dot{C} + P\dot{\bar{C}})$ is included in the action; (*ii*) the ghost correction (3.2) is included in $\mathcal{H}_{\mu}$; (*iii*) the functions $N^{\mu}$ are taken to be independent of time. The sum over histories is then carried over all functions $g_{ij}(x,\tau)$, $\pi^{ij}(x,\tau)$, $C(x,\tau)$, $\bar{C}(x,\tau)$, $\bar{P}(x,\tau)$, $P(x,\tau)$ which are such that $g_{ij}(x,\tau_2) = g_{ij}(2)$, $g_{ij}(x,\tau_1) = g_{ij}(1)$, $C(x,\tau_1) = \bar{C}(x,\tau_1) = C(x,\tau_2) = \bar{C}(x,\tau_2) = 0$. The functions $N^{\mu}(x)$ are kept fixed.

In the classical limit the dominating contribution comes from the trajectory which satisfies the equations of motion obtained from the Hamiltonian,

$$\int d^3x\, N^\mu(x)\, \mathcal{H}^{\text{eff}}_\mu(x), \tag{4.1}$$

with the above specified boundary conditions for $g_{ij}$, C and $\bar{C}$. That trajectory is such that C, $\bar{C}$, P and $\bar{P}$ are equal to zero for all times whereas $g_{ij}$ and $\pi^{ij}$ satisfy the dynamical Einstein equations

$$^{(4)}G_{ik} = 0 \tag{4.2}$$

in the four-dimensional coordinate system in which $\dot{N}^\mu = 0$. At this stage the initial value equations $\mathcal{H}_\mu = 0$ which are equivalent to

$$^{(4)}G_{\perp\mu} = 0 \tag{4.3}$$

are not enforced and hence the dynamical equations (4.2) will have, in the generic case, a unique solution $g_{ij}(x,\tau)$, $\pi^{ij}(x,\tau)$ such that $g_{ij}$ takes prescribed values at $\tau_1$ and $\tau_2$, for any fixed $N^\mu(x)$.

It is appropriate to point out here that there is no inconsistency in assuming (4.2) to hold while not requiring (4.3), provided the four-dimensional coordinate system is fixed. In fact one may obtain (4.3) from (4.2) only if one assumes the latter to hold for every four-dimensional coordinate system. A similar analysis carried out directly in terms of the Hamiltonian (Teitelboim 1973, 1980a) shows that the dynamical equations imply the initial value constraints provided one requires compatibility between the equations obtained for arbitrary $N^\mu(x,\tau)$. That requirement cannot be enforced here since the $N^\mu$ are fixed to be time independent.

Second, one may imagine inserting the solution of the dynamical equations into the action (2.1) to express that functional in terms of $g_{ij}(2)$, $g_{ij}(1)$ and $T^\mu$. Last, after that is done one also approximates the integral over $T^\mu(x)$ by stationary phase. This implies that the dominating contribution comes from those $T^\mu(x)$ which extremize the classical action keeping $g_{ij}\ (x,\tau_2)$ and $g_{ij}\ (x,\tau_1)$ fixed.

The condition for the action to be an extremum under arbitrary variations of the time independent $T^\mu(x)$ is

$$\int_{\tau_1}^{\tau_2} d\tau\ \mathcal{H}_\mu\ (x,\tau) = 0 \tag{4.4}$$

Here it is understood that the arguments $g_{ij}$, $\pi^{ij}$ of $\mathcal{H}_\mu$ are expressed in terms of $g_{ij}(1)$, $g_{ij}(2)$, $N^\mu$ and the time $\tau$ through the solution of the equations of motion (4.2) which has the boundary values $g_{ij}(1)$, $g_{ij}(2)$ for the given $N^\mu$.

Now, one observes that the time derivative of the $\mathcal{H}'_\mu$s is a linear combination of the $\mathcal{H}_\mu$'s themselves at all space points. This is so because the $\mathcal{H}_\mu$'s obey a closed set of Poisson bracket relations (Teitelboim 1973). That is, one has

$$\dot{\mathcal{H}}_\mu\ (x) = \int d^3x'd^3x''\ \kappa_{\mu\nu}{}^\rho(x,x';x'')\ N^\nu(x')\ \mathcal{H}_\rho\ (x''), \tag{4.5}$$

or, in a more compact notation

$$\dot{\mathcal{H}}_\mu = \kappa_{\mu\nu'}{}^{\rho''}\ N^{\nu'}\ \mathcal{H}_{\rho''} \tag{4.6}$$

which is of the form

$$\dot{\mathcal{H}}_\mu = \Omega_\mu{}^{\rho'}(t)\ \mathcal{H}_{\rho'} \tag{4.7}$$

Equation (4.7) is of the Schroedinger type with $\mathcal{H}_\mu$ being the state vector and $i\Omega_\mu{}^{\rho'}$ (t) being the Hamiltonian--which is explicitly time dependent because $\kappa_{\perp\perp}{}^{i}$ contains $g_{ij}$. The solution of (4.7) is

$$\mathcal{H}_\mu\ (\tau) = U_\mu{}^{\rho'}\ (\tau,\tau_1)\ \mathcal{H}_{\rho'}\ (\tau_1), \tag{4.8}$$

where $U_\mu{}^{\rho'}$ is the corresponding evolution operator. Hence (4.4) implies

$$\left(\int_{\tau_1}^{\tau_2} U_\mu{}^{\rho'}\ (\tau,\tau_0)d\tau\right)\ \mathcal{H}_{\rho'}\ (\tau_0) = 0 \tag{4.9}$$

for any $\tau_0$ in between $\tau_1$ and $\tau_2$. However, the operator $\int U_\mu{}^{\rho'}\ (\tau,\tau_1)\ d\tau$ is invertible. Indeed, it has a power expansion of the form

$$(\tau_2-\tau_1)\ \delta_\mu{}^{\rho'} + \alpha_\mu{}^{\rho'}\ (\tau_2-\tau_1)^2+\ldots\ldots. \tag{4.10}$$

so that unless the coefficients $\alpha_\mu{}^{\rho'}$ ... are carefully arranged by means of a special selection of $g_{ij}(2)$ and $g_{ij}(1)$ to cancel the zeroth order term in (4.10) (in which case one may just change $\tau_o$) the inverse will exist. Thus one obtains from (4.9) the initial value equations

$$\mathcal{H}\rho\ (\tau_o) = 0 \tag{4.11}$$

for all $\tau_o$.

Equations (4.11) are known to generically determine the functions $N^\mu(x)$ ("thin sandwich theorem"). More precisely, if we set, from the equations of motion,

$$\pi_{ij} = (2N^\perp)^{-1}\left[\dot{g}_{ij} - N_{(i|j)} - (g^{kl}\dot{g}_{kl}-2N^k{}_{|k})g_{ij}\right] \tag{4.12}$$

then (4.11) can be solved to express the infinitesimal separation $\delta T^\mu = N^\mu\delta\tau$ as a function of $g_{ij}(1) = g_{ij}$ and $g_{ij}(2) = g_{ij} + \dot{g}_{ij}\ \delta\tau$.

This is, then, how the thin sandwich theorem (determined $N^\mu$) appears in the classical limit from exactly the opposite situation in the quantum regime (all $N^\mu$ equally allowed and present at once).

The determination of $T^\mu$ from (4.11) is not so straightforward where the sandwich is not thin. In that case the function $g_{ij}\ (x,\tau)$ appearing in $\mathcal{H}_\rho$ must be first expressed in terms of $g_{ij}(1)$, $g_{ij}(2)$ and $T^\mu$ itself by means of the equations of motion. The dependence of $\mathcal{H}_\rho$ on $N^\mu$ is not, then, just the one visible in (4.12), but it is more complicated since $N^\mu$ is also contained implicitly in $g_{ij}$ and $\dot{g}_{ij}$. However, one still expects that equations (4.11) will determine $N^\mu(x)$ in that case, although this problem appears not to have been studied.

It is interesting to notice in this context that the equations for $N^\mu$ obtained by demanding (4.11) are identical for all times due to the fact that $\mathcal{H}_\rho = 0$ is preserved by the equations of motion (Eq. 4.8). In other words the knowledge of the infinitesimal separation $N^\mu\delta\tau$ between two consecutive slices is sufficient to determine the total separation $T^\mu = N^\mu\ (\tau_2-\tau_1)$ between the initial and final surfaces. In this sense the use of the proper time gauge $N^\mu = 0$ reduces the thick sandwich problem to the thin one by dividing the thick sandwich into many thin ones of the same thickness.

## 5 INTERSECTING SURFACES: CAUSALITY VERSUS SMOOTHNESS

There is one central feature of the quantum amplitude which has played no role in the discussion so far, namely the fact that one integrates in (3.1) over positive proper times only.

This restriction on the range of integration for $T^{\perp}(x)$ may be considered as the way in which causality is built into quantum gravity. It says that each history contributing to the path integral is sliced so that the spatial sections do not intersect at any point, that is, the future is not mixed with the past. A similar restriction for a single relativistic particle distinguishes the causal Green function--or Feynman propagator--from other possibilities.

However, in the context of the thin sandwich problem one gives the two boundary data $g_{ij}$ and $g_{ij} + \delta g_{ij}$ arbitrarily and it may very well happen that when embedded into a four-dimensional spacetime the two surfaces which have $g_{ij}$ and $g_{ij} + \delta g_{ij}$ as intrinsic metrics will intersect each other. Indeed, to create that situation it is sufficient to proceed the other way around by picking two neighboring slices which mutually intersect within a spacetime which solves Einstein's equations, calculate their intrinsic metrics and use the resulting $g_{ij}$ and $g_{ij} + \delta g_{ij}$ as data for the thin sandwich problem.

If the two surfaces intersect $N^{\perp}$ goes through zero at the crossing points. Furthermore, if one demands that $N^{\perp}$ be continuously differentiable then it will change signs at the crossing point if its derivative does not happen accidentally to vanish there too. This means that, for crossing surfaces, a continuously differentiable solution $N^{\perp}$ of the thin sandwich problem will necessarily lie outside the range of $N^{\perp}$'s allowed in the quantum amplitude.

Now, if one demands continuous differentiability, the function $N^{\perp}$ which solves the thin sandwich problem is unique up to an overall sign. However, if the assumption of continuous differentiability is dropped one has the freedom of adjusting the sign of $N^{\perp}$ in every region where $N^{\perp}$ is different from zero. (Two points belong to a different region if every line joining them goes through a point where $N^{\perp} = 0$).

The question then arises of which solution is the one which

correctly approximates the quantum mechanical amplitude in the semiclassical limit. The answer is that one must take that solution which is within the range of integration for $N^{\perp}$, i.e. one must adjust the independent signs so that $N^{\perp} = |N^{\perp}| \geq 0$.

As already indicated, such a non-negative $N^{\perp}$ will in general not be continuously differentiable. This feature has an interesting implication, namely that the extrinsic geometry of the surfaces (governed by $\pi^{ij}$) becomes not smooth.

In fact, if one gives $g_{ij}$ and $\dot{g}_{ij}$ to be say, continuously differentiable, then $\pi^{ij}$ will in general turn out to be discontinuous: every time that it has been necessary to change the relative sign between two adjoining regions to achieve $N^{\perp}=|N^{\perp}|$ from the continuously differentiable solution, then $\pi_{ij}$ given by (4.12) will tend to values with opposite sign and equal (in general non-zero) magnitude if one approaches the crossing point from each of the two regions.

What has happened? Indeed, suppose, as previously indicated, that one had constructed the boundary data by selecting two smooth intersecting hyper-surfaces within a smooth four-dimensional spacetime which solves Einstein's equations. How can it be that the solution of the thin-sandwich problem selected by quantum mechanics implies that the surfaces are not smooth?

The explanation is simple. As illustrated in Fig. 2, quantum mechanics tells us to keep the same set of points within the same spacetime but to group them differently. That is, we are instructed to exchange regions from surface 1 with regions from surface 2 so that after the exchange is made past and future touch at the points where $N^{\perp}=0$ but do not intersect. However, after the exchange, the geometry of the surfaces is in general not smooth. In this way causality is preserved at the price of sacrificing smoothness.

The author would like to thank Dr. Marc Henneaux for enlightening discussions. This work was supported in part by the U.S. National Science Foundation under Grant No. PHY-8011432 to the University of Texas.

Figure 2. Two neighboring surfaces looked at classically and quantum mechanically. (a) Shows two smooth surfaces A (solid line) and B( broken line), which intersect each other at the points P and Q. The function $N^{\perp}$ is zero at P and Q. These two points divide both surfaces in three regions labeled $A_1$, $A_2$, $A_3$ and $B_1$, $B_2$, $B_3$ respectively. In regions 1 and 3 $N^{\perp}$ is positive (A lies in the future of B) whereas it is negative in region 2 (A lies in the past of B).

(b) Shows the surfaces A' (solid line) and B' (broken line) obtained from A and B through the exchange of regions $A_2$ and $B_2$. The surface A' is formed by $A_1$, $B_2$ and $A_3$ whereas B' is formed by $B_1$, $A_2$ and $B_3$. The surfaces A' and B' touch at P and Q but do not intersect each other; A' never reaches the past of B and the function $N^{\perp}$ is now positive in all three regions. However, the surfaces A' and B' are not geometrically smooth (the extrinsic curvature is singular at P and Q).

According to quantum mechanics the action assigned to the propagation from the smooth configuration B to the smooth configuration A is not that evaluated with $N^{\perp}$ corresponding to diagram (a) but rather the one obtained with the positive $N^{\perp}$ of diagram (b).

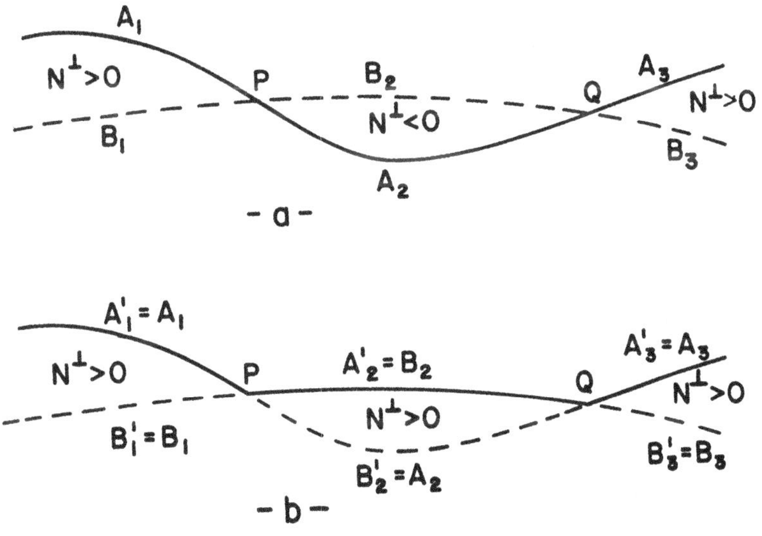

References

Arnowitt, R. *et al.* (1962) The Dynamics of General Relativity. In Gravitation, an Introduction to Current Research, ed. L. Witten. New York: Wiley.

Baierlein, R.F. *et al.* (1962). Three Dimensional Geometry as a Carrier of Information About Time. Phys. Rev. 126, p. 1864.

Dirac, D.A.M. (1958). The Theory of Gravitation in Hamiltonian Form. Proc. Roy. Soc. London, A246, p. 333.

Feynman, R.P. (1948). Space-time Approach to Non-relativistic Quantum Mechanics. Rev. Mod. Phys. 20, p. 267.

Teitelboim, C. (1973). How Commutators of Constraints Reflect the Space-time Structure. Ann. Phys. (N.Y.), 79, p. 592.

Teitelboim, C. (1980a). The Hamiltonian Structure of Spacetime. In Einstein Centenary volume, ed. A. Held. New York: Plenum.

Teitelboim, C. (1980b) Proper Time Approach to the Quantization of the Gravitational Field. Phys. Lett. 96b, p. 77.

Teitelboim, C. (1981). Quantum Mechanics of the Gravitational Field. Austin-Princeton preprint,to appear in Proceedings of the Second Chilean Symposium of Theoretical Physics. Santiago: Universidad de Santiago Press.

Teitelboim, C. (1982a). The Propagation Amplitude in Quantum Theory of Gravitation. In preparation.

Teitelboim, C. (1982b). Proper Time Approach to the Quantum Theory of Gravitation in Asymptotically Flat Space. In preparation.

Wheeler, J.A. (1964). Geometrodynamics and the Issue of the Final State. In Relativity, Groups and Topology: Les Houches 1963, eds. B. DeWitt and C. DeWitt. New York: Gordon and Breach.

# θ-STATES INDUCED BY THE DIFFEOMORPHISM GROUP IN CANONICALLY QUANTIZED GRAVITY

C.J. Isham
Blackett Laboratory, Imperial College, London SW7 2BZ

## 1. INTRODUCTION

Studies of the effects of spatial or spacetime topology in quantum gravity are an interesting part of the whole quantization programme. In particular, investigations of this type seem a natural prerequisite to the attainment of one of the ultimate goals - to subject the topology itself to some sort of quantization. There are several possible approaches to the problem and in this paper we will concentrate on developing an analogue of the vacuum θ-structure that arises in Yang-Mills theory. The decomposition of the vector space of quantum states into θ-sectors is a direct consequence of specific topological properties of the local gauge group. These, in turn, reflect certain aspects of the topological structure of space and/or spacetime which are thereby coded into the quantum field theory.

A priori, two candidates for the gauge group in quantum gravity are the group of diffeomorphisms and the local tangent space group of triad or tetrad rotations - the choice between three-dimensional space or four-dimensional spacetime being determined primarily by whether a canonical or covariant quantization scheme is to be employed. In a Yang-Mills theory, canonically quantized on a compact, orientable, three-manifold $\Sigma$, the gauge group is the set of differentiable functions from $\Sigma$ into the internal symmetry group G and the n-vacua are labelled by the set of connected components of this group, whilst the θ-vacua correspond to the characters (homomorphisms into U1) of this set. We will show that the diffeomorphism group (which in general is also disconnected) plays an analogous role in quantum gravity. To do so it is expedient to rederive the Yang-Mills results in a way which emphasizes the non simple connectedness of the infinite dimensional physical configuration space Q of the theory. The key homotopy group is $\pi_1(Q)$ and the equivalence with the conventional approach is confirmed by proving that $\pi_1(Q)$ is isomorphic to the group of components of the gauge group.

A natural analogue of Q in quantum gravity is Wheeler's superspace - the set of equivalence classes of riemannian metrics defined on the underlying three-manifold of physical space, in which two metrics are identified if one may be reached from the other by an action of the diffeomorphism group i.e. if they are isometric. The Yang-Mills results may be duplicated in this framework and $\pi_1(Q)$ computed in terms of the homotopical properties of the diffeomorphism group. However a subtlety enters that is peculiar to quantum gravity and concerns the way in which time appears in the canonical form of general relativity. One of the functional degrees of freedom in Q can be regarded as an intrinsic time label and must be factored out in order to yield the genuine physical configuration space $\tilde{Q}$. The new group of interest is $\pi_1(\tilde{Q})$ but for reasons given in the text the techniques successfully employed to compute $\pi_1(Q)$, break down here.

The plan of the paper is as follows. In Section 2 the θ- and n-vacua of Yang-Mills theory are introduced via the configuration space Q and are shown to be equivalent to those obtained by more conventional means. Section 3 contains the discussion of the analogous treatment of canonically quantized gravity whilst some specific examples of disconnected diffeomorphism groups are exhibited in section 4. The paper concludes with a short discussion of the gravitational analogues of instantons and tunnelling.

The treatment throughout is deliberately pedagogical and problems such as the precise choice of function space topologies are neglected. Fortunately, powerful theorems of Palais guarantee that $\pi_1(Q)$ is independent of this choice within the class of topologies considered in the references cited in the text. A previous discussion of the θ-states in quantum gravity may be found in Isham 1981a.

## 2. θ-VACUA IN YANG-MILLS THEORY

One possible starting point for a study of topological sectors in a canonically quantized Yang-Mills theory lies in the classical, static, zero-energy solutions to the field equations; it being supposed that the quantum vacuum state functionals are peaked around such field configurations. (Jackiw & Rebbi 1976;Callen et al.1976). It is convenient to choose the gauge in which the Yang-Mills potential $\vec{A}_\mu(x)$ satisfies $\vec{A}_o=0$. Note that time independent gauge transformations on $\vec{A}_i(x)$ (i=1,2,3) are still permitted. The energy of a static field in Minkowski space is

$$E = \int_{R^3} \vec{F}_{ij}.\vec{F}^{ij}d^3\underline{x} \tag{2.1}$$

and the field strength $\vec{F}_{ij}(\underline{x})$ of a zero energy configuration vanishes:

$$\vec{F}_{ij}(\underline{x}) = 0 \tag{2.2}$$

This implies that $\vec{A}_i(\underline{x})$ is 'pure gauge', viz

$$\vec{A}_i(\underline{x}) = \Omega(\underline{x})\partial_i\Omega^{-1}(\underline{x}) \tag{2.3}$$

where $\Omega$ is some function from physical three-space ($R^3$) into the internal symmetry group G. The boundary conditions frequently imposed on $\vec{A}_i(\underline{x})$ are such that this potential is naturally defined on the three-sphere ($S^3$) obtained by identifying the boundary of $R^3$ to a point $\underline{x}_\infty$. Under these circumstances $\Omega$ becomes a differentiable function from $S^3$ into G which maps $\underline{x}_\infty$ to the identity element. Thus the gauge group $\mathcal{G}$ is the set $G_*^{S^3}$ of all such functions (conveniently equipped with a compact-open topology) and is disconnected, as is the group O(n) for example. The components of $\mathcal{G}$ form a discrete group labelled by the homotopy set $\pi_o(G_*^{S^3}) \equiv [S^3,G]_*$ where, in general, $[X,Y]_*$ denotes the set of homotopy classes of continuous maps (preserving a fixed basepoint such as $\underline{x}_\infty$) from the topological space X into Y. Now $[S^3,G]_*$ is just the third homotopy group of G and is isomorphic to the integers Z for any simple, non abelian, lie group G. Hence, for such a G, the gauge group $G_*^{S^3}$ has a countable number of components (whereas, for example, O(n) has only two).

The significance of this feature within a canonical quantization scheme is discussed in Jackiw (1980). There exists a generator of gauge transformations (depending only on the spatial coordinates $\underline{x}$) and this should annihilate physical states $|\psi\rangle$.

$$\hat{J}^i(\underline{x})|\psi\rangle = 0 \quad i=1....\text{dim}.G \tag{2.4}$$

In so far as all elements in the connected component of the identity gauge function can be reached by iterating "infinitesimal transformations", eqn (2.4) implies that $|\psi\rangle$ is invariant under the action of such members of G. However nothing can be inferred concerning the action of the disconnected parts of the gauge group and it is this feature which allows the existence of n-vacua $|n\rangle$. These approximate vacuum states (approximate in the sense that tunnelling is ignored) are labelled by the integers and transform under the action $T_\Omega$ of a gauge function as

$$T_\Omega |n\rangle = |n + [\Omega]\rangle \tag{2.5}$$

where $\Omega$ denotes the homotopy class of $\Omega$ as an element of $\pi_3(G)=Z$. The state functional $\langle\vec{A}_i|n\rangle$ is presumed to peak around a field configuration of the type in (2.3) with the homotopy class of $\Omega$ being the integer n.

The states cannot be regarded as the true vacua because i) as shown by (2.5) they are not invariant under "large" gauge transformations (i.e. those arising from gauge functions that are not homotopic to the identity) and ii) quantum tunnelling may occur between them. Both defects are eliminated by defining the θ-vacua:

$$|\theta\rangle \equiv \sum_{n=-\infty}^{\infty} e^{-in\theta}|n\rangle , \qquad 0 \leq \theta < 2\pi \tag{2.6}$$

which are invariant up to a physically irrelevant phase factor and which diagonalise the S-matrix.

The above discussion may be generalised to the case where the canonical Yang-Mills theory is defined on an arbitrary, compact, orientable three-space $\Sigma$ (rather than $S^3$). The gauge group is defined as the space $G_*^\Sigma$ of all differentiable, base point preserving, maps of $\Sigma$ into G. The homotopy set $[\Sigma,G]_*$ may be computed using cohomological methods and shown to be equal to $Z \oplus \mathrm{Hom}\,(\pi_1(\Sigma),\pi_1(G))$ or to some specific subgroup of this group (Isham 1981b). Hence the "n"-vacua are now labelled by homomorphisms of $\pi_1(\Sigma)$ into $\pi_1(G)$ in addition to the integers Z. When $\Sigma$ is simply connected (i.e. $\Sigma = S^3$) these are the same integers as those encountered above in the guise of $\pi_3(G)$. The phase factors in (2.6) become elements of the character group of $[\Sigma,G]_*$ (i.e. homomorphisms from $[\Sigma,G]_*$ into U1) and the θ-and (n,h)-vacua are related by

$$|\theta\rangle = \sum_{(n,h)} \theta(n,h)|n,h\rangle \tag{2.7}$$

where θ and (n,h) belong respectively to Hom $([\Sigma,G]_*, U1)$ and

$Z \oplus \mathrm{Hom}\,(\pi_1(\Sigma), \pi_1(G))$. It should be emphasised that the entire Hilbert space of states (and not just the set of vacua) splits into sectors with the θ- and (n,h)-states being related by (2.7). In fact, when $\pi_1(\Sigma) \neq 0$, the definition of a vacuum state is subject to a complication arising from the existence of solutions to $\vec{F}_{ij} = 0$ other than the pure gauge fields of (2.3) which can enhance and/or contract the set of naive (n,h)-vacua. This phenomenon is discussed in Isham and Kunstatter (1981a,b) and in Kunstatter's article in this volume.

Various problems arise in the development of an analogous picture for canonically quantized gravity. Two natural candidates for playing the role of the Yang-Mills gauge group are the group of diffeomorphisms of the three-space Σ and the tangent space group of local triad rotations. For almost all Σ both these groups, together with a number of subgroups, are disconnected and hence any of them might lead to the existence of a θ-vacuum structure. It is not clear a priori on which one attention should be focussed. The definition of a "vacuum" state also involves difficulties closely related to the subtleties connected with the notion of time in the canonical form of general relativity. For these and related reasons it is profitable to rederive the Yang-Mills results in a way that is more adaptable to the gravitational case.

Let $\mathcal{A}$ denote the set of all Yang-Mills potentials, i.e. connections in a principal G-bundle ξ over Σ. The different bundles are classified by the elements of the cohomology group $H^2(\Sigma;\pi_1(G))$ but, for simplicity, we will restrict ourselves to the situation where ξ is the trivial product Σ×G in which case the gauge group $\mathcal{G}$ is just the function space $G_*^\Sigma$ (as above). Physical results should be invariant under the action of $\mathcal{G}$ on $\mathcal{A}$:

$$\vec{A}_i \to \vec{A}_i = \Omega A_i \Omega^{-1} + \Omega\partial_i \Omega^{-1} \tag{2.8}$$

and correspondingly it is natural to regard two potentials as being equivalent if one may be obtained from the other by a gauge transformation; i.e. if they lie on the same orbit of the $\mathcal{G}$-action on $\mathcal{A}$. The set $Q \equiv \mathcal{A}/\mathcal{G}$ of all such equivalence classes may be viewed as the true, physical configuration space of the system and canonical quantum states might (naively) be identified with functionals defined on Q. However, for reasons to be explained below, $\pi_1(Q)$ is non trivial, and hence we are involved with the quantum theory of a system whose configuration space is not simply connected. This subject has been treated at length in the literature (Schulman 1971; Laidlaw & DeWitt 1971; Dowker 1972) and the possibility of thus regarding Yang-Mills theory is discussed in Dowker (1980). In particular it is shown that quantization is not unique and that for example, the possible two point Green's functions are labelled by the character group $\mathrm{Hom}(\pi_1(Q),U1)$ of $\pi_1(Q)$ and may be expressed in the form

$$K_\chi(q,q') = \sum_{[p]} \chi[p] \quad K_{[p]}(q,q');\chi\varepsilon\mathrm{Hom}(\pi_1(Q),U1) \tag{2.9}$$

where, in a path integral context, $K_{[p]}(q,q')$ is the propagator obtained by summing overall paths from q to q' that are homotopic to the path p. The sum in (2.9) is over representatives of each of the homotopy classes of paths between q and q' and this set can be labelled by the elements of $\pi_1(Q)$.

In an operator/state vector approach we may proceed as follows. Rather than considering states as complex valued functionals on Q let us contemplate the possibility that, as we move about Q, the complex numbers "twist around". In other words we generalise the notion of a functional to include cross-sections of complex line bundles (i.e. vector bundles whose fibres are copies of the complex numbers) over Q. At a more heuristic level this is equivalent to "gauging" the U1 quantum mechanical phase group. In particular the functional derivative $-i\hbar\frac{\delta}{\delta\phi_a(\underline{x})}$ which represents the momentum $\hat{\pi}^a(\underline{x})$ conjugate to the true dynamical variables $\phi_a(\underline{x})$, must be replaced by a covariant derivative:

$$\hat{\pi}^a(\underline{x}) = -i\hbar\frac{\delta}{\delta\phi_a(\underline{x})} - \mathfrak{A}^a_{\underline{x}}[\phi] \tag{2.10}$$

where the connection $\mathfrak{A}^a_{\underline{x}}[\phi]$ transforms in an appropriate way under U1 gauge transformations.

This appearance of $\mathfrak{A}^a_{\underline{x}}[\phi]$ could change dramatically the dynamics and we would like to minimize its effects. An obvious restriction is to require the associated curvature two-form

$$\mathfrak{F}^{a,b}_{\underline{x},\underline{y}}[\phi] := \frac{\delta\mathfrak{A}^a_{\underline{x}}[\phi]}{\delta\phi_b(\underline{y})} - \frac{\delta\mathfrak{A}^b_{\underline{y}}[\phi]}{\delta\phi_a(\underline{x})} \tag{2.11}$$

to vanish which allows $\mathfrak{A}$ to be locally gauged away. This condition significantly restricts the topological class of the line bundle (the real Chern class vanishes) and the resulting bundles and connections are classified, up to small gauge transformations, by elements of the group $\mathrm{Hom}(\pi_1(Q),U1)$ (Milnor 1957; Kostant 1970). In this way we recover the labelling scheme of (2.9). Note that because $\pi_1(Q) \neq 0$, the vanishing of $\mathfrak{F}^{a,b}_{\underline{x},\underline{y}}$ does not completely eliminate $\mathfrak{A}^a_{\underline{x}}[\phi]$ and there is a residual "Bohm-Aharanov" type of effect. Indeed the map of closed curves in Q, into U1 given by :

$$\Gamma \to e^{i\int_\Gamma \mathfrak{A}} \tag{2.12}$$

precisely represents the particular homomorphism from $\pi_1(Q)$ into U1 that is being employed. It should also be observed that in writing expressions like (2.10) - (2.12) it is tacitly assumed that Q has the structure of a (infinite dimensional) differentiable manifold. However as things stand this is false and Q is actually a stratified space (Fischer 1967, Singer 1978). It is highly desirable for quantum field theoretical purposes that these strata are removed and we will show shortly how this is to be achieved.

The concept of θ - and (n,h)-states will be regained in this new approach if the equality $\pi_1(Q) = \pi_o(\mathcal{G}) = [\Sigma,G]_*$ can be demonstrated. Indeed elementary bundle theory shows that cross-sections of a line bundle over Q are in bijective correspondence with functionals on $\mathcal{A}$ transforming in the correct way (e.g. see Isham 1981a).

(Note however that in the present approach the primary construct is the θ-state whilst an (n,h)-state is defined using the inverse transformation to that employed in (2.7)). The desired relation between $\pi_1(Q)$ and $\dot{\pi}_o(\mathcal{G})$ ensues from the following considerations. Suppose that the $\mathcal{G}$-action on $\mathcal{A}$ were free, i.e. $\vec{A}_i = \vec{A}'_i$ in (2.8) implies that $\Omega = \mathbb{1}$. Then each $\mathcal{G}$-orbit would be a copy of $\mathcal{G}$ and there would be a chance that $\mathcal{A}$ was a principal $\mathcal{G}$-bundle over Q (denoted $\mathcal{G} \to \mathcal{A} \to Q$). We would then have the long exact sequence (Steenrod 1951).

$$\to \pi_2(Q) \to \pi_1(\mathcal{G}) \to \pi_1(\mathcal{A}) \to \pi_1(Q) \to \pi_o(\mathcal{G}) \to \pi_o(\mathcal{A}) \tag{2.13}$$

relating the homotopy groups of $\mathcal{G}$, $\mathcal{A}$ and Q. However $\mathcal{A}$ is an affine space ($A^{(1)}, A^{(2)}$ are in $\mathcal{A}$ then so is $\lambda A^{(1)} + (1-\lambda)A^{(2)}$ for all λ) and is hence contractible. This implies $\pi_o(\mathcal{A}) = \pi_1(\mathcal{A}) = 0$ and then (2.13) shows that

$\pi_1(Q)$ and $\pi_0(\mathcal{G})$ are indeed isomorphic.

Now the choice above of the gauge group $\mathcal{G}$, as the set of base point preserving maps, does give a free action and the fibre bundle character of $\mathcal{A}$ was proved in Singer (1978), Narisimhan & Ramadas (1979) and Mitter and Viallet (1981) in the context of discussions of the Gribov effect. If non base point preserving gauge transformations are to be admitted then a free action may be obtained by factoring G by its centre C(G) and removing the remaining non fixed points from $\mathcal{A}$ to give the space $\mathcal{A}_r$ of 'regular' connections. By these means two possible models

$$Q_r \equiv \mathcal{A}_r/(G/C(G))^\Sigma \quad \text{and} \quad Q_* \equiv \mathcal{A}/G_*^\Sigma \tag{2.14}$$

are obtained for the physical configuration space of the Yang-Mills system and

$$\pi_1(Q_r) = [\Sigma, G/C(G)] \quad ; \quad \pi_1(Q_*) = [\Sigma, G]_* \tag{2.15}$$

where $(G/C(G))^\Sigma$ and $[\Sigma, G/C(G]$ are respectively the set of non basepoint preserving maps from $\Sigma$ into $G/C(G)$ and the free homotopy classes of such maps.

Both $\mathcal{A}_r$ and $\mathcal{A}$ can be given the structure of infinite dimensional manifolds in various ways as can the associated gauge groups. As a consequence the bundle base spaces $Q_r$ and $Q_*$ are also differentiable manifolds and hence restricting $\mathcal{A}$ and/or $\mathcal{G}$ in this way also solves the strata-removing problem alluded to earlier. The exact topology to be assigned to these spaces is likely to be determined by the appropriate theory of the canonical commutation relations of the dynamical field variables. Fortunately Palais' results (Palais 1966,1968) show that $\pi_1(Q)$, and hence the θ-structure, is unaffected by these considerations.

## 3. θ-STATES IN QUANTUM GRAVITY

In the canonical version of general relativity the basic field variables are a riemannian metric $g_{ij}$ on the three-manifold $\Sigma$ and its canonical conjugate $\pi^{k\ell}$ (for a recent review of the canonical formalism see Kuchar 1981). These are subject to the first class constraints

$$H_\perp(g \;, \pi\;) = 0 \tag{3.1}$$

$$H_i(\;g\;, \;\pi) = 0 \qquad i = 1,2,3 \tag{3.2}$$

where $H_\perp$ generates deformations normal to $\Sigma$ in the enveloping spacetime and $H_i$ are the generators of diffeomorphisms of $\Sigma$. In the quantum theory of gravity these constraints may either be solved classically (in principle - not in practice!) after imposing a gauge (which renders them second class) or they can be implemented as operator constraints on the allowed state vectors:

$$H_{\perp}(\hat{g},\hat{\pi})|\psi\rangle = 0 \tag{3.3}$$

$$H_i(\hat{g},\hat{\pi})|\psi\rangle = 0 \qquad i = 1,2,3. \tag{3.4}$$

Let us concentrate first on (3.4). This is the gravitational analogue of (2.4) and suggests that "n"-states (i.e. states that are not invariant under large diffeomorphisms) may occur if the diffeomorphism group of $\Sigma$ (which replaces $G^{\Sigma}$) is disconnected. To confirm this and in analogy with the Yang-Mills theory discussed in section 2, let Riem $\Sigma$ denote the set of all riemannian metrics on $\Sigma$. Consider the action of the diffeomorphism group Diff $\Sigma$ of $\Sigma$ on Riem $\Sigma$ in which $\phi \varepsilon$Diff $\Sigma$ takes $g\varepsilon$ Riem $\Sigma$ into the pullback $\phi^* g$. Superspace (Wheeler 1965,1968) is the space of equivalence classes $S(\Sigma) \equiv$ Riem $\Sigma$/Diff $\Sigma$ and, as in section 2, we can argue that quantum states are cross-sections of a complex line bundle over $S(\Sigma)$ whose connection has a vanishing curvature. This structure is to be augmented in some way with the constraint (3.1) which ultimately describes the time evolution. The full superspace $S(\Sigma)$ is a stratified manifold (Fischer 1967) and, for the same reasons as those in section 2, we desire to adjust Riem $\Sigma$ and/or Diff $\Sigma$, in some physically acceptable way, so that the resulting quotient space is a genuine, infinite dimensional, differentiable manifold.

In this superspace approach, the θ-states associated with Diff $\Sigma$ will be classified by elements of the abelian group Hom$(\pi_1(S(\Sigma)),U1)$ and the main problem is to compute this group in terms of the topological properties of the diffeomorphism group. In analogy with the Yang-Mills theory this could be achieved if Riem $\Sigma$(which, like $\mathcal{A}$, is contractible) were a principal Diff $\Sigma$-bundle over $S(\Sigma)$ in which case the exact homotopy sequence would yield

$$\pi_1(S(\Sigma)) = \pi_0(\mathrm{Diff}\Sigma) \tag{3.5}$$

However the action of Diff $\Sigma$ on Riem $\Sigma$ is not free but possesses fixed points which are metrics with isometries. One possibility is to restrict Riem $\Sigma$ to the subspace Riem$_r$ $\Sigma$ of metrics that have no such invariances -

rather as $\mathcal{A}$ was restricted to $\mathcal{A}_r$. However this is incompatible with the action principle in the sense that there certainly exist static solutions to Einstein's equations that possess an isometry group and hence do not lie in $\text{Riem}_r\,\Sigma$. An alternative is to seek an analogue of the other Yang-Mills procedure in which the gauge group was restricted to a freely acting subgroup by admitting only basepoint preserving maps.

There is one obvious way of achieving this. Let $\text{Diff}_*\,\Sigma$ denote the subgroup of diffeomorphisms that map some fixed basepoint $\underline{x}_o$ in $\Sigma$ into itself and also induce the identity transformation on the tangent space at that point. (Note that if $\underline{x}_o$ is the point at infinity of a one point compactification $\Sigma$ of a non-compact three-space $\Sigma'$ then $\text{Diff}_*\,\Sigma$ corresponds to diffeomorphisms on $\Sigma'$ that produce no asymptotic rotations or translations). It is well known that no element of $\text{Diff}_*\Sigma$ can be the isometry group of a metric. Furthermore, the work of Ebin (1970) and Fischer (1967) shows that Riem $\Sigma$ is a principal $\text{Diff}_*\,\Sigma$-bundle over $S_*(\Sigma) \equiv \text{Riem}\,\Sigma/\text{Diff}_*$. Hence, since Riem $\Sigma$ is contractible, we can deduce that

$$\pi_1(S_*(\Sigma)) = \pi_o\,(\text{Diff}_*\Sigma) \tag{3.6}$$

with the corresponding $\theta$-states being labelled by $\text{Hom}(\pi_o(\text{Diff}_*\Sigma),U1)$. Note that the configuration space $S_*(\Sigma)$ is precisely what DeWitt has called extended superspace and that, as desired, it is a strata free infinite dimensional manifold.

It should be observed that Gribov phenomena may arise in this scheme. These occur if an attempt is made to choose a global gauge for the diffeomorphism group, for example, as would be necessary if the constraints $H_i = 0$ were to be solved before quantization. Mathematically, this is equivalent to constructing a cross-section of the principal $\text{Diff}_*$ $\Sigma$-bundle over $S_*(\Sigma)$ and is only possible if the bundle is trivial. This in turn gives

$$0 = \pi_i(\text{Riem}\,\Sigma) = \pi_i(S_*(\Sigma)) \times \pi_i\,(\text{Diff}_*\Sigma),\ \text{for all } i \geq o. \tag{3.7}$$

and hence the nonvanishing of any homotopy group of $\text{Diff}_*\,\Sigma$ is sufficient to ensure that no such cross-section exists and that a corresponding Gribov effect occurs.

Note that $\text{Diff}_*\Sigma$ is a subgroup of the group $\text{Diff}^+\Sigma$ of orientation preserving diffeomorphisms of $\Sigma$. These two groups are related via the fibration

$$\mathrm{Diff}_{*}\,\Sigma \rightarrow \mathrm{Diff}^{+}\Sigma \rightarrow F^{+}(\Sigma) \tag{3.8}$$

where $F^{+}(\Sigma)$ is the bundle of oriented frames. This is diffeomorphic to $\Sigma \times SO3$ if $\Sigma$ is parallelizable (eg. if it admits spinors).

By use of the techniques above we have obtained results in quantum gravity that are entirely analogous to those pertaining to Yang-Mills theory. However there is a subtlety in the former case which conceivably should be taken into account. Superspace $S(\Sigma)$ has roughly speaking three degrees of freedom per space point whereas the true, dynamical configuration space would be expected to possess one less. The extra function corresponds to an intrinsic time whose handling is intimately connected with the attitude adopted to the $H_{\perp} = 0$ constraint. In the operator formalism (3.3), an identification of the intrinsic time converts $\hat{H}_{\perp}|\psi\rangle = 0$ into a multitime evolution equation and it is sometimes argued that the physical Hilbert space should consist of functionals defined on what is left of superspace after factoring out the extra degree of freedom. This attitude also allows the possibility of choosing prior to quantization, a time gauge in the form $T = T(g,\pi)$, for some function T of the canonical variables, and then solving $H_{\perp} = 0$ classically for the canonical conjugate $P(g,\pi)$ of T. The action term $\int_{\Sigma\times R} \pi^{ij}\dot{g}_{ij}\, d^4x$ then acquires a contribution $\int_{\Sigma\times R} P(g,\pi)\, d^4x$ and the effective Hamiltonian for the remaining dynamical degrees of freedom is $-\int_{\Sigma} P(g,\pi) d^3\underline{x}$. (see for example Hansen et. al., 1976). For example, the choice $t = {}^2/_3(\det g)^{-\frac{1}{2}}\, \pi^{k}{}_{k}$ (York 1972) gives $P = -(\det g)^{\frac{1}{2}}$ and the dynamical variables may be chosen to be $\tilde{g}_{ij} \equiv g_{ij}/(\det g)^{\frac{1}{3}}$ and $\tilde{\pi}^{ik} = \pi^{ik} - \frac{1}{3} g^{ik}\pi^{\ell}{}_{\ell}$ with the effective $H_{eff} = \int_{\Sigma}\{(\det g)^{\frac{1}{2}} - N_i H^i\} \equiv \pm$ volume of $\Sigma - \int N_i H^i$ where the $N_i$ are Lagrange multipliers and $(\det g)^{\frac{1}{2}}$ is obtained as a function of the dynamical variables by solving (3.1).

The new canonical variables $\tilde{g}_{ij}$ constitute a space $W(\Sigma)$ of tensor densities that is isomorphic to the factor space $\mathrm{Riem}\,\Sigma/P$ where the multiplicative group P of positive, differentiable, functions $\Omega$ on $\Sigma$ acts on Riem $\Sigma$ by $g_{ij}(\underline{x}) \rightarrow \Omega(\underline{x})\, g_{ij}(\underline{x})$ (Fischer & Marsden 1977). The diffeomorphism group acts on Riem $\Sigma/P$ by $[g] \rightarrow [\phi^{*}g]$ and the resulting quotient space $S(\Sigma) \equiv W(\Sigma)/\mathrm{Diff}\,\Sigma$ has only two degrees of freedom per space point. This conformal superspace is the true configuration space and the associated θ-structure derives from the group Hom $(\pi_1(\tilde{S}(\Sigma)), U1)$. If we were to proceed in a way analogous to that employed for discussing $\mathcal{A}^{\Sigma}/G_*$ and Riem $\Sigma$/Diff $\Sigma$ we would **note** that Riem $\Sigma$ is a principal P-bundle over $W(\Sigma)$. Furthermore P is contractible and hence the homotopy exact sequence yields $\pi_i(W(\Sigma)) = 0$ for all i. The next step would be to show that $W(\Sigma)$ is a principal Diff $\Sigma$-bundle over $\tilde{S}(\Sigma)$ and hence that

$$\pi_1(\tilde{S}(\Sigma)) = \pi_o(\text{Diff}\Sigma) \tag{3.9}$$

However the Diff $\Sigma$-action is once again not free. The fixed points correspond to metrics with conformal Killing vectors and, unlike the earlier case, it is unclear how any or all of the spaces, Riem $\Sigma$, P and Diff $\Sigma$ should be restricted in a way which removes these fixed points whilst being physically acceptable. Robbed of (3.9) there is no obvious way of computing $\pi_1(\tilde{S}(\Sigma))$ and the programme grinds to a halt. Thus for the moment we must restrict our attention to the earlier scheme employing the extended superspace $S_*(\Sigma)$.

## 4. SOME EXAMPLES OF DIFF$_*\Sigma$

Homotopically the simplest three-manifold is the three-sphere $S^3$. Cerf (1959) shows that

$$\pi_i(\text{Diff}^+S^3) = \pi_i(SO4), \text{ for all } i \geqslant 0. \tag{4.1}$$

which, together with the homotopy exact sequence of (3.8) shows that $\pi_1(S_*(\Sigma)) = 0$ and hence there is no θ-structure. Another simple example is $\Sigma = S^1xS^2$ and the diffeomorphism group of this space is briefly discussed in Gluck (1961). He shows that

$$\pi_o(\text{Diff}^+S^1xS^2) = Z_2 \oplus Z_2 \tag{4.2}$$

which implies that Diff$_*S^1xS^2$ is disconnected and hence there exist both θ-sectors and Gribov effects. The multitorus $S^1xS^1xS^1$ was investigated by Waldhausen (1968) and Hatcher (1976) and $\pi_o(\text{Diff}^+S^1xS^1xS^1)$ shown to be the orientation preserving subgroup of the group of automorphisms of $Z \oplus Z \oplus Z$, which is again a finite group. A typical "large" diffeomorphism (i.e. one not homotopic to the identity) is expressed in terms of the three angles $\alpha_1, \alpha_2, \alpha_3$ as $\phi(\alpha_1,\alpha_2,\alpha_3) = (\pm\alpha_{p(1)}, \pm\alpha_{p(2)}, \pm\alpha_{p(3)})$ for some permutation p of (1,2,3). By suitable deformations, the orientation preserving maps of this type can be shown to belong to $Diff_*(S^1xS^1xS^1)$ so that again we have θ-sectors and the Gribov phenomenon.

In general the computation of $Diff_*\Sigma$ is very nontrivial and there is no simple algorithm of the type that, in the Yang-Mills case, related $\pi_o(G_*^{\Sigma})$ to $Z\oplus$ Hom $(\pi_1(\Sigma),\pi_1(G))$. Fortunately, some recent mathematical results are of considerable help. If $\Sigma_1$ and $\Sigma_2$ are two compact, orientable, three-manifolds, their topological sum $\Sigma_1 * \Sigma_2$ is obtained by cutting a three-ball from each manifold and identifying the two-sphere boundaries with an orientation reversing diffeomorphism. The resulting three-manifold is itself compact and orientable and its homotopical properties are closely related to those of $\Sigma_1$ and $\Sigma_2$. In particular $\pi_1(\Sigma_1 * \Sigma_2)$ is equal to the free product of $\pi_1(\Sigma_1)$ and $\pi_1(\Sigma_2)$. De Sa and Rourke (1979) have considered the unique decomposition of any three-manifold $\Sigma$ into a topological sum of copies of $S^1xS^2$ and certain basic three-manifolds that are $P^2$-irreducible (see for example Hempel, 1976). They show how $\pi_o(Diff_*\Sigma)$ may be related to the $\pi_o$ of the summands and the latter are known from the work of Gluck (1961) and Hatcher (1976). Thus, in principle, the necessary information on $\pi_o(Diff_*\Sigma)$ is available and it is clear that this group is almost always non trivial.

## 5. CONCLUSIONS

We have seen that the disconnectedness of the diffeomorphism group of $\Sigma$ produces θ-sectors in the canonical quantization of gravity. This is an exact analogue of the process at work in the Yang-Mills theory and it is natural to enquire if any of the well known Yang-Mills phenomena associated with the θ-structure have analogues in the present case. The most important is the notion of a diffeomorphism group (DG-) instanton. This would be a four-metric, solving Einstein's equations of motion, with the property of interpolating between two three-metrics on $\Sigma$ that were related by a large diffeomorphism. Depending on precisely how time t is introduced it might be possible to analytically continue t to -it and hence study DG-instantons on a riemannian, rather than pseudo-riemannian, four-space. Physically these would provide the semi-classical approximation to quantum tunnelling between the states associated with the three-metrics.

The term 'gravitational instanton' has already been employed by S.W. Hawking (1978,1979) to describe riemannian solutions to the

Einstein equations. These are classified according to various types of boundary conditions and none of them seem to be related to the DG-instantons considered here.

In the Yang-Mills theory, CP violating effects arise when the "n" in the $e^{-in\theta}$ coefficient of the generating functional in the n-sector ( cf.2.6) is represented by an addition to the action of a suitable multiple of $\int \vec{F}_\wedge \vec{F}$. The formal gravitational analogue of this effect is described in Deser et al. (1980) but it seems unrelated to the diffeomorphism group structure as there is no obvious way of representing the elements of $\pi_0(\mathrm{Diff}_*\Sigma)$ by a differential form integrated over $\Sigma$. Similarly it is as yet unclear if the inclusion of massless fermions will produce an analogue to the suppression of tunnelling which arises from the existence of zero energy modes of the Dirac equation in a Yang-Mills instanton background.

A number of extensions of the diffeomorphism group results are possible. A particularly interesting one would be a topological study of supergravity. This would involve understanding the global nature of the supergauge group $\mathcal{G}$ and would be particularly important if the structure of $\mathcal{G}$ was such that $\pi_0(\mathcal{G})$ was not simply the zero'th homotopy set of the underlying diffeomorphism group of $\Sigma$.

## REFERENCES

Callen C.G., Dashen R.F. & Gross D.J. (1976). The structure of the gauge theory vacuum. Phys. Letts. 63B, 334-340.

Cerf. J. (1959). Groupes d'automorphisms et groupes de diffeomorphismes des varietes compactes de dimension 3. Bull. Soc. Math. France 87, 319-329.

Deser, S., Duff M.J. & Isham C.J. (1980). Gravitationally induced CP effects. Phys. Letts. B93, 419-423.

Dowker J.S. (1972) Quantum mechanics and field theory on multiply connected and on homogeneous spaces. Journ. Phys. A5, 936-943.

Dowker J.S. (1980). Selected topics in topology and quantum field theory. University of Austin preprint.

Ebin D.G. (1970). The manifold of riemannian metrics. In Proc. Symp. Pure Math. 15, pp.11-37: American Mathematical Society.

Fischer A.E. (1967). The theory of superspace. In Relativity, eds. M. Carmeli, S. Fickler and L. Witten pp. 303-357: Plenum.

Fischer A.E. & Marsden, J.E. (1977). The manifold of conformally equivalent metrics. Can. J. Math. XXIX, 193-209.

Gluck H. (1961). The embedding of two-spheres in the four-sphere. Bull. Am. Math. Soc. 67, 586-589.

Hanson A.J., Regge T.& Teitelboim C. (1976). Constrained hamiltonian systems: Accademia Nazionale dei Lincei Rome.

Hatcher, A. (1976). Homeomorphisms of sufficiently large $P^2$-irreducible 3-manifolds. Topology 15, 343-347.

Hawking S.W. (1978). Spacetime foam. Nuc. Phys. B144, 349-362.

Hawking S.W. (1979). The path-integral approach to quantum gravity. In General Relativity - an Einstein centenary survey, eds. S.W. Hawking and W. Israel pp.746-789. Cambridge, Cambridge University Press.

Hempel J. (1976). 3-Manifolds. Princeton University Press.

Isham C.J. (1981a). Topological θ -sectors in canonically quantized gravity. Phys. Letts. B. 106, 188-192.

Isham C.J. (1981b). Vacuum tunnelling in static spacetimes. In essays in honour of Wolfgang Yourgrau, ed. A. van der Merwe : Plenum Press, to appear.

Isham C.J. & Kunstatter G. (1981a). Yang-Mills canonical vacuum structure in a general three-space. Phys. Letts. B102, 417-420.

Isham C.J. & Kunstatter G. (1981b). Spatial topology and Yang-Mills vacua. To appear in Journ. Math. Phys.

Jackiw R. & Rebbi C. (1976). Vacuum periodicity in a Yang-Mills quantum theory. Phys. Rev. Letts. 37, 172-175.

Jackiw R. (1980). Introduction to the Yang-Mills quantum theory. Rev. Mod. Phys. 52, 661.

Kostant B. (1970). Quantization and unitary representations. In lecture note in mathematics. Vol. 170 New York: Springer Verlag.

Kuchar K. (1981). Canonical methods of quantization. In Quantum Gravity - a second Oxford symposium. C.J. Isham, R. Penrose and D.W. Sciama. Oxford; Oxford University Press.

Laidlaw M. & DeWitt C. (1971). Feynman functional integrals for systems of indistinguishable particles. Phys. Rev. D3, 1375-1379.

Milnor J. (1957). On the existence of a connection with curvature zero. Comm. Math. Helv. 32, 215-223.

Mitter P.K. & Viallet C.M. (1981). On the bundle of connections and the gauge orbit manifold in Yang-Mills theory. Comm. Math, Phys. 79, 457-472.

Narasimhan M.S. & Ramadas T.R. (1979). Geometry of SU2 gauge fields. Comm. Math. Phys. 67, 121-136.

Palais, R.S. (1966). Homotopy theory of infinite dimensional manifolds. Topology 5, 1-16.

Palais, R.S. (1968) . Foundations of global non-linear analysis. New York: Benjamin.

de Sa E.C. & Rourke C. (1979). The homotopy type of homeomorphisms of 3-manifolds. Bull. Amer. Math. Soc. 1, 251-254.

Schulman L.S. (1971). Approximate topologies. Jour. Math. Phys. 12,304-308.

Singer I. (1978). Some remarks on the Gribov ambiguity. Comm. Math. Phys. 60, 7-12.

Steenrod N. (1951). The topology of fibre bundles. Princeton, New Jersey; Princeton University Press.

Waldhausen F. (1968). On irreducible 3-manifolds which are sufficiently large. Ann. Math. 87, 56-88.

Wheeler J.A. (1965). In Relativity, groups and topology, eds. C. De Witt and B.S. DeWitt, pp.317-522. London: Blackie and Son Ltd.

Wheeler J.A. (1968). In Battelle rencontres 1967. eds. C. DeWitt and J.A. Wheeler, New York, Benjamin.

York J.W. (1972). Role of conformal three geometry in the dynamics of gravitation. Phys. Rev. Lett. 28, 1082-1085.

# STRONG COUPLING QUANTUM GRAVITY: AN INTRODUCTION

Martin Pilati
Imperial College
The Blackett Laboratory
London SW7 2BZ

## 1. INTRODUCTION

In recent years strong coupling quantum field theory has excited much interest. The problem that has been of most interest is that of strong coupling Yang-Mills as applied to the strong interactions. In this paper a strong coupling limit of general relativity will be considered.

In the covariant approach to quantum gravity the quantity $Gp^2$ is an important dimensionless quantity. When this quantity is large the corresponding distance scale is small. Strong coupling corresponds to short distances, and in quantizing the strong coupling limit we are investigating the quantum behaviour of the gravitational field at distances smaller than the Planck length. The weak coupling, covariant approach applies to large distances, and breaks down at short distances. By studying the strong coupling limit we are looking at an aspect of the quantized gravitational field complimentary to that normally considered. In studying strong coupling gravity we take the strong coupling limit of general relativity and not of one of the numerous modified theories of gravity that claim to be appropriate for short distances.

The most fruitful approach for strong coupling Yang-Mills has been to put the theory on a lattice. This is not appropriate for the strong coupling limit of general relativity. For Yang-Mills strong coupling corresponds to the large distance behaviour of the theory; so the particular form of the lattice should not affect the results. The opposite is true of general relativity. Another fact that makes lattice gauge theory possible is the compactness of Yang-Mills gauge groups. The gauge group of general relativity, the diffeomorphism group, is not compact. Finally, forcing gravity onto a lattice destroys the gauge invariance since the diffeomorphism group represents a non-internal gauge symmetry. Strong coupling gravity must be quantized in the continuum.

General relativity is a theory of a self-interacting gauge field. As a result, the strong coupling limit is expected to be a non-linear field theory. Our desire is to find an exact quantization of this theory. A perturbation expansion in 1/G will then be made about this exact solution. Exact quantizations of non-linear field theories are rather rare, but the strong coupling limit of general relativity has a property that makes an exact solution possible. The dynamics of this limit are such that the fields at separate spatial points are uncoupled. Information cannot propagate spatially, i.e. the limit is ultralocal. The techniques for quantizing self-coupled ultralocal field theories have been developed by Klauder (1970a,b,c, 1971,1973), and here we apply his methods to general relativity.

We start the process of quantization with the Hamiltonian formulation of general relativity. The canonical variables are the spatial metric $g_{ij}$ $(i,j = 1,2,3)$ and its conjugate $\pi^{ij}$ with the Hamiltonian being given by

$$H = \int d^3x \, (N(x)\,\mathcal{H}_\perp(x) + N^i(x)\,\mathcal{H}_i(x)) \;, \tag{1.1}$$

$$\mathcal{H}_\perp = \kappa g^{-\frac{1}{2}} G_{ijk\ell}\pi^{ij}\,\pi^{k\ell} - \kappa^{-1}\, g^{\frac{1}{2}}R = 0, \tag{1.2}$$

$$\mathcal{H}_i = -2\,\pi^j_{i|j} = 0, \tag{1.3}$$

where $\kappa = \dfrac{16\pi G}{c^3}$, $|$ denotes covariant derivative in $g_{ij}$, $R$ is the three dimensional curvature scalar, and

$$G_{ijk\ell} = \tfrac{1}{2}\,(g_{ik}g_{j\ell} + g_{i\ell}g_{jk} - g_{ij}\,g_{k\ell}). \tag{1.4}$$

The generators $\mathcal{H}_\perp$ and $\mathcal{H}_i$ satisfy the bracket relations

$$\begin{aligned}
\{\mathcal{H}_\perp(x)\,,\mathcal{H}_\perp(x')\} &= (g^{rs}(x)\mathcal{H}_s(x) + g^{rs}(x')\mathcal{H}_s(x'))\,\delta_{,r}(x,x')\;,\\
\{\mathcal{H}_\perp(x)\,,\mathcal{H}_j(x')\} &= \mathcal{H}_\perp(x')\;\delta_{,j}(x,x')\;,\\
\{\mathcal{H}_i(x),\mathcal{H}_j(x')\} &= \mathcal{H}_i(x')\;\delta_{,j}(x,x') - \mathcal{H}_j(x)\;\delta_{,i}(x',x).
\end{aligned} \tag{1.5}$$

We take the $G \to \infty$ limit of the equations of motion and ask for the Hamiltonian that generates them. The answer is essentially the same as (1.1), the only change being the replacement of $\mathcal{H}_\perp$(1.2) by

$$\mathcal{H}_o = g^{-\frac{1}{2}} G_{ijk\ell} \pi^{ij} \pi^{k\ell} = 0. \tag{1.6}$$

The brackets of $\mathcal{H}_o$ and $\mathcal{H}_i$ are the same as in (1.5) except that

$$\{\mathcal{H}_o(x), \mathcal{H}_o(x')\} = 0. \tag{1.7}$$

The dynamics of the full theory are generated by $\mathcal{H}_\perp$(1.2). Notice that in passing from $\mathcal{H}_\perp$ to $\mathcal{H}_o$ the spatial derivatives in $\mathcal{H}_\perp$ have been dropped. The terms that dynamically couple different spatial points have been eliminated. The light cones have collapsed to lines. This is what is meant by "ultralocal". The combined actions of $\mathcal{H}_o$ and $\mathcal{H}_i$ determine the type of four dimensional manifold (Teitelboim, 1973) that is associated with the strong coupling theory. In particular the bracket (1.7) implies that the four dimensional manifold is not Riemannian; there is no four dimensional metric. The geometry for the ultralocal theory has been worked out in detail (Henneaux, 1979a) and is found to be that of a manifold with degenerate metric (not surprisingly, since by taking the limit the timelike directions have been squeezed into null-lines). The Lagrangian associated with the strong coupling Hamiltonian has been shown to be generally covariant (Henneaux, 1979a).

The dynamics of the strong coupling theory are generated by $\mathcal{H}_o$ while $\mathcal{H}_i$ generates coordinate transformations on three dimensional spacelike hypersurfaces, i.e. gauge transformations. The coupling constant $G$ does not appear in either (1.3) or the brackets (1.5) involving $\mathcal{H}_i$; so $\mathcal{H}_i$ is unaffected by taking the $G \to \infty$ limit. This is what one expects since $\mathcal{H}_i$ is concerned with the coordinate transformation properties of fields defined on spacelike surfaces. This should not involve the coupling constant. Our goal is to, eventually, quantize the Hamiltonian with dynamics generated by $\mathcal{H}_o$ and gauge invariance generated by $\mathcal{H}_i$, and then to include as a perturbation the $g^{\frac{1}{2}}R$ term dropped from (1.2). This program for quantizing the gravitational field was first proposed by Isham (1976). For a general review of canonical quantization see (Kuchař, 1981).

The development given in this paper of the above quantization scheme will, to a large extent, be purely formal. At this stage the appropriate physical questions to ask of this theory are not known, but it is hoped that as the perturbation theory is developed its physical content will become clearer.

In this paper the label x will refer to a point in a coordinate patch on a compact, three dimensional manifold. By making this restriction the inclusion of various surface terms in the Hamiltonian is avoided. It is also true that the treatments of gauge invariance and the concepts of time for the compact and non-compact cases are quite different. Notice that the choice of compact manifold precludes the use of standard S-matrix theory. This will cause us no problem.

## 2. THE METRIC AS A QUANTUM OPERATOR

The metric $g_{ij}$, classically, has signature (+,+,+), i.e. it is not an arbitrary field. When we quantize, $g_{ij}$ becomes an operator and this classical restriction must be incorporated into the quantum theory.

Most of the work in quantum gravity has been in a weak coupling perturbation theory in which perturbations are made about a fixed, unquantized, background metric of the proper signature. In this case one might feel justified in ignoring any subtleties with the restriction on $g_{ij}$. This is not the sort of perturbation theory that we want to do. Rather than butcher the metric into quantized and unquantized parts we wish to quantize it whole. The classical restriction on the signature is incorporated into the quantum theory by requiring that the operators $g_{ij}$ have positive definite spectrum.

Usually one requires that the canonically conjugate operators $g_{ij}$, $\pi^{ij}$ be self-adjoint. This cannot be done while maintaining the above restriction on the spectrum of $g_{ij}$ (Klauder 1970b, Isham 1976). Following Klauder we will represent the variables $\pi^i_j$ (classically given by $g_{j\ell}\,\pi^{\ell i}$) and $g_{ij}$ as self-adjoint. These variables satisfy the commutation relations

$$\left[g_{ij}(x),\ \pi^k_\ell(x')\right] = \frac{i}{2}\ (g_{\ell i}\ \ \delta^k_j + g_{j\ell}\delta^k_i)\delta(x,x'), \tag{2.1}$$

$$\left[\pi^i_j(x), \pi^k_\ell(x')\right] = \frac{i}{2}\ (\ \pi^i_\ell\ \delta^k_j - \pi^k_j\ \ \delta^i_\ell)\ \delta(x,x'). \tag{2.2}$$

To see why this choice of variables is consistent with the restriction on $g_{ij}$ compute

$$e^{i\int\lambda_k^{\ell}(x')\,\pi_{\ell}^{k}(x')dx'}\, g_{ij}(x)\, e^{-i\int\lambda_k^{\ell}(x')\,\pi_{\ell}^{k}(x')\,dx'} = e^{-\frac{1}{2}\lambda_{\cdot i}^{k}(x)}\, g_{k\ell}(x)\, e^{-\frac{1}{2}\lambda_{\cdot j}^{\ell}(x)}, \qquad (2.3)$$

where the $\lambda_{\cdot i}^{k}(x)$ are c-number elements of a suitable function space. It is not difficult to see from this that the action of $\pi_{\cdot j}^{i}$ on $g_{ij}$ respects the restriction on the spectrum (Pilati 1980,1981). Equation (2.3) is to be contrasted with the similar expression for the action of $\pi^{ij}$ on $g_{ij}$ i.e. $g_{ij} \to g_{ij} + \lambda_{ij}$ with $\lambda_{ij}$ some more or less arbitrary function.

In addition to being consistent with the positive definite spectrum of $g_{ij}$ the variables $\pi_{\cdot j}^{i}$ have the following attractive properties

(1) Their brackets (2.2) are those of the generators of GL(3, ℝ). This enables us to use known properties of this group and especially of the symmetric space SL(3,ℝ)/SO(3).

(2) They are Killing vectors for the DeWitt-Misner metric $G_{ijk\ell}$(1.4) that is naturally used as a metric for the configuration space (DeWitt 1967, Misner 1972).

(3) The Hamiltonian $\mathcal{H}_o$ takes a particularly simple form when expressed in terms of the $\pi_{\cdot j}^{i}$'s. This form enables us to see the close connection between $\mathcal{H}_o$ and the Casimir operator on SL(3, ℝ).

(4) The natural variables to use for any ultralocal quantum field theory (e.g. the scalar field) satisfy commutators analogous to (2.1) and (2.2) (Klauder 1971). Thus our desire to quantize the ultralocal limit of gravity is intimately connected with the special nature of the metric field.

These properties will be discussed more fully in the remainder of this paper.

Another way to guarantee the positivity of $g_{ij}$ is to use triads (i.e. $e_i^a$ satisfying $e_i^a e_{ja} = g_{ij}$, $g^{ij} e_{ia} e_{jb} = \delta_{ab}$) (Isham 1976). Here again, however, it is natural to use variables analogous to $\pi_{\cdot j}^{i}$. This approach has much to recommend it, and work along these lines is currently in progress (Isham 1981).

## 3. ULTRALOCAL QUANTIZATION

We want to find the exact quantum field theory appropriate to the ultralocal Hamiltonian (1.6). A quantum field theory is a representation on a Hilbert space of the commutation relations of the fields (see e.g. Araki 1960). For each set of commutation relations there are an infinite number of inequivalent representations. The type of representation that is applicable to ultralocal fields is the exponential representation (Klauder 1970a,b,c, 1971, 1973). Rather than give a detailed derivation of the properties of this type of representation we will content ourselves with a statement of the results.

The most important feature of a theory with ultralocal dynamics is the statistical independence of the fields evaluated at different spatial points. For a scalar field one thus expects the field $\phi(x)$ to in some sense reduce to a parameter $\lambda$. In addition higher powers, $\phi^p(x)$, of the field should be represented by $\lambda^p$.

We take a Fock space; i.e. a fiducial vector $|0\rangle$ annihilated by operators $A(x,\lambda)$ with the Hilbert space spanned by the vectors resulting from the action of powers of $A^\dagger$ on $|0\rangle$. The operators $A$, $A^\dagger$ satisfy the commutation relations

$$[A(x,\lambda),\ A^\dagger(x',\lambda')] = \delta(x,x')\ \delta(\lambda,\lambda'). \tag{3.1}$$

We consider the operators

$$B(x,\lambda) = A(x,\lambda) + C(\lambda) \quad , \tag{3.2}$$

where $C(\lambda)$ is some function of $\lambda$, and form the (non-Fock) representation

$$\phi(x) = \int d\lambda\ B^\dagger(x,\lambda)\ \lambda B(x,\lambda) \tag{3.3}$$

for the field $\phi(x)$. The close relation between $\lambda$ and $\phi$ is obvious from (3.3). To define $\phi^2(x)$ first look at

$$\phi(x)\phi(x') = \delta(x,x')\int d\lambda B^\dagger(x,\lambda)\lambda^2 B(x,\lambda) + \int d\lambda d\lambda' B^\dagger(x,\lambda)B^\dagger(x',\lambda')\lambda\lambda' B(x,\lambda)B(x',\lambda')$$

and take $x \to x'$. Keeping the most singular part we naturally define

$$(\phi^2(x))_r = \int d\lambda B^\dagger(x,\lambda)\ \lambda^2 B(x,\lambda), \tag{3.4}$$

where $(\phi^2(x))_r = Z^{-1}\phi^2(x)$, $Z=\delta(o)$. Thus $\phi^2$ is represented by $\lambda^2$. A Fock representation for $\phi$ would not have this property. This argument can be extended to higher powers of the field, and it can be made rigorous within the context of the operator product expansion (Hegerfeldt and Klauder 1974). That this representation is the appropriate one for ultralocal fields is derived from first principles in the references quoted above.

It is natural to represent $\pi$ by

$$\pi = -i\int d\lambda B^{\dagger} \frac{\partial}{\partial\lambda} B, \tag{3.5}$$

and this is the best expression for it. The problem is that for the most desirable choices of $C(\lambda)$ (3.5) is not a self-adjoint operator. This will not concern us in this paper.

The Hamiltonian $\pi^2+U(\phi)$ is naturally represented by

$$\mathcal{H} = \int d\lambda B^{\dagger} h(\lambda) B, \tag{3.6}$$

with

$$h(\lambda) = -\frac{\partial^2}{\partial\lambda^2} + V(\lambda), \tag{3.7}$$

where $V(\lambda)$ is chosen such that

$$\mathcal{H}|0\rangle = 0. \tag{3.8}$$

This is the analog of normal ordering the harmonic oscillator Hamiltonian. Equation (3.8) is equivalent to $h(\lambda)\, C(\lambda) = 0$, and, with this, $\mathcal{H}$ is a self-adjoint operator. This provides a connection between $C(\lambda)$ and $V(\lambda)$ where $V(\lambda)$ is to differ from $U(\lambda)$ by renormalization terms. The quantized scalar field theory has been converted into a one dimensional problem in the space of $\lambda$.

To give the explicit expression for the inner product construct the overcomplete set of states (coherent states) satisfying

$$A|\psi\rangle = \psi(x,\lambda)|\psi\rangle \tag{3.9}$$

The innerproduct is given by

$$\langle\psi'|\psi\rangle = \exp\left(-\tfrac{1}{2}\,||\psi'||^2 - \tfrac{1}{2}\,||\psi||^2 + (\psi',\psi)\right) \tag{3.10}$$

where ( , ) is the innerproduct on the Hilbert space d of the eigenfunctions in (3.9). In this way to every vector in d is associated a vector in a complete set of state for the Hilbert space H of the field theory. The space d is the space of states associated to the finite dimensional Hamiltonian (3.7).

We wish to apply the above type of representation to the metric field; so we represent the metric operator as

$$g_{ij}(x) = \int dk_{mn}\ B^{\dagger}(x,k_{k\ell})\ k_{ij}B(x,k_{k\ell}), \tag{3.11}$$

where $k_{ij}$ is a symmetric, 3x3, positive definite matrix, and

$$\left[B(x,k_{ij}),\ B^{\dagger}(x',k'_{ij})\right] = \delta(x,x')\ \delta(k_{ij},k'_{ij}). \tag{3.12}$$

The construction of the Hilbert space of states is a straightforward generalization of that given above for the scalar field. The $\pi^i_{\cdot j}$ introduced in section 2 are represented by

$$\pi^i_{\cdot j} = -\frac{i}{2}\int dk_{mn}\ B^{\dagger}(\rho^{\dagger i}_{\cdot j} - \rho^i_{\cdot j})B \tag{3.13}$$

with

$$\rho^i_{\cdot j} = k_{jm}\ \frac{\partial}{\partial k_{mi}}\ . \tag{3.14}$$

The expression (3.13) is self-adjoint for a much wider class of $C(k_{ij})$ than is the similar expression for $\pi^{ij}$, where $C(k_{ij})$ is determined by

$$B|0\rangle = C(k_{ij})|0\rangle\ . \tag{3.15}$$

The manifold of the $k_{ij}$ will obviously play an important role in what follows.

We now digress a little to discuss the classical expression for $\mathcal{H}_o$. In terms of the variables $\pi^i_{\cdot j}$, $\mathcal{H}_o$ is given by

$$\mathcal{H}_o = g^{-\frac{1}{2}}(\pi^i_{\cdot j}\ \pi^j_{\cdot i} - \tfrac{1}{2}\ \pi^2) \tag{3.16}$$

where

$$\pi = \pi^i_{\cdot i}\ . \tag{3.17}$$

It has been found convenient (Misner 1972, Pilati 1980, Teitelboim 1980, 1981) to drop the $g^{-\frac{1}{2}}$ in (3.16) and we do this here. In addition we will rewrite (3.16) in terms of the variables

$$P^i_{\cdot j} = \pi^i_{\cdot j} - \frac{1}{3}\ \pi\ \delta^i_{\cdot j}, \tag{3.18}$$

which satisfy $P^i_{\cdot i}=0$ and

$$\left[P^i_{\cdot j}(x), P^k_{\cdot \ell}(x')\right] = \frac{i}{2}\ (P^i_{\cdot \ell}\ \delta^k_{\cdot j} - P^k\delta^i_{\cdot \ell}). \tag{3.19}$$

The $P^i_{\cdot j}$ correspond to generators of SL(3,R). We finally obtain the expression

$$\mathcal{H}_o = P^i_{\cdot j}P^j_{\cdot i} - \frac{1}{6}\pi^2. \tag{3.20}$$

Notice that $P^i_{.j} P^j_{.i}$ commutes with all of the generators $P^k_\ell$, and corresponds to the Casimir operator for SL(3,$\mathbb{R}$).

Now we turn to the space D of $k_{ij}$. The natural metric to put on this space (DeWitt 1967) is the analog of (1.4) i.e.

$$G_{ijk\ell} = \frac{1}{2}\,(k_{ik}k_{j\ell} + k_{i\ell}k_{jk} - k_{ij}k_{k\ell}) \qquad (3.21)$$

which has the signature (−+++++). The $\rho^i_{.j}$ (3.14) are vectors on D, and $\rho = \rho^i_{.i}$ is easily seen to be timelike. It can be checked that the $\rho^i_{.j}$ are Killing vectors for the metric (3.21).

The coordinate conjugate to $\rho$ is

$$\tau = \frac{1}{3}\,\ell n(\det k_{ij}) \qquad (3.22)$$

(i.e. $\rho = \frac{\partial}{\partial\tau}$) and, when expressed in the (nonholonomic) basis $\frac{\partial}{\partial\tau}, \frac{\partial}{\partial\tilde{k}_{ij}}$, with

$$\tilde{k}_{ij} = e^{-\tau} k_{ij} \,, \qquad (3.23)$$

the metric takes the form

$$\begin{pmatrix} -1/6 & 0 \\ 0 & \tilde{G}_{ijk\ell} \end{pmatrix} , \qquad (3.24)$$

where

$$\tilde{G}_{ijk\ell} = \frac{1}{2}(\tilde{k}_{ik}\,\tilde{k}_{j\ell} + \tilde{k}_{i\ell}\tilde{k}_{jk} - \frac{2}{3}\,\tilde{k}_{ij}\,\tilde{k}_{k\ell}). \qquad (3.25)$$

From (3.24) it follows that $\rho$ is a timelike, hypersurface orthogonal Killing vector (Kuchař 1981). The variable $\tau$ is called an intrinsic time.

The $\tau$, $\frac{\partial}{\partial\tau}$ part of the above basis has a trivial structure, but the space $\tilde{D}$ of the $\tilde{k}_{ij}$ (i.e. those $k_{ij}$ with determinant equal to 1) is much more interesting. It is the symmetric space SL(3,$\mathbb{R}$)/SO(3) (Helgason 1968) and the metric (3.25) is the natural SL(3,$\mathbb{R}$) invariant metric defined on $\tilde{D}$. This space will be discussed in more detail in the next section.

If we take

$$\tilde{\rho}^i_j = \rho^i_j - \frac{1}{3}\,\delta^i_j\,\rho \qquad (3.26)$$

then the expression for the operator $\mathcal{H}_o$ is given by

$$\mathcal{H}_o = \int dk_{mn}\, B^\dagger(\tilde{\rho}^{i\dagger}_{\cdot j}\, \tilde{\rho}^{j}_{\cdot i} - \frac{1}{6}\, \frac{\partial^2}{\partial\tau^2})B. \qquad (3.27)$$

The representation of the Casimir operation $\tilde{\rho}^{i\dagger}_{\cdot j}\ \tilde{\rho}^{j}_{\cdot i}$ on $\tilde{D}$ is just the Laplace-Beltrami operator for $\tilde{D}$ (Helgason 1962) i.e. $\mathcal{H}_o$ has the form

$$\mathcal{H}_o = \int dk_{mn} B^\dagger\ (\Delta_{\tilde{D}} - \frac{1}{6}\frac{\partial^2}{\partial\tau^2})B \qquad (3.28)$$

The expression (3.28) (and in particular $\Delta_{\tilde{D}} - \frac{1}{6}\ \frac{\partial^2}{\partial\tau^2}$ ) is what we will concentrate on for the remainder of the paper. We should state that (3.28) will normally require renormalization, but we will ignore this here.

Throughout this paper the gauge freedom in the theory has been and will be ignored. This means that the quantum theory as stated is not entirely correct. The gauge modes will have to be cancelled through some "probability eating" mechanism when the full quantum theory is finally obtained. For weak coupling Yang-Mills this is done by introducing ghosts. In that case one is helped by the fact that for zero coupling the theory is a set of uncoupled vector gauge fields for which a gauge condition exists that does not require ghosts. The ghosts then appear in an essential way only in perturbation theory. An analog of this has not yet been found for strong coupling gravity although there has been some speculation (Pilati 1981). Here we will write purely formal expressions that contain all of the components of the metric. These expressions and the techniques that accompany them will be an integral part of the complete theory.

## 4. DYNAMICS

The discussion of the previous section suggests the direction in which the quantum theory should be developed. The fact that $\rho$ is a timelike, hypersurface orthogonal Killing vector suggests working in terms of positive and negative (with respect to $\rho$) frequency states. The definition (3.22) of $\tau$ implies that positive frequency corresponds to a geometry that is expanding, and conversely for negative frequency (Misner 1972, Teitelboim 1980, 1981). We decompose solutions to the wave equation

$$(\Delta_{\tilde{D}} - \frac{1}{6}\ \frac{\partial^2}{\partial\tau^2})\psi\ (\tilde{k}_{ij},\tau) = 0 \qquad (4.1)$$

into positive and negative frequency parts. The equation (4.1) represents propagation on the light cone of the metric $G_{ijk\ell}$ on D. It is also the equation governing the quantum cosmological model for a Kasner type universe (Misner 1972).

Our eventual goal is to construct a scattering theory in which the solutions to (4.1), which are elements of some Hilbert space, represent the asymptotic states. For such a scattering theory we need a label of time such that, in this time, states evolve from $-\infty$ to $+\infty$. This clearly cannot be $\tau$ since we know classically that expanding universes can turn around and contract, i.e. $\tau$ goes from $+\infty$ to $-\infty$.

The usual Hilbert space for an equation of Klein-Gordon type has an innerproduct which integrates over the spatial variables (in this case $\tilde{k}_{ij}$) the time component of the Klein-Gordon current. In expressions (3.11), (3.13), (3.27) and (3.28) we have integrated over $\tau$ as well as $\tilde{k}_{ij}$, and it is difficult to avoid this (Pilati 1981).

A way to solve these two problems was given for the Klein-Gordon equation by Stückelberg (1941). It involves working with the equation

$$(\Delta_{\tilde{D}} - \frac{1}{6}\frac{\partial^2}{\partial\tau^2})\psi = -i\frac{\partial}{\partial\theta}\psi \tag{4.2}$$

which gives the evolution in $\theta$. The Hilbert space consists of square integrable functions of $\tilde{k}_{ij}$ and $\tau$. The results are equivalent to those of the standard treatment with the Klein-Gordon innerproduct. If we analytically continue $\theta \to -i\theta$ and $\tau \to -i\tau$ (4.2) becomes the heat equation

$$(\Delta_{\tilde{D}} + \frac{1}{6}\frac{\partial^2}{\partial\tau^2})\psi = \frac{\partial}{\partial\theta}\psi. \tag{4.3}$$

A path integral formulation of (4.2) that gives an equivalent, but more physical, formulation of the results of heat equation techniques can be found in (Teitelboim 1980,1981).

To use the wave equation (4.3) for further study we need expressions for plane waves and for the Fourier decomposition in terms of them on D. The $\tau$ part of the decomposition is trivial; so we will concentrate on $\tilde{D}$. The Riemannian manifold $\tilde{D}$ is curved and, in general, Fourier analysis cannot be done on curved spaces. However, it can be done on symmetric spaces. To proceed further we need a short summary of the geometry of the symmetric space $\tilde{D}$. The discussion given here is a drastically abbreviated version of that given in (Helgason 1968). Note that the plane wave states (propagating on D, not on spacetime) will be the "free" (asymptotic) states in the perturbation theory.

The space $\tilde{D}$ of 3x3, symmetric, positive definite matrice of determinant 1 is acted upon by the group SL(3,R), the action being given by

$$\gamma \to Q\, \gamma Q^t; \quad Q \epsilon SL(3,\mathbb{R}),\ \gamma \epsilon \tilde{D}. \tag{4.4}$$

The group acts transitively on $\tilde{D}$ and, since SO(3) is the maximal subgroup that leaves the unit element $(\delta_{ij})$ of $\tilde{D}$ invariant, $\tilde{D}$ is given by $SL(3,\mathbb{R})/SO(3)$.

The Lie algebra $sl(3,\mathbb{R})$ of $SL(3,\mathbb{R})$ is given by the 3x3, traceless matrices. If $m \epsilon sl(3,\mathbb{R})$ then it has a unique decomposition of the form

$$m = k+h+\ell \tag{4.5}$$

where $k \,\epsilon\, so(3)$,

$h \,\epsilon\, a :=$ the set of matrices of the form $\begin{pmatrix} r_1 & o & o \\ o & r_2 & o \\ o & o & -(r_1+r_2) \end{pmatrix}$,

$\ell \,\epsilon\, n :=$ the set of matrices of the form $\begin{pmatrix} o & n_1 & n_2 \\ o & o & n_3 \\ o & o & o \end{pmatrix}$.

The Iwasawa decomposition states that each element Q of $SL(3,\mathbb{R})$ can be uniquely decomposed in the form

$$Q = KAN, \tag{4.6}$$

$K \epsilon SO(3)$, $A \epsilon e^a$, and $N \epsilon e^n$. The group $e^a$ is a maximal abelian subgroup of $SL(3,\mathbb{R})$ and $e^n$ is a unipotent subgroup. All points of $\tilde{D}$ can be written in the form

$$\gamma = Q\, \mathbb{1}\, Q = N\, A\, A^t N^t. \tag{4.7}$$

Additionally, we introduce the following subgroup of SO(3)

$$M = \{k \epsilon\, SO(3) \mid Ad(k)\, h = h \ \text{ for all } h \epsilon a\}, \tag{4.8}$$

i.e. the centralizer of a in SO(3). This group is discrete and has four elements (Pilati 1981). Finally we note that the Cartan form for SL(3,$\mathbb{R}$) induces an SL(3,$\mathbb{R}$) invariant metric on $\tilde{D}$, and, as previously mentioned, this metric is (3.25).

To describe a plane wave one needs to have the notion of the direction in which it is travelling. In $\mathbb{R}^2$ the directions in which plane waves travel can be labelled by elements of SO(2), the boundary of $\mathbb{R}^2$. For $\tilde{D}$ (which is noncompact) the boundary B can similarly be taken as SO(3)/M (Helgason 1968), and the elements of B label the directions of the plane waves.

One also needs to know what a plane is. The action of SO(3) on $\tilde{D}$ generates spheres. These spheres should approach planes as the radius becomes infinite. The planes so obtained are generated by the action of groups conjugate to the group $e^n$ and are called horocycles. Thus a plane wave is given by an eigenfunction of $\Delta_{\tilde{D}}$ that is constant on one of these planes. The horocycle given by the action of the group $e^n$ on the origin is denoted by $\xi_o$. Any horocycle $\xi$ can be written as

$$\xi = kh \cdot \xi_o \tag{4.9}$$

where $h \varepsilon e^a$ and the coset $kM \varepsilon SO(3)/M$ are unique. The element b of the boundary corresponding to k is called the normal to the horocycle and h the complex distance from the origin to $\xi$. Given a point $\gamma \varepsilon D$ and a direction b there is a unique horocycle normal to b passing through $\gamma$. Denote by $H(\gamma,b)$ the element of a such that $be^{H(\gamma,b)} \cdot \xi_o$ gives this horocycle.

A plane wave travelling in the direction b is

$$e^{\lambda_c(H(\gamma,b))} \tag{4.10}$$

where $\lambda_c : a \to \mathbf{C}$ is linear. Given an arbitrary function $f(\gamma)$ of compact support the following Fourier transform can be made

$$\hat{f}(\lambda,b) = \int_{\tilde{D}} f(\gamma)\, e^{(-i\lambda+\rho')(H(\gamma,b))}\, d\gamma, \tag{4.11}$$

where $\lambda \varepsilon a^*$ the dual of a, and $\rho'$ is a specific element of $a^*$. The expansion of $f(\gamma)$ in terms of plane waves is

$$f(\gamma) = \int_{a^*} \int_B \hat{f}(\lambda,b)\; e^{(i\lambda+\rho')(H(\gamma,b))} |c(\lambda)|^{-2}\, d\lambda db, \tag{4.12}$$

where $c(\lambda)$ is a known expression in terms of gamma functions.

To complete the prerequisites for perturbation theory we need the propagator for functions on $\tilde{D}$ (again, we ignore the simple $\tau$ dependence). Using the plane wave decomposition (4.12) and the fact that plane waves are constant on horocycles we need only know how to propagate functions that are constant on horocycles. This problem has been solved by Karpelevič (1967) by finding the heat kernel solution to (4.3). The heat kernel is

$$P(\theta,H_1(\gamma_1,b),\ H_2(\gamma_2,b)) = \frac{1}{4\pi\theta}\exp-\frac{(H_1-H_2-2\theta\rho')^2}{4\theta} \quad . \tag{4.13}$$

Expressions applicable to $\tilde{D}$ are obtained by integrating $\theta$ from 0 to $\infty$.

The above provides the skeleton around which the perturbation theory in 1/G will be constructed. The ultralocal techniques of Klauder show us that the quantum field theory can be reformulated as a finite dimensional theory on D. Amazingly, the Hamiltonian $\mathcal{H}_0$ reduces to the Laplace-Beltrami operator on $\tilde{D}$. Free states are identified with plane waves on D, and the special geometrical structure of D allows explicit expressions for the plane waves to be determined. There is also an explicit expression for the propagators.

What we lack now is a method for doing the perturbation theory about the ultralocal limit. There is no reason in principle why we cannot establish such a perturbation theory and work on this is currently in progress.

## 5. DISCUSSION

The classical ultralocal theory along with the perturbations about it are known as velocity dominated cosmology (Belinsky et al. 1970, Eardley et al 1972, Henneaux 1979b). The solution to the ultralocal theory has the behaviour of an independent Kasner universe at each spatial point. Velocity dominated cosmology has been particularly useful in studying the behaviour of the gravitational field near singularities (as one might expect of a strong coupling theory), because, typically, it is near singularities that the "$\pi^2$" in $\mathcal{H}_\perp$ dominates "R". Similarly the plane wave solutions of the quantum theory as given above can be considered as independent quantum Kasner models (Misner 1972) at each spatial point. The solution approaches the singularity as the intrinsic time $\tau \to -\infty$. We would consider incoming, asymptotic states (positive frequency) to be coming out of the singularity, and interaction with the potential R can either scatter them into other positive frequency states or into negative frequency states. Even classically there is mixing between positive and negative frequencies, i.e. expanding solutions can stop and then contract.

As was mentioned before, strong coupling gravity corresponds to short distances. The discussion of the previous paragraph was about the connection between strong coupling and cosmology. The reconciliation of these contradictory aspects of strong coupling gravity will be crucial to the full understanding of the formalism that we are developing. At present we are not even close to obtaining this reconciliation although there is one suggestive result from earlier studies of quantum cosmological models. This is the "quantum puff" solution (Misner 1972) which consists of a cosmology that expands from the initial singularity to the Planck volume and then contracts. Is this a unit of "spacetime foam "?

We do not wish to imply that this approach to quantum gravity will solve all (or any) of the problems associated with the standard covariant methods. These covariant methods describe large distances and run into trouble at the Planck length. We might similarly expect the strong coupling approach to be good at short distance, but breakdown at large scales. That this might be true can be seen from the classical solution known as the Gowdy $T^3$(Gowdy 1974). This starts off as a velocity dominated solution, i.e. its initial state is one that might be described by one of our asymptotic states, but it continues expanding indefinitely. Its final state is one with gravitational radiation on a fixed background, a state in which "$\pi^2$" no longer dominates "R" and ultralocality does not apply.

The primary fruitfulness of the strong coupling approach to quantized general relativity will be the understanding it gives of gravity at short distances, and not its viability as a complete quantum theory of gravity. The formalism presented in this paper indicates that the behaviour of gravity at short distances will be simple and elegant. We hope that this view of the calm inside the wild, boiling structure that exists at the Planck length will contribute substantially toward the attainment of a complete theory of the "quantum structure of spacetime".

ACKNOWLEDGEMENTS.

The author would like to thank R. Beig, M. Henneaux, C. Isham, J. Klauder, K. Kuchař, and C. Teitelboim for helpful discussions.

REFERENCES.

Araki, H. (1960). Hamiltonian Formalism and the Canonical Commutation Relations in Quantum Field Theory. J. Math. Phys, 1, 492.

Belinsky, V.A., Khalatnikov, I.M. & Lifschitz, E.M. (1970). Oscillatory Approach to a Singular Point in the Relativistic Cosmology. Adv. Phys., 19, 525.

DeWitt, B.S. (1967). Quantum Theory Gravity. I. The Canonical Theory. Phys. Rev. 160, 1113.

Eardley, D., Liang, E. & Sachs, R. (1972). Velocity-Dominated Singularities in Irrotational Dust Cosmologies. J. Math. Phys. 13, 99.

Gowdy, R.H. (1974). Vacuum Spacetimes with Two-Barameter Spacelike Isometry Groups and Compact Invariant Hypersurfaces: Topologies and Boundary Conditions. Ann. Phys. (N.Y.) 83,203.

Hegerfeldt, G.C. & Klauder, J.R. (1974). Field Product Renormalization and the Wilson-Zimmerman Expansion in a Class of Model Field Theory. Il Nuovo Cimento, 19, 153.

Helgason, S. (1962). Differential Geometry and Symmetric Spaces, New York: Academic Press.

Helgason, S. (1968). Lie Groups and Symmetric Spaces. In Battelle Rencontres, eds. C.M. De Witt & J.A. Wheeler, pp.1-74. New York: Benjamin.

Henneaux, M. (1979a). Zero Hamiltonian Signature Spacetimes. Bull.Soc. Math. Belg. 31, 49.

Henneaux, M. (1979b). unpublished.

Isham, C.J. (1976). Some Quantum Field Theory Aspects of the Superspace Quantization of General Relativity. Proc. Roy. Soc. Lond. A. 351, 209.

Isham, C.J. (1981). Work in progress.

Karpelevič, F.I. (1967). The Geometry of Geodesics and the Eigenfunctions of the Beltrami-Laplace Operator on Symmetric Spaces. In Transactions of the Moscow Mathematical Society - 1965, pp. 51-199. Providence, R.I.: American Mathematical Society.

Klauder, J.R. (1970a). Exponential Hilbert Space: Fock Space Revisited. J. Math. Phys., 11, 609.

Klauder, J.R. (1970b). Soluble Models of Quantum Gravitation. In Relativity, eds. M.S. Carmeli, S.I. Fickler & L. Witten, pp. 1-17. New York: Plenum Press.

Klauder, J.R. (1970c). Ultralocal Scalar Field Models. Comm. Math. Phys., 18, 307.

Klauder, J.R. (1971). Ultralocal Quantum Field Theory. Acta. Phys. Austr., Suppl. VIII, 227.

Klauder, J.R. (1973). Functional Techniques and Their Application in Quantum Field Theory. In Mathematical Methods in Theoretical Physics, Vol. 14B, ed. W.E. Brittin, pp.329-421. Boulder Colorado: Colorado Associated University Press.

Kuchař, K. (1981). Canonical Methods of Quantization. In Quantum Gravity - A second Oxford Symposium,eds: C.J. Isham, R. Penrose & D.W. Sciama. Oxford: Oxford University Press.

Misner, C.W. (1972). Minisuperspace. In Magic without Magic, ed. J.R. Klauder, pp. 441-473. San Francisco: W.H. Freeman.

Pilati, M. (1980). Ph.D thesis. Princeton University, unpublished.

Pilati, M. (1981). Strong Coupling Quantum Gravity I: Solution in a Particular Gauge. Imperial College Preprint.

Stückelberg, E.C.G. (1941). La Mécanique du Point Matériel en Théorie de Relavité et en Théorie des Quantas. Helv. Phys. Acta 15, 23.

Teitelboim, C. (1973). How Commutators of Constraints Reflect the Spacetime Structure. Ann. Phys. (N.Y.), 79, 542.

Teitelboim, C. (1980). Proper Time Approach to the Quantization of the Gravitational Field. Phys. Letts., 96B, 77.

Teitelboim, C. (1981). Quantum Mechanics of the Gravitational Field. Univ. of Texas preprint.

# QUANTIZING FOURTH ORDER GRAVITY THEORIES

S.M. Christensen
Institute for Field Physics
Department of Physics and Astronomy
University of North Carolina
Chapel Hill, NC 27514, USA

## INTRODUCTION

Standard quantum gravity theory is based on a classical Lagrangian constructed from at most two derivatives of the metric tensor. It is well known that such theories are particularly difficult to study due in large part to the tedious calculations involved. Why then should one be interested in quantizing a theory with a Lagrangian made from as many as four metric derivatives? Won't such a theory be a technical mess? The answer is yes, but it also has been found to have many interesting aspects.

We motivate these calculations in several ways:

(1) Fourth order theories are usually renormalizable (Stelle 1977), but contain unphysical particles at the tree level. It is hoped that quantum corrections will eliminate these particles without sacrificing renormalizability.

(2) Recent interest in induced gravity and 1/N expansions in gravity [See for example Adler (1980); Tomboulis (1980); Hasslacher & Mottola (1981).] point toward a need to study fourth order gravity.

(3) One-loop calculations in fourth order theories were carried out several years ago (Julve & Tonin 1978; Salam & Strathdee 1978) and more complete results on this subject have been given in a very interesting recent paper by Fradkin & Tseytlin (1981). In a series of papers (Christensen 1981a; Christensen & Fulling 1982a,b; Christensen 1982; Christensen & York 1982) we plan to check and extend these results using somewhat different and more general methods.

(4) The fourth order theory provides an interesting physical example of a system that requires study of differential operators of the form

$$-[A^{\alpha\beta}]^{i}{}_{j}\nabla_{\alpha}\nabla_{\beta}+[B^{\alpha}]^{i}{}_{j}\nabla_{\alpha}+[C]^{i}{}_{j}$$

and

$$\Box^2\delta^i{}_j+[A^{\alpha\beta}]^i{}_j\nabla_\alpha\nabla_\beta+[B^\alpha]^i{}_j\nabla_\alpha+[C]^i{}_j ,$$

where $\nabla_\alpha$ is the usual covariant derivative and the abstract indices i,j can represent any combination of spinorial or tensorial indices.

The fourth order gravity theories provide a "laboratory" for investigating nearly every pathological feature that can occur in a quantum field theory without being hopelessly non-renormalizable. Here we will present an overview of the important features of the one-loop calculations involved in quantizing fourth order gravity.

## ACTIONS, FIELD EQUATIONS, PARTICLE CONTENT

For simplicity we will consider a compact manifold without boundaries. The metric on the manifold will have signature (+,+,+,+) and we use MTW (Misner et al 1973) Riemann tensor conventions. Our fourth order action will be

$$S=\int g^{1/2}[-\gamma(R-2\Lambda)+\alpha R^2+\beta R_{\mu\nu}R^{\mu\nu}]d^4x+\varepsilon\chi , \tag{1}$$

where $\alpha$, $\beta$, $\gamma$ and $\varepsilon$ are coupling constants ($\gamma>0$ always) and $\Lambda$ is the cosmological constant. We include a topological term built from the topological invariant Euler number

$$\chi=\frac{1}{32\pi^2}\int g^{1/2}(R_{\mu\nu\sigma\tau}R^{\mu\nu\sigma\tau}-4R_{\mu\nu}R^{\mu\nu}+R^2)d^4x . \tag{2}$$

This is done because $\chi$ counterterms are nearly always produced at the one-loop level. We must have a $\chi$ term in the original action or, since $\chi$ is not necessarily zero, the theory won't be renormalizable. In any case this $\chi$ term does not affect the field equations. The first functional derivative of $\chi$ is zero, that is,

$$\frac{\delta\chi}{\delta g_{\alpha\beta}}=0 . \tag{3}$$

The fourth order field equations are

$$\frac{1}{2}\alpha R^2 g^{\alpha\beta}+\frac{1}{2}\beta R_{\mu\nu}R^{\mu\nu}g^{\alpha\beta}-2\alpha RR^{\alpha\beta}-2\beta R_{\rho\tau}R^{\alpha\rho\beta\tau}-(2\alpha+\frac{1}{2}\beta)\Box Rg^{\alpha\beta}-\beta\Box R^{\alpha\beta}$$

$$+(2\alpha+\beta)R_{;}{}^{\alpha\beta}-\frac{1}{2}\gamma Rg^{\alpha\beta}+\gamma R^{\alpha\beta}+\gamma\Lambda g^{\alpha\beta}=0\ , \tag{4}$$

which when traced on $\alpha$ and $\beta$ give

$$-2(3\alpha+\beta)\Box R-\gamma R+4\gamma\Lambda=0\ . \tag{5}$$

If we linearize the theory we can obtain the particle content (Stelle 1977). The theory has eight total degrees of freedom. Two degrees represent a massless spin-2 particle, a graviton. Another degree is a spin-0 particle with mass

$$m_0 = \left\{\frac{-\gamma}{2(3\alpha+\beta)}\right\}^{1/2}\ . \tag{6}$$

The final five degrees of freedom are tied up in a spin-2 particle with mass

$$m_2= \left\{\frac{\gamma}{\beta}\right\}^{1/2}\ . \tag{7}$$

It is at this point that we run into one of the problems with fourth order theories. If $\beta>0$, then the massive spin-2 particle is a tachyon and if $\beta<0$, it is a negative definite linearized energy particle called a ghost. Similarly if $3\alpha+\beta>0$, the massive scalar is a real particle, but if $3\alpha+\beta<0$, the scalar is a tachyon. Finally, if either $\beta=0$ or $3\alpha+\beta=0$, the theory loses its nice renormalizability property. Our goal is to see if renormalization of the coupling constants will eliminate the ghost/tachyon problem.

## THE FUNCTIONAL INTEGRAL, GAUGE GHOSTS

The general form for the functional integral for the generating functional Z is given by (DeWitt 1965)

$$Z=N(\det\gamma_{\alpha\beta})^{1/2}\int e^{-(\frac{1}{2}\phi^i S_{,ij}\phi^j+\frac{1}{2}\phi^i P_i{}^\alpha\gamma_{\alpha\beta}P_j{}^\beta\phi^j)}\det F^\alpha{}_\beta d\phi^i\ , \tag{8}$$

where $\phi^i$ is some quantum field, N is a normalization constant and a comma represents functional differentiation with respect to $\phi^i$. The argument of the exponential will give one-loop results since we have written only terms quadratic in $\phi^i$.

If S is invariant under the gauge transformation

$$\delta\phi^i = Q^i_{\ \alpha}\delta\xi^\alpha , \tag{9}$$

where $Q^i_{\ \alpha}$ is in general some functional of the $\phi^i$'s and $\delta\xi^\alpha$ is an infinitesimal gauge group element, then $S_{,ij}$ is a singular operator. We remedy this by demanding

$$P_i^{\ \alpha}\phi^i = \zeta^\alpha , \tag{10}$$

which is implemented by adding a "gauge fixing term"

$$\frac{1}{2}\phi^i P_i^{\ \alpha}\gamma_{\alpha\beta}P_j^{\ \beta}\phi^j \tag{11}$$

in the exponential. The operator

$$F_{ij} = S_{,ij} + P_i^{\ \alpha}\gamma_{\alpha\beta}P_j^{\ \beta} \tag{12}$$

will be non-singular with appropriate choices of $P_i^{\ \alpha}$ and $\gamma_{\alpha\beta}$. We compensate for the breaking of gauge invariance with the determinants of $\gamma_{\alpha\beta}$ and $F^\alpha_{\ \beta}$, where

$$F^\alpha_{\ \beta} = P_i^{\ \alpha}Q^i_{\ \beta} . \tag{13}$$

Standard functional integration then gives

$$Z_{(1)} = N\frac{(\det\gamma_{\alpha\beta})^{1/2}[(\det F^\alpha_{\ \beta})^{1/2}]^2}{(\det F_{ij})^{1/2}} \tag{14}$$

for the one-loop part of the generating functional. The one-loop counterterms $\Delta W_{(1)}$ are defined to be the divergent parts of the one-loop effective action

$$W_{(1)} = \frac{1}{2}(\ln\det F_{ij} - \ln\det\gamma_{\alpha\beta} - 2\ln\det F^\alpha_{\ \beta}). \tag{15}$$

The $\ln \det F^{\alpha}{}_{\beta}$ term represents the two "gauge ghosts" while the $\ln \det\gamma_{\alpha\beta}$ is the "third ghost" contribution. [In usual second order gravity theory $\ln \det\gamma_{\alpha\beta}$ is a constant and is absorbed into a discarded normalization constant term. The gauge ghosts remain.]

## A SIMPLE FOURTH ORDER EXAMPLE

The action (1) is invariant under the infinitesimal gauge transformation

$$\delta g_{\alpha\beta}=\delta\xi_{\alpha;\beta}+\delta\xi_{\beta;\alpha}\ , \tag{16}$$

where $\delta\xi_{\alpha}$ is an infinitesimal vector. We will have to pick a gauge fixing term. We start this process by computing $\frac{1}{2}\phi^{i}S_{,ij}\phi^{j}$, where $\phi^{i}=h_{\alpha\beta}$ is a symmetric tensor quantum field propagating on a classical background manifold with mètric $g_{\alpha\beta}$. This is a very lengthy undertaking and we will not present the details here (Christensen 1981a). The final result (with all abstract indices written out) has over sixty terms. We can look at the simpler $R_{\mu\nu}=0$ (which forces $\Lambda=0$) case. We find that

$$\frac{1}{2}\phi^{i}S_{,ij}\phi^{j}=\frac{1}{2}\iint h^{\alpha\beta}\frac{\delta^{2}S}{\delta g_{\alpha\beta}\delta g_{\gamma'\delta'}}\,h^{\gamma'\delta'}d^{4}xd^{4}x'$$

$$=\frac{1}{2}\int g^{1/2}\Big\{\frac{1}{2}\beta h^{\alpha\beta}\Box^{2}h_{\alpha\beta}+\beta A^{\mu}\Box A_{\mu}+(2\alpha+\frac{1}{2}\beta)h\Box^{2}h-(4\alpha+\beta)\Box hA_{\mu;}{}^{\mu}$$

$$+(2\alpha+\beta)A_{\mu;}{}^{\mu}A_{\nu;}{}^{\nu}-\frac{1}{2}\gamma h^{\alpha\beta}\Box h_{\alpha\beta}-\gamma hA_{\mu;}{}^{\mu}-\gamma A^{\mu}A_{\mu}+\frac{1}{2}\gamma h\Box h$$

$$+2\beta h^{\alpha\beta}R_{\alpha}{}^{\mu}{}_{\beta}{}^{\nu}\Box h_{\mu\nu}-\gamma h^{\alpha\beta}R_{\alpha}{}^{\mu}{}_{\beta}{}^{\nu}h_{\mu\nu}+2\beta h^{\alpha\beta}R_{\alpha}{}^{\rho}{}_{\beta}{}^{\tau}R_{\rho}{}^{\mu}{}_{\tau}{}^{\nu}h_{\mu\nu}\Big\}\,d^{4}x\ , \tag{17}$$

where $h=h_{\alpha}{}^{\alpha}$, $A_{\mu}=h_{\mu\nu;}{}^{\nu}$ and primes on indices indicate evaluation at the point x' rather that the point x. The Riemann tensors and covariant derivatives in (17) are constructed from $g_{\alpha\beta}$.

There are many gauge fixing terms we could choose. However, there is one choice that will make calculations easier. Consider the choice

$$\frac{1}{2}\int g^{1/2}(A^{\mu}-\eta h_{;}{}^{\mu})(\xi_{1}\Box g_{\mu\nu}+\xi_{2}\nabla_{(\mu}\nabla_{\nu)}+\xi_{3}\gamma g_{\mu\nu})(A^{\nu}-\eta h_{;}{}^{\nu})d^{4}x, \tag{18}$$

where $\eta$, $\xi_1$, $\xi_2$ and $\xi_3$ are constants to be determined. Multiplying out the integrand in (18) we find that if we pick

$$\eta=1+\frac{\beta}{4\alpha}\ ,\quad \xi_1=-\beta\ ,\quad \xi_2=2\alpha+\beta \text{ and } \xi_3=1\ , \tag{19}$$

then we can diagonalize the $h_{\alpha\beta}$ terms. This means we only have $\bar{h}^{\alpha\beta}\Box^2\bar{h}_{\alpha\beta}$ ($\bar{h}_{\alpha\beta}=h_{\alpha\beta}-\frac{1}{2}hg_{\alpha\beta}$) and $h\Box^2 h$ in the four derivative terms. We will simplify the calculation even further by choosing the special case $\beta=-2\alpha$. This allows us to write the term

$$\frac{1}{2}\phi^i S_{,ij}\phi^j+\frac{1}{2}\phi^i P_i{}^\alpha \gamma_{\alpha\beta} P_j{}^\beta \phi^j$$

as

$$\frac{1}{2}\int g^{1/2}\Big\{\frac{1}{2}\beta\bar{h}^{\alpha\beta}(-\Box\delta_{\alpha\beta}{}^{\gamma\delta}-2R_\alpha{}^\gamma{}_\beta{}^\delta)(-\Box\delta_{\gamma\delta}{}^{\mu\nu}-2R_\gamma{}^\mu{}_\delta{}^\nu+\frac{\gamma}{\beta}\delta_{\gamma\delta}{}^{\mu\nu})\bar{h}_{\mu\nu}$$
$$+\frac{3}{8}\beta h(-\Box)(-\Box-\frac{\gamma}{\beta})h\Big\}d^4x\ . \tag{20}$$

In this special case it is easy to see that

$$\gamma_{\alpha\beta}=-\Box g_{\alpha\beta}+\frac{\gamma}{\beta}g_{\alpha\beta} \tag{21}$$

and

$$F^\alpha{}_\beta=-\Box\delta^\alpha{}_\beta\ . \tag{22}$$

Looking at (20)-(22) we see that the operators $-\Box$, $-\Box\delta^\alpha{}_\beta$ and $-\Box\delta_{\alpha\beta}{}^{\gamma\delta}-2R_\alpha{}^\gamma{}_\beta{}^\delta$ represent the massless graviton. The operator $-\Box-\frac{\gamma}{\beta}$ is the one for the massive scalar while $-\Box\delta_{\alpha\beta}{}^{\gamma\delta}-2R_\alpha{}^\gamma{}_\beta{}^\delta+\frac{\gamma}{\beta}\delta_{\alpha\beta}{}^{\gamma\delta}$ and $-\Box g_{\alpha\beta}+\frac{\gamma}{\beta}g_{\alpha\beta}$ go together to give the massive spin-2 contribution.

We note that each operator is of the form

$$-\Box\delta^i{}_j+X^i{}_j\ . \tag{23}$$

Using the usual algorithm (DeWitt 1965; Christensen 1975; Gilkey 1975) we find the one-loop counterterms via

$$\Delta W_{(1)}=\frac{1}{n-4}\Big\{A_2[\text{graviton}]+A_2[\text{massive spin-2}]$$
$$+A_2[\text{massive spin-0}]\Big\} \quad . \tag{24}$$

where

$$A_2[\cdots]=\int g^{1/2}\frac{1}{(4\pi)^2}\, a_2{}^i{}_i[\cdots]d^4x \tag{25}$$

and 1/(n-4) is the dimensional regularization parameter. The coefficients $a_2{}^i{}_j[\cdots]$ come from

$$a_2{}^i{}_j[\text{graviton}]=a_2{}^i{}_j[-\Box\delta_{\alpha\beta}{}^{\gamma\delta}-2R_\alpha{}^\gamma{}_\beta{}^\delta]+a_2{}^i{}_j[-\Box]-2a_2{}^i{}_j[-\Box\,\delta^\alpha{}_\beta],$$

$$a_2{}^i{}_j[\text{massive spin-2}]=a_2{}^i{}_j[-\Box\,\delta_{\alpha\beta}{}^{\gamma\delta}-2R_\alpha{}^\gamma{}_\beta{}^\delta+\frac{\gamma}{\beta}\delta_{\alpha\beta}{}^{\gamma\delta}]$$
$$-a_2{}^i{}_j[-\Box\,\delta^\alpha{}_\beta+\frac{\gamma}{\beta}\delta^\alpha{}_\beta],$$

$$a_2{}^i{}_j[\text{massive spin-0}]=a_2{}^i{}_j[-\Box-\frac{\gamma}{\beta}] \quad , \tag{26}$$

and the general formula

$$a_2{}^i{}_j[-\Box\,\delta^i{}_j+X^i{}_j]=\frac{1}{180}\Big\{(R_{\mu\nu\sigma\tau}R^{\mu\nu\sigma\tau}-R_{\mu\nu}R^{\mu\nu}+\frac{5}{2}R^2$$
$$+6\Box R)\delta^i{}_j-30RX^i{}_j+90X^i{}_kX^k{}_j$$
$$+15[Y^{\mu\nu}]^i{}_k[Y_{\mu\nu}]^k{}_j-30\Box X^i{}_j\Big\} \quad , \tag{27}$$

where $[Y_{\mu\nu}]^i{}_j$ is defined from the Ricci identity

$$[\nabla_\mu,\nabla_\nu]\phi^i=[Y_{\mu\nu}]^i{}_j\phi^j \quad . \tag{28}$$

[The -2 and -1 factors that appear in front of the $a_2{}^i{}_j$ terms in (26) are from (15). Gauge and third ghost contributions always appear with signs opposite to the $F_{ij}$ terms.]

Putting together the formulae (23)-(28) gives, after some algebra,

$$\Delta W_{(1)}=\frac{1}{n-4}\left\{\frac{413}{90}\chi+3\,\frac{\gamma^2}{\beta^2}\,V\right\}\;. \qquad (29)$$

where

$$V=\int g^{1/2}d^4x \qquad (30)$$

and $\chi$ is the Euler number. This result agrees with Fradkin & Tseytlin (1981). We see that the coupling constant $\varepsilon$ is renormalized. The V term is a constant that reintroduces a cosmological constant at the one-loop level.

## THE GENERAL CASE

If we want to relax the $R_{\mu\nu}=0$ and $\beta=-2\alpha$ restrictions of the previous example, we run into two problems. The third ghost operator (with $\eta=1+\beta/4\alpha$) is

$$Y_{\alpha\beta}=-\Box g_{\alpha\beta}+\frac{2\eta-1}{2\eta-2}\nabla_{(\alpha}\nabla_{\beta)}+\frac{1}{2}\,\frac{2\eta+1}{2\eta-2}R_{\alpha\beta}+\frac{\gamma}{\beta}g_{\alpha\beta}\;, \qquad (31)$$

the gauge ghost operator is

$$F_{\alpha\beta}=-\Box g_{\alpha\beta}+(2\eta-1)\nabla_{(\alpha}\nabla_{\beta)}-\frac{1}{2}(2\eta+1)R_{\alpha\beta} \qquad (32)$$

and the fourth order operator $F_{ij}$ does not split into the product of two second order operators, but has a very complicated form

$$\Box^2\delta_{\alpha\beta}{}^{\gamma\delta}+[A^{\mu\nu}]_{\alpha\beta}{}^{\gamma\delta}\nabla_\mu\nabla_\nu+[B^\mu]_{\alpha\beta}{}^{\gamma\delta}\nabla_\mu+[C]_{\alpha\beta}{}^{\gamma\delta}\;. \qquad (33)$$

The $[A^{\mu\nu}]$, $[B^\mu]$ and $[C]$ are constructed from the Riemann tensor and its derivatives. Clearly the usual $-\Box\delta^i{}_j+X^i{}_j$ one-loop algorithm will not work in these new cases.

Searching the literature we find very few useful discussions of the operators in (31)-(33). Methods for calculating counterterms in fourth order theories are also poorly developed. Gilkey (1980) does give many valuable clues however.

## DIMENSIONAL REGULARIZATION, FOURTH ORDER COUNTERTERMS

To determine the divergent parts of $W_{(1)}$ for a fourth order theory we proceed in the usual fashion. [See for example the work of DeWitt (1975; 1979).] The one-loop effective action $W_{(1)}$ can be written as an n-dimensional integral

$$W_{(1)}=\frac{1}{2}\int d^nxg^{1/2}\int_0^\infty \frac{1}{is}K^i{}_i(s)ds \tag{34}$$

where

$$K^i{}_i(s)=\delta^j{}_i \lim_{x'\to x} K^i{}_{j'}(s) \quad . \tag{35}$$

The bi-tensor/spinor $K^i{}_{j'}$ is the so-called heat kernel and satisfies

$$-i\,\frac{\partial}{\partial s}K^i{}_{j'}=F^i{}_kK^k{}_{j'}\ , \tag{36}$$

where $F^i{}_k$ is the operator in (33) say.

Without loss of generality we can consider

$$-i\frac{\partial}{\partial s}K=FK\ . \tag{37}$$

We have suppressed all indices and F can be any operator of even order of a form similar to (33). [See Gilkey (1980) for further details.] Let u=2v be the order of F. From Gilkey (1980) we conclude that

$$K(x,x',s)=-iE(x,x',s)\frac{1}{(4\pi)^{n/2}}\,e^{-i\mu^2 s}\sum_{k=0}^{\infty}a_{k,n,u}(x,x')(is)^{\frac{(2k-n)}{u}}\ . \tag{38}$$

E(x,x',s) is some bi-tensor/spinor with the property that

$$\lim_{x'\to x}E(x,x',s)=I\ . \tag{39}$$

In this case I is the identity tensor/spinor. In (38) the $a_{k,n,u}(x,x')$ coefficients must be determined by some recursion relation technique. The arbitrary "mass" $\mu$ is a renormalization device that "cuts off" the integral at the upper end. If we are concerned with ultraviolet divergences we take this mass to zero in the end to obtain the counter-terms.

From (34), (38) and

$$W_{(1)}=\int d^n x g^{1/2} L_{(1)} , \tag{40}$$

we obtain the effective Lagrangian

$$\begin{aligned} L_{(1)} &= -\frac{1}{2(4\pi)^{n/2}} \operatorname{tr} \int_0^\infty e^{-i\mu^2 s} \sum_{k=0}^{\infty} \lim_{x'\to x} a_{k,n,u}(x,x')(is)^{\frac{2k-n}{u}-1} ds \\ &= -\frac{1}{2(4\pi)^{n/2}} \sum_{k=0}^{\infty} \operatorname{tr} \lim_{x'\to x} a_{k,n,u}(x,x')(\mu^2)^{(n-k)/u} \Gamma\left(\frac{2k-n}{u}\right). \end{aligned} \tag{41}$$

The symbol tr means "trace over the suppressed indices on $a_{k,n,u}$". If we expand around n=4 and take $\mu\to 0$ we can isolate the divergent part of $L_{(1)}$

$$\Delta L_{(1)} = \frac{1}{(4\pi)^2} v \frac{1}{n-4} \operatorname{tr} \lim_{x'\to x} a_{2,4,u}(x,x') . \tag{42}$$

The one-loop counterterms are then

$$\Delta W_{(1)} = \frac{1}{(4\pi)^2} v \frac{1}{n-4} \int \operatorname{tr} \lim_{x'\to x} a_{2,4,u}(x,x') g^{1/2} d^4 x . \tag{43}$$

This formula is the same as (24) except for the factor v. For second order operators we have v=1, so the results of (24) and (43) match. We still must determine an algorithm for finding the

$$\operatorname{tr} \lim_{x'\to x} a_{2,4,u}(x,x')$$

and we shall investigate this next.

<u>FOURTH ORDER OPERATOR ALGORITHM</u>

The standard procedures for determining the coincidence limit

$$\lim_{x'\to x} a_{2,4,u}{}^{i}{}_{j'}$$

come from substituting (38) into (37) and obtaining recursion relations for the $a_{k,n,u}{}^{i}{}_{j}$'s. Unfortunately we do not know the properties of $E(x,x',s)$ well enough to do this. [This subject will be treated in Christensen & Fulling (1982a).] However, we can use the results given in Gilkey (1980). If we start with the operator

$$\Box^2 \delta^i{}_j + [A^{\mu\nu}]^i{}_j \nabla_\mu \nabla_\nu + [B^\mu]^i{}_j \nabla_\mu + [C]^i{}_j \ , \tag{44}$$

which is of order four, then we see that $[A^{\mu\nu}]^i{}_j$ is of order two, $[B^\mu]^i{}_j$ is of order three and $[C]^i{}_j$ has order four. From past experience we know that $\lim_{x'\to x} a_{2,4,u}(x,x')$ will be of fourth order and will be a linear combination of the fourth order invariants constructed from $R_{\mu\nu\sigma\tau}$, $[Y^{\mu\nu}]^i{}_j$, $[A^{\mu\nu}]^i{}_j$, $[B^\mu]^i{}_j$, $[C]^i{}_j$ and $\nabla_\mu$. So we must have

$$\begin{aligned}\lim_{x'\to x} a_{2,4,u}{}^{i}{}_{j}{}' = \frac{1}{180}\Big\{ & aR_{\mu\nu\sigma\tau}R^{\mu\nu\sigma\tau}\delta^i{}_j + bR_{\mu\nu}R^{\mu\nu}\delta^i{}_j + cR^2\delta^i{}_j \\ & + e[Y_{\mu\nu}]^i{}_k[Y^{\mu\nu}]^k{}_j + f[B^\mu]^i{}_{j;\mu} + g[C]^i{}_j \\ & + h[A^{\mu\nu}]^i{}_{j;\mu\nu} + i[A_\mu{}^\mu]^i{}_{j;\nu}{}^\nu + d\Box R\delta^i{}_j \\ & + j[A^{\mu\nu}]^i{}_k[A_{\mu\nu}]^k{}_j + kR_{\mu\nu}[A^{\mu\nu}]^i{}_j \\ & + \ell R[A_\mu{}^\mu]^i{}_j + m[A_\mu{}^\mu]^i{}_k[A_\nu{}^\nu]^k{}_j \Big\} \ .\end{aligned} \tag{45}$$

Some of the constants a, b,..., m can be determined easily. For example, consider two second order operators

$$-\Box\delta^i{}_j + X^i{}_j \quad \text{and} \quad -\Box\delta^i{}_j + Y^i{}_j \ , \tag{46}$$

where

$$X^i{}_{j;\mu} = Y^i{}_{j;\mu} = 0 \quad \text{and} \quad X^i{}_k Y^k{}_j = Y^i{}_k X^k{}_j \ . \tag{47}$$

If we multiply these two operators together we get a fourth order operator like (44) with

$$[A^{\mu\nu}]^i{}_j = -(X^i{}_j + Y^i{}_j)g^{\mu\nu} \ , \quad [B^\mu]^i{}_j = 0 \quad \text{and} \quad [C]^i{}_j = X^i{}_k Y^k{}_j \ . \tag{48}$$

Because this fourth order operator is formed as the product of two second order operators in (46) we have

$$a_{2,4,4}{}^{i}{}_{j}[\Box^{2}\delta^{i}{}_{j}+[A^{\mu\nu}]^{i}{}_{j}\nabla_{\mu}\nabla_{\nu}+[B^{\mu}]^{i}{}_{j}\nabla_{\mu}+[C]^{i}{}_{j}]$$
$$=\frac{1}{2}\left\{a_{2,4,2}{}^{i}{}_{j}[-\Box\delta^{i}{}_{j}+X^{i}{}_{j}]+a_{2,4,2}[-\Box\delta^{i}{}_{j}+Y^{i}{}_{j}]\right\} \quad . \tag{49}$$

Using (27), (45) and (49) we can match terms to obtain

$$a=1,\ b=-1,\ c=5/2,\ d=6 \text{ and } e=15, \tag{50}$$

as well as

$$g+8j+32m=0,$$
$$4j+16m=45$$

and

$$-k-4\ell=-15. \tag{51}$$

The first two equations in (51) simply

$$g=-90, \tag{52}$$

This method does not give j, k, $\ell$ or m. These have been found by Gilkey (1980) using other techniques. They are

$$j=15/4,\ k=15,\ \ell=-15/2 \text{ and } m=15/8\ . \tag{53}$$

Also we see that f, h and i are undetermined. As far as we know now, these constants can only be found with a heat kernel algorithm (Christensen & Fulling 1982a).

SECOND ORDER OPERATOR ALGORITHM

The operators (31) and (32) are of the general form

$$-[A^{\mu\nu}]^{i}{}_{j}\nabla_{\mu}\nabla_{\nu}+[B^{\mu}]^{i}{}_{j}\nabla_{\mu}+[C]^{i}{}_{j}\ . \tag{54}$$

Can we find a heat kernel algorithm for (54)? Consider the "simple" case where

$$[A^{\mu\nu}]^i{}_j = A^{\mu\nu}\delta^i{}_j \ . \tag{55}$$

We assume $A^{\mu\nu}$ is symmetric and has an inverse $A^{-1}_{\mu\nu}$. Now pretend that $A^{-1}_{\mu\nu}$ is a new metric on the manifold and define the "twiddled" Christoffel symbol

$$\tilde{\Gamma}^\alpha{}_{\beta\gamma} = \frac{1}{2}A^{\alpha\delta}(A^{-1}_{\beta\delta,\gamma} + A^{-1}_{\gamma\delta,\beta} - A^{-1}_{\beta\gamma,\delta}) \ . \tag{56}$$

Using (56) we can eliminate the $[B^\mu]$ term by defining a twiddled covariant derivative and rewriting (54) as

$$-A^{\mu\nu}\tilde{\nabla}_\mu\tilde{\nabla}_\nu\delta^i{}_j + \tilde{C}^i{}_j \ , \tag{57}$$

where $\tilde{C}^i{}_j$ is a complicated combination of $A^{\mu\nu}$, $[B^\mu]^i{}_j$, $[C]^i{}_j$ and their derivatives. We see that (57) is now of the form (23) and so we can use the usual algorithm (57) to find

$$a_2{}^i{}_j[-\tilde{\Box}\delta^i{}_j + \tilde{C}^i{}_j] = \frac{1}{180}\Big\{(\tilde{R}_{\mu\nu\sigma\tau}\tilde{R}^{\mu\nu\sigma\tau} - \tilde{R}_{\mu\nu}\tilde{R}^{\mu\nu} + \frac{5}{2}\tilde{R}^2 + 6\tilde{\Box}\tilde{R})\delta^i{}_j$$
$$-30\tilde{R}\tilde{C}^i{}_j + 90\tilde{C}^i{}_k\tilde{C}^k{}_j + 15[\tilde{Y}^{\mu\nu}]^i{}_k[\tilde{Y}_{\mu\nu}]^k{}_j - 30\tilde{\Box}\tilde{C}^i{}_j\Big\} \tag{58}$$

where $\tilde{R}_{\mu\nu\sigma\tau}$ is constructed from $\tilde{\Gamma}^\alpha{}_{\beta\gamma}$. To obtain $a_2{}^i{}_j$ in terms of "detwiddled" $R_{\mu\nu\sigma\tau}$'s, $A^{\mu\nu}$, $[B^\mu]^i{}_j$, etc. is annoyingly tedious since, for example,

$$\tilde{R}^\alpha{}_{\beta\gamma\delta} = R^\alpha{}_{\beta\gamma\delta} + \frac{1}{4}A^{-1}_{\beta\mu}A^{-1}_{\gamma\nu}A^{-1}_{\varepsilon\lambda}A^{\alpha\kappa}A^{\mu\varepsilon}{}_{;\delta}A^{\nu\lambda}{}_{;\kappa} + 27 \text{ other terms.} \tag{59}$$

Clearly producing $\tilde{R}_{\mu\nu\sigma\tau}\tilde{R}^{\mu\nu\sigma\tau}$ is a major undertaking. Details of this "simple" calculation will be presented elsewhere (Christensen 1981b).

As the discussion above indicates, an $a_2{}^i{}_j$ algorithm for the operator (54) will be difficult to obtain. There may be hope however. Looking at (31) and (32) we see that $[A^{\mu\nu}]^i{}_{j;\alpha}=0$ and $[B^\mu]^i{}_j=0$. This means that many terms that might normally appear in $a_2{}^i{}_j$ will be absent and a simple algorithm may exist. Some progress has been made on this and will be discussed in Christensen & Fulling (1982b). [See Fradkin & Tseytlin (1982) for some interesting comments on this problem.]

Once the fourth and second order algorithms are found we will be able to obtain the most general form for $\Delta W_{(1)}$. The renormalization of $\alpha$, $\beta$, $\gamma$ and $\varepsilon$ will then be known. An analysis of this should shed considerable light on whether or not quantum corrections will solve the ghost/tachyon problem. We plan to present our results on this in a future paper (Christensen 1982).

## BOUNDARIES

Finally we will look briefly at what happens if we allow manifolds M with boundaries $\partial M$. First we look at the situation in second order gravity. Beginning with the action

$$S=-\gamma\int_M g^{1/2} R d^4x , \qquad (60)$$

we can find its first variation under $g_{\alpha\beta}\to g_{\alpha\beta}+\delta g_{\alpha\beta}$

$$\delta S=-\gamma\int_M g^{1/2}(\tfrac{1}{2}Rg^{\alpha\beta}-R^{\alpha\beta})\,\delta g_{\alpha\beta}d^4x$$
$$+\gamma\int_{\partial M} g^{\alpha\beta}(n^{\rho}\nabla_{\rho}\delta g_{\alpha\beta})\tilde{\gamma}^{1/2}d^3x-\int_{\partial M}(Kn^{\alpha}n^{\beta}-K^{\alpha\beta}-n^{\alpha}a^{\beta})\,\delta g_{\alpha\beta}\tilde{\gamma}^{1/2}$$
$$\times\ d^3x , \qquad (61)$$

where $n^{\alpha}$ is the unit normal on $\partial M$, $a^{\alpha}=n^{\rho}\nabla_{\rho}n^{\alpha}$, $\tilde{\gamma}_{ij}$ is the induced metric on $\partial M$, $\tilde{\gamma}=\det\tilde{\gamma}_{ij}$, $K_{\alpha\beta}$ is the second fundamental form and $K=K_{\alpha}{}^{\alpha}$. For stationarity of the action to be equivalent to the usual field equations

$$\tfrac{1}{2}Rg^{\alpha\beta}-R^{\alpha\beta}=0 .$$

we can make a number of possible choices of boundary conditions. For example, if we want

$$\delta g_{\alpha\beta}\big|_{\partial M}=0,\quad (n^{\rho}\nabla_{\rho}\delta g_{\alpha\beta})\big|_{\partial M}\neq 0$$

and no other restrictions, we see that the second boundary term in (61) will not vanish unless $g^{\alpha\beta}|_{\partial M}=0$, which is impossible. We get around this by modifying (60) with a boundary action (York 1972)

$$S=-\gamma\int_M g^{1/2}Rd^4x-\gamma\int_{\partial M}\gamma^{1/2}2Kd^3x\ , \tag{62}$$

The variation of this is

$$\delta S=-\gamma\int_M g^{1/2}(\tfrac{1}{2}Rg^{\alpha\beta}-R^{\alpha\beta})\delta g_{\alpha\beta}d^4x\ ,$$

as we want.

Moving on to fourth order theory we can ask what happens if we demand

$$\delta g_{\alpha\beta}|_{\partial M}=0,\quad (n^\rho\nabla_\rho\delta g_{\alpha\beta})|_{\partial M}\neq 0$$

and no other restrictions? Bunch (1981) has found that stationarity of the fourth order action with these boundary conditions requires the curvature squared terms in S to be some multiple of the Euler number $\chi$ and so the theory is not really fourth order. Thus we must put on some restriction like $R|_{\partial M}=0$ in order to obtain stationarity of the action. We see that the problems of boundaries are not clearly understood. For example, York has shown that the proper definition of "$\partial M$" in expressions such as (62) is not as simple as one might imagine if M has a non-trivial topology. This can lead to important physical effects. [See York (1981) and Christensen & York (1982).]

## CONCLUSION

Fourth order gravity theories are rich with interesting and complicated questions. The mathematical techniques needed to study such theories at only the one-loop level are only just beginning to be developed in full. Therefore knowledge of any physical content is still tentative.

## ACKNOWLEDGEMENTS

Thanks are due to B. DeWitt, M. Duff, S. Fulling, C. Isham, R. Jackiw, K. Stelle, P. van Nieuwenhuizen and J. York for numerous useful discussions. This work was supported by the National Science

Foundation and the Bahnson Trust Fund at the University of North Carolina at Chapel Hill.

REFERENCES

Adler, S.L. (1980). A formula for the induced gravitational constant. Phys. Lett., 95B, 241.

Bunch, T.S. (1981). Surface terms in higher derivative gravity. J. Math. Phys. A: Math. Gen., 14, L139.

Christensen, S.M. (1975). Covariant coordinate space methods for calculations in the quantum theory of gravity. Dissertation, University of Texas at Austin.

Christensen, S.M. (1981a). Quantizing fourth order gravity theories. I: The functional integral. Paper in preparation.

Christensen, S.M. (1981b). A simple extension of the one-loop counterterm algorithm in gravity theory. Paper in preparation.

Christensen, S.M. (1982). Quantizing fourth order gravity theories. IV: The beta function. Work in progress.

Christensen, S.M. & Fulling, S.A. (1982a). Quantizing fourth order gravity theories. II: Heat kernel methods for fourth order operators. Work in progress.

Christensen, S.M. & Fulling, S.A. (1982b). Quantizing fourth order gravity theories. III: The ghosts. Work in progress.

Christensen, S.M. & York, J.W. (1982). Quantizing fourth order gravity theories. V: The boundary problem. Work in progress.

DeWitt, B.S. (1965). Dynamical theory of groups and fields. New York, New York: Gordon and Breach.

DeWitt, B.S. (1975). Quantum field theory in curved spacetimes. Phys. Rep., 19C, 295.

DeWitt, B.S. (1979). Quantum gravity: The new synthesis. In General Relativity, an Einstein centenary survey, ed. S.W. Hawking & W. Israel, Cambridge, England: Cambridge.

Fradkin, E.S. & Tseytlin, A.A. (1981). Higher derivative quantum gravity: One-loop counterterms and asymptotic freedom. P.N. Lebedev Physical Institute report.

Gilkey, P.B. (1975). The spectral geometry of a Riemannian manifold. J. Duff. Geo., 10, 601.

Gilkey, P.B. (1980). The spectral geometry of the higher order Laplacian. Duke Math. J., 47, 511.

Hasslacher, B. & Mottola, E. (1981). Asymptotically free quantum gravity and black holes. Phys. Lett., 99B, 221.

Julve, J. & Tonin, M. (1978). Quantum gravity with higher derivatives. Nuovo Cim., 46B, 137.

Misner, C., Thorne, K. & Wheeler, J. (1973). Gravitation. San Francisco, California: Freeman.

Salam, A. & Strathdee, J. (1978). Remarks on high-energy stability and renormalizability of gravity theory. Phys. Rev., D18, 4480.

Stelle, K.S. (1977). Renormalization of higher-derivative quantum gravity. Phys. Rev., D16, 953.

Tomboulis, E. (1980). Renormalizability and asymptotic freedom in quantum gravity. Phys. Lett., 97B, 77.

York, J.W. (1972). Role of conformal three-geometry in the dynamics of gravitation. Phys. Rev. Lett., 28, 1082.

York. J.W. (1981). Boundaries in quantum gravity. Paper in preparation.

# GREEN'S FUNCTIONS, STATES AND RENORMALISATION

M.R. Brown
Department of Astrophysics, South Parks Road, Oxford OX1 3RQ

A.C. Ottewill
Department of Astrophysics, South Parks Road, Oxford OX1 3RQ

## 1 INTRODUCTION

In this article we shall examine the significance that is to be attached to different operator orderings of free quantum field theories in curved space-time. By so doing we hope to cast some light on the now standard, albeit adhoc, renormalisation of such theories. We shall argue that, as in flat space, these theories should be rendered finite by normal ordering with respect to a local, geometrical vacuum state.

First let us introduce the key elements of the field theory that we shall require later. We use the real scalar field as an example. It has the classical action

$$S(\phi) \equiv -\tfrac{1}{2}\int d^4x\, g^{\frac{1}{2}}\{\phi_{;c}\phi^{;c} + \xi R\phi^2 + m^2\phi^2\} ,$$

and is conformally invariant when $m=0$ and $\xi=1/6$. The energy-momentum tensor is defined by

$$T^{ab}(\phi) \equiv 2g^{-\frac{1}{2}}\frac{\delta}{\delta g_{ab}} S(\phi) .$$

The quantum theory is constructed in the standard way: Corresponding to a complete set of solutions to the field equations, $\{u_k(x)\}$, the u-vacuum $|u\rangle$, is defined by

$$\hat{a}_k|u\rangle = 0 \ ; \ \langle u|u\rangle = 1 ,$$

where the field operator $\hat{\phi}$ is represented by

$$\hat{\phi}(x) = \sum_k \hat{a}_k u_k(x) + \hat{a}_k^\dagger u_k^*(x) ,$$

and the creation and annihilation operators, $\hat{a}_k^\dagger$ and $\hat{a}_k$, satisfy the usual commutation relations. The state $|u\rangle$ is determined by a knowledge of the Feynman Green's function, $G_u(x,x')$, where

$$G_u(x,x') \equiv i\langle u|T(\hat{\phi}(x)\hat{\phi}(x'))|u\rangle,$$

T denotes time-ordering and

$$(\Box - \xi R - m^2)G_u(x,x') = -g^{-\frac{1}{2}}(x)\delta^4(x-x') \ . \tag{1.1}$$

We shall also need to make reference to the following formal relationships: The bare effective action, W[u], is given by the equation

$$e^{iW(u)} \propto \{\det G_u(x,x')\}^{\frac{1}{2}} , \tag{1.2}$$

or, equivalently, by the equation

$$2g^{-\frac{1}{2}}\frac{\delta}{\delta g_{ab}} \ W(u) = \langle u|T^{ab}(\hat{\phi})|u\rangle. \tag{1.3}$$

The essential difficulty with the quantum theory is in giving a preferred meaning to expectation values of, for example, the energy-momentum tensor operator of equation (1.3). $T^{ab}(\hat{\phi})$ is a quadratic function of the field operators that has as many different values as there are distinct operator orderings. Undaunted, renormalisation theory seeks to limit the ambiguity in the variously infinite, undefined expression $\langle u|T^{ab}(\hat{\phi})|u\rangle$ to the numerical coefficients of a few local curvature terms. We should like to ask: "Can one do any better?".

When space-time is flat the answer to this question is yes. All currently favoured methods of renormalisation agree with the result that can be stated as follows: In flat space-time, the renormalised expectation value of the energy-momentum tensor in the state $|u\rangle$ is given by the expression

$$T^{ab}_R(u) = \langle u|{}^*_*T^{ab}(\hat{\phi}){}^*_*|u\rangle - \langle M|{}^*_*T^{ab}(\hat{\phi}){}^*_*|M\rangle , \tag{1.4}$$

where $|M\rangle$ is the Minkowski vacuum state and $({}^*_*)$ denotes any given ordering. The equation holds for any ordering because $T^{ab}(\hat{\phi})$ is a quadratic function of $\hat{\phi}$ and one ordering differs from another by, at most, a c-number.

Notice that equation (1.4) has a meaning that is independent of any method of regularisation. Indeed, by choosing $({}^*_*)$ to be normal ordering with respect to $|M\rangle$ (conventionally denoted by $(:)$), we are able to represent the renormalised stress tensor as a single finite expectation value of a well defined operator, viz.,

$$T_R^{ab}(u) = \langle u|:T^{ab}(\hat{\phi}):|u\rangle \quad . \tag{1.5}$$

Notice also that equation (1.4) is consistent with the definition of the renormalised effective action,

$$W_R(u) \equiv W(u) - W(M) \quad . \tag{1.6}$$

Equation (1.6) makes it apparent that what we have been calling renormalisation here should be more accurately be described as a normalisation of the theory; no coupling constants or wave function is being adjusted, rather, the effective action is normalised to be zero for the state $|M\rangle$. For the remainder of this paper we shall be concerned to find the curved space analogue of this process of normalisation.

We begin by examining the rôle played by the Minkowski vacuum in equation (1.4). We shall attempt to describe it in a way that offers a natural generalisation to curved space-times. First let us consider some examples; here, for simplicity, we shall work with the massless theory.

If our flat space is Minkowski space itself then $|M\rangle$ is the natural global vacuum state and $T_R^{ab}(M)$ is identically zero. This is not too revealing.

Suppose next we take Minkowski space but identify the points $(t,x,y,z)$ and $(t,x+a,y,z)$ for all $(t,x,y,z)$. By so doing we change the topology from $R^4$ to $R^3 \times S^1$. The natural global vacuum for this flat space has a Feynman Green's function, $G_u$, that is periodic in the coordinate x (with period a). Dowker & Critchley (1976) express this function as the sum

$$G_u(x,x') = \frac{i}{4\pi^2} \sum_{-\infty}^{+\infty} \{(x-x'+na)^2 - s^2 + i\varepsilon\}^{-1} \quad ,$$

$$= \frac{i}{8\pi as} \{\cot\pi(x-s)/a - \cot\pi(x+s)/a\} \quad ,$$

where $s^2 \equiv (t-t')^2-(y-y')^2-(z-z')^2$. The function $G_M(x,x')$, the Feynman Green's function for the Minkowski vacuum, is necessarily a solution to equation (1.1) for this space-time but now it clearly has nothing to do with the global properties of the space. In fact, it respects only the local symmetries of the space-time; it corresponds exactly to the n=0 term of equation (1.6). We can usefully make this picture a little more precise: The geodesic distance between two points in this space, x and x', is a many valued function corresponding to the freedom to wind around the $S^1$ any number of times. The geodesic distance that takes n turns is $\sqrt{2\sigma_n(x,x')}$ , where

$$2\sigma_n = -(t-t')^2 + (x-x'+na)^2 + (y-y')^2 + (z-z')^2 . \tag{1.7}$$

Equation (1.6) represents $G_u(x,x')$ as a sum of contributions from these distinct paths. $G_M(x,x')$ is the direct (n=0) geodesic contribution. Dowker & Critchley (1976) give the renormalised stress tensor for this space in the state $|u\rangle$ as

$$T_R{}^a{}_b(u) = \frac{\pi^2}{90a^4} \operatorname{diag}(1,-3,1,1)^a{}_b .$$

These examples are sufficient to exhibit many of the properties that $G_M(x,x')$ enjoys in any flat space: It is a local, geometrical, symmetric solution to equation (1.1) that is a function of the direct geodesic alone. (We say that a Green's function is geometrical if it is a function of geometrical objects alone.) Loosely speaking, for any flat space, the function $G_M(x,x')$ defines a preferred, geometrical vacuum, $|M\rangle$ , that is appropriate for a description of local physics. It is with respect to this vacuum that the operators of the free quantum field theory are to be normal ordered. In the very special case of Minkowski space this local vacuum coincides with the global vacuum of the space.

So much for flat space. In the next section we shall see how far it is possible to define an analogous local, geometrical Green's function when space-time is curved. We shall adopt the conventions of Hawking & Ellis (1973).

## 2 LOCAL GREEN'S FUNCTIONS IN CURVED SPACE-TIME

First let us consider a simple example: The Einstein static universe has the topology $R \times S^3$ and its line element can be written as

$$ds^2 = -dt^2 + a^2(d\chi^2 + \sin^2\chi(d\theta^2 + \sin^2\theta d\phi^2)) ,$$

where $-\infty < t < \infty$ ; $0 \leqslant \chi,\theta \leqslant \pi$ ; $0 \leqslant \phi < 2\pi$ .

Dowker (1971) has shown that the global Feynman Green's function for the massless, conformally coupled scalar field on this space can be written as a sum over contributions from distinct geodesics in the space-time, viz.,

$$G(t,\underset{\sim}{x};t',\underset{\sim}{x}') = \frac{i}{8a\pi^2}\sum_{-\infty}^{+\infty} \frac{(s+2n\pi a)}{\sin(s/a)} (\sigma_n + i\varepsilon)^{-1} \quad , \tag{2.1}$$

where $s(\underset{\sim}{x},\underset{\sim}{x}')$ is the shortest geodesic distance on the spatial section, $S^3$, and

$$2\sigma_n \equiv -(t-t')^2 + (s+2n\pi a)^2 \ .$$

$2\sigma_n$ is the square of the geodesic distance along the $n^{th}$ geodesic joining x and x'. For any curved space we define

$$\Delta(x,x') \equiv -g^{-\frac{1}{2}}(x)g^{-\frac{1}{2}}(x')\det(-\sigma_{;ab'}) \ , \tag{2.2}$$

where $2\sigma$ is the square of the distance along a geodesic joining x and x'. $\Delta$ is the usual biscalar constructed from the VanVleck-Morette determinant. For the Einstein static universe we define $\Delta_n$ to be given by equation (2.2) with $\sigma$ replaced by $\sigma_n$. One can then rewrite equation (2.1) as

$$G(x,x') = \frac{i}{8\pi^2}\sum_{-\infty}^{+\infty} \Delta_n^{\frac{1}{2}}(\sigma_n + i\varepsilon)^{-1} \ . \tag{2.3}$$

The n=0 term in this sum is of special interest and we denote it by $G_o$. It is clearly analogous to the n=0 term in the sum (1.6) in that it respects the local symmetries of the space and is a function of the direct geodesic distance ($\sigma_o$) alone. Moreover, $G_o$ satisfies the equation

$$(\Box - \tfrac{1}{6}R)G_o(x,x') = -g^{-\frac{1}{2}}(x)\delta^4(x-x') \ ,$$

at least for x in some open neighbourhood of x'. $G_o$ is not a suitable candidate for a global Green's function for the space as it does not possess the required periodicity properties and blows up on the line $\Delta_o = \infty$ , which corresponds to $s=\pi$ . (For a general space, the locus of points, x, for which $\Delta(x,x')$ is infinite is the caustic surface corresponding to x'. Equivalently, this surface can be described as the envelope of geodesics emanating from x'. Normally this will be a three dimensional surface but when, for example, the space has symmetries it may be of lower dimension.)

$G_o$ is an example, for the special case of the Einstein static universe, of what we shall call a local Green's function, $G_L$. In this section, it is our intention to give a definition of $G_L$ for general space-times. First we must make the concepts of "locality" and "direct geodesics" more precise: $G_L(x,x')$ is described as local in the sense that, for a given point x', it is a symmetric solution to equation (1.1) for x in an open neighbourhood of x', but it need not exist for all points, x, in the space. Moreover, although we shall demand that $G_L$ respects the local symmetries of the space-time, we shall make no such demands with respect to global symmetries such as periodicity. In order to incorporate "direct geodesics" we shall, in general, require that x and x' lie in a "simple region", N. (For an explanation of the terminology see Penrose (1972).) This ensures that there exists a unique geodesic joining x and x' which lies entirely in N. We shall call this unique geodesic the direct geodesic. It is worth noting that if two points in the space (x and y), not necessarily lying in a simple region, can be connected by a path then, by covering the path with simple regions, we can analytically continue $G_L$ from a simple region containing x to a simple region containing y. This is possible even though $G_L(x,y)$ itself may be infinite.

The physical significance of the local Feynman Green's function is essentially the same whether space-time is curved or flat. For definiteness, consider again the example of the Einstein static universe: Suppose there is an observer in this space who follows some arbitrary timelike trajectory and performs experiments inside a laboratory which has dimensions very much less than the radius of the universe. His concept of classical physics is straightforward. The presence of his laboratory restricts classical particles from reaching the boundary of the space or from following paths that wrap themselves around the space. The only path which a freely falling classical particle can follow between two points in the laboratory, that is not entirely dependent on properties of the laboratory, is the direct geodesic between the points. (We assume the observer's laboratory to be contained within a simple region, N.) The quantum mechanical particles corresponding to this classical picture are precisely those whose vacuum is defined by the function $G_O$, the direct geodesic or local Feynman Green's function. If our observer wishes to allow for his motion and the presence of his laboratory he may define an appropriate observer dependent vacuum state. (This corresponds to the philosophy underlying the theory of particle detectors. See, for example, Unruh (1976), Grove & Ottewill (1981).) However, if he wishes to discount these effects he will naturally define the observer independent, local, geometrical vacuum described by $G_O$.

Now suppose this Einstein static universe is in its natural global ground state, that defined by the Green's function of equation (2.1). Then our local observer who normal orders his stress tensor with respect to the local, geometrical vacuum will discover that the expectation value of this operator in the global ground state is non-zero. In fact, it will have the same value as the renormalised Einstein static stress tensor that was calculated by Dowker & Critchley (1976). Their renormalisation ansatz precisely corresponds to normal ordering with respect to the vacuum defined by $G_o$. We shall have more to say about renormalisation later. We now return to our attempt to give a definition of $G_L$ for general space-times.

An excellent intuitive picture is that provided by path integration. (For a recent review see, for example, the article by Hawking (1979).) In this approach one introduces a proper time parameter, w, to write any solution to equation (1.1) in the form

$$G_u(x,x') = \int dw \, \exp-(im^2w + \varepsilon/w)F(x,x',w) \; . \tag{2.4}$$

Then, following the ansatz due to Feynman, the propagator, $F(x,x',w)$, is expressed as a path integral:

$$F(x,x',w) = \int D(x(u)) \, \exp\{\tfrac{i}{4}\int_0^w du \; g(\dot{x},\dot{x})\} \quad , \tag{2.5}$$

where dot denotes differentiation with respect to u and the paths, x(u), are restricted by the boundary conditions $x(0)=x$, $x(w)=x'$. These boundary conditions are insufficient to give a precise meaning to the path integral: one must further specify the measure on the space of paths. It is not known how to do this in general. However, by formal manipulations of equations (2.4) and (2.5) one can deduce that F satisfies a Schrödinger equation. Moreover, one can give a stationary phase analysis of the integral: a full discussion of this is given in the review article by DeWitt-Morette et al. (1979). The stationary phase "points" are just the geodesics joining x to x'. If the integral is approximated by keeping only the quadratic fluctuations about these paths this yields the standard WKB approximation:

$$F(x,x',w) \sim w^{-2}\sum \Delta^{\frac{1}{2}}(x,x') \; \exp i(\sigma/2w) \; , \tag{2.6}$$

where the sum is taken over all geodesics joining x and x', $2\sigma$ is the square of the geodesic distance along a given geodesic and $\Delta$ is given by equation (2.2). For each geodesic one can proceed to calculate the higher order terms in the approximation. (The integral can be reduced to a sum of Feynman diagrams; this is described, for example, by D'Eath (1981).) It is known that the resulting series for each geodesic separately satisfies the Schrödinger equation satisfied by F. This provides us with at least a formal definition for $G_L$: For x and x' belonging to N, $G_L(x,x')$ has the Feynman perturbation expansion that is obtained from equations (2.4) and (2.5) by expanding the integrand to all orders about the direct geodesic alone.

An equivalent proper time representation of this function has been given by DeWitt (1965) as

$$G_L = \frac{\Delta^{\frac{1}{2}}(x,x')}{16\pi^2} \int \frac{dw}{w^2} \Lambda(x,x',w) \exp i\{(\sigma+i\varepsilon)/2w - m^2 w\} , \qquad (2.7)$$

where $\Lambda(x,x',w)$ satisfies the equation

$$\frac{\partial\Lambda}{\partial w} + w^{-1}\sigma^{;a}\Lambda_{;a} = i\Delta^{-\frac{1}{2}}(\Box - \xi R)(\Delta^{\frac{1}{2}}\Lambda) , \qquad (2.8)$$

with the boundary condition $\Lambda(x,x',0)=1$. Equation (2.8) is solved by the power series

$$\Lambda(x,x',w) = \sum_{n=o}^{\infty} a_n(x,x')(iw)^n , \qquad (2.9)$$

where the coefficients , $a_n$, are determined by the recursion relations

$$a_o(x,x') = 1 ;$$

$$\sigma^{;c}a_{n;c} + na_n = \Delta^{-\frac{1}{2}}(\Box - \xi R)(\Delta^{\frac{1}{2}}a_{n-1}) , \quad (n>0) \qquad (2.10)$$

together with the requirement that they be regular as $x \to x'$. Equations (2.7)-(2.10) provide us with a workable definition for $G_L$. (There are known to be problems with these equations as a representation for a global Green's function, but these need not concern us for our local Green's function.) In section 3 we shall establish that this definition is consistent with the choice of the massless Green's function $G_o$ of the Einstein static universe. It is clearly consistent with the flat space examples.

For massive theories equation (2.7) can be developed as the series

$$G_L = \frac{-\Delta^{\frac{1}{2}}}{8\pi^2} \sum_{n=0}^{\infty} a_n \left(-\frac{\partial}{\partial m^2}\right)^n \frac{m^2}{(-2m^2\sigma)^{\frac{1}{2}}} H_1^2\{(-2m^2\sigma)^{\frac{1}{2}}\} , \qquad (2.11)$$

where $H_1^2$ is a Hankel function of the second kind of order 1. This equation determines the behaviour of $G_L$ for x close to x'.

For massless theories it is not so straightforward to obtain this small distance behaviour; there is no natural dimensionless expansion parameter. However, we would expect that the massless local Green's function will have a Hadamard series expansion. This series was introduced by Hadamard (1923) and later developed by DeWitt & Brehme (1960): A massless scalar Green's function satisfying the equation

$$(\Box - \tfrac{1}{6}R)G(x,x') = -g^{-\frac{1}{2}}(x)\delta^4(x-x') , \qquad (2.12)$$

and having the appropriate singularity structure as $x \to x'$ has the form

$$G(x,x') = \frac{i\Delta^{\frac{1}{2}}}{8\pi^2} \{(\sigma+i\varepsilon)^{-1} + v\ln(\sigma+i\varepsilon) + w\} , \qquad (2.13)$$

where

$$v(x,x') = \sum_{n=0}^{\infty} v_n(x,x')\sigma^n \qquad (2.14)$$

and

$$w(x,x') = \sum_{n=0}^{\infty} w_n(x,x')\sigma^n . \qquad (2.15)$$

The requirement that equation (2.12) be satisfied yields differential recursion relations for the coefficients $v_n$ and $w_n$. Hadamard has shown that when the metric is analytic the series (2.13) and (2.14) converge uniformly inside the region where $\sigma$ is single valued. The recursion relations suffice to determine the function v(x,x') uniquely; however $w_0(x,x')$ is arbitrary, corresponding to the freedom to add to G any solution of the homogeneous equation. The condition that G be symmetric in x and x' restricts the biscalar $w_0$ to some extent. (The function v(x,x') is known to be symmetric.) We can write the condition that w(x,x') be symmetric entirely in terms of constraints on $w_0$. These constraints relate the coincidence limits of covariant derivatives of $w_0$ and are obtained by differentiating the sum $\sum[w_n(x,x') - w_n(x',x)]\sigma^n$, using the recursion relations and taking coincidence limits. There are an infinite number of these constraints, the first two can be written

$$[w_o{}^{;c}]_{;c} - \tfrac{1}{2}\Box[w_o] = 0 \, , \tag{2.16}$$

$$[w_{o;c}{}^{a}]_{;a} - \tfrac{1}{4}[\Box w_o]_{;c} - \tfrac{1}{2}\Box[w_{o;c}] - \tfrac{1}{2}R_{ca}[w_o{}^{;a}] + \tfrac{1}{6}R_{ca}[w_o]^{;a} = \tfrac{1}{2}[v_1]_{;c} \tag{2.17}$$

where we define $[f(x,x')]$ to be the limit of $f(x,x')$ as $x \to x'$. The next constraint relates the tensor $[w_o^{;abc}]$ to $[w_o^{;ab}]$, $[w_o^{;a}]$ and $[w_o]$, and so on. However, only equations (2.16) and (2.17) are relevant to the calculation of effective stress tensors.

The representation of the massless Green's function provided by equation (2.12) is much better suited to our needs than the previous expressions. We should like to derive from equation (2.7) an expression for $w_o^L$, i.e. that particular biscalar that defines the massless $G_L$. At present we can only do this for certain space-times and these we shall discuss in the next section.

DeWitt introduced the representation (2.7) in order to discuss the divergences of quantum field theory in curved space-time, and it is to the subject of renormalisation that we now turn. In particular, we wish to contrast renormalisation in curved space-time with what we have described as normalisation. Now that we have, in equation (2.7), a local geometrical Green's function, $G_L$, we can define the normalised expectation value of the stress tensor operator in the state $|u\rangle$ to be

$$T_N^{ab}(u) \equiv \langle u|{\ast}T^{ab}(\hat{\phi}){\ast}|u\rangle - \langle L|{\ast}T^{ab}(\hat{\phi}){\ast}|L\rangle \tag{2.18}$$

where the local vacuum, $|L\rangle$, is defined implicitly by the equation

$$G_L(x,x') \equiv i\langle L|T(\hat{\phi}(x)\hat{\phi}(x'))|L\rangle \, , \tag{2.19}$$

and, as in equation (1.4), $({\ast})$ denotes any operator ordering. One also has the normalised effective action defined by

$$2g^{-\frac{1}{2}}\frac{\delta}{\delta g_{ab}}W_N(u) = T_N^{ab}(u) \, , \tag{2.20}$$

or, equivalently,

$$\exp iW_N(u) = \{\det(G_u G_L^{-1})\}^{\frac{1}{2}} \tag{2.21}$$

In flat space-times normalisation is the same as renormalisation. In some curved space-times the same is true, the Einstein static universe is an example. However, in general, these procedures are inequivalent.

If we denote by $W_R[u]$ the standard renormalised effective action obtained from equation (1.2), the interdependence of normalisation and renormalisation is described by the equation

$$W_N(u) = W_R(u) - W_R(L) \ . \tag{2.22}$$

There is, of course, a similar equation for stress tensors, viz.,

$$T_N^{ab}(u) = T_R^{ab}(u) - T_R^{ab}(L) \ . \tag{2.23}$$

A few years ago Adler et al. (1977) proposed a method of renormalisation that required the subtraction of a local Green's function. They chose the solution to equation (2.12) for which $w_o(x,x')=0$. This Green's function is not symmetric: $w_o$ does not satisfy equation (2.17) unless $[v_1]$ is constant. $[v_1]$ is proportional to $a_2(x,x)$, a function that is commonly known as the trace anomaly. (See, for example, Duff (1977).) Wald (1978) subsequently showed that this lack of symmetry gave rise to a stress tensor that was not conserved and so the renormalisation scheme of Adler et al. was abandoned. (We note, in passing, that our definition of $G_L$ is manifestly symmetric.) Wald (1977) had previously given a set of four axioms in an attempt to define the renormalised stress tensor from general principles. These axioms limit the ambiguity in the definition of renormalised stress tensors to a conserved local curvature tensor. We might now ask whether the process that we describe as normalisation, when regarded as a method of renormalisation, is consistent with Wald's axioms. That it is may be seen from equation (2.23) when we take $|u\rangle$ to be the global vacuum for the space-time. $T_N^{ab}(L)$ is, by construction, a conserved local curvature tensor. Thus if the standard method of renormalisation ("R") satisfies Wald's axioms, and we suppose it does, then normalisation also satisfies these axioms.

We note that normalisation preserves all local symmetries of the field theory. In particular, it preserves conformal invariance: the trace of the normalised stress tensor vanishes for conformally invariant theories. However, the local Green's functions for conformally related spaces are not, in general, related by a simple scaling law. This point is discussed in more detail by Brown et al. (1981). An example is provided by the Einstein static universe and Minkowski space. The local Green's functions are $G_o$ and $G_M$, respectively. These functions are not related by the conformal factor that maps Minkowski space into the Einstein static universe; instead it is the functions $G_M$ and the global Einstein static

Green's function (given by equation (2.1)) that are so related.

We conclude this section with a few remarks concerning the underlying philosophy of this approach: Although normalisation can be regarded as just another version of renormalisation we are not arguing that it should be. We take the view that free quantum field theories are theories of operators and states and that the physical content of these theories is that described by well defined expectation values of well defined operators. Free quantum field theory in curved space-time is as good (or as bad) as free quantum field theory in flat space-time. The former has a rather more complicated structure but it is qualitatively not very different. In both curved and flat space-time it is necessary to do something about the problem of operator ordering. Standard renormalisation theory essentially ignores this problem. In this article we have tried not to ignore it while, at the same time, giving a geometrical determination of the vacuum stress-energy density of a space-time. The result is the normalised expectation value of the stress tensor operator, $T_N^{ab}(u)$. It is normalised to be zero in the local vacuum state $|L\rangle$ : $T_N^{ab}(L)=0$.

## 3 EXAMPLES

The examples that we shall give are all symmetric space-times. These are defined by the property that $R_{abcd;e}=0$ . We first derive some results concerning the geodesic structure of such space-times.

Differentiation of the equation

$$\sigma_{;a}\sigma^{;a} = 2\sigma \tag{3.1}$$

yields a differential equation for $\sigma_{;bc}$, viz.,

$$\sigma^{;a}(\sigma_{;bc})_{;a} + \sigma_{;ba}\sigma^{;a}{}_{c} + \Gamma_{bc} = \sigma_{;bc} \quad , \tag{3.2}$$

where $\Gamma_{ac} = R_{abcd}\sigma^{;b}\sigma^{;d}$ . For symmetric spaces equation (3.2) has the solution

$$\sigma_{;bc} = (2\sigma)^{-1}\sigma_{;b}\sigma_{;c} + \Gamma^{\frac{1}{2}}{}_{ba}(\cot\Gamma^{\frac{1}{2}})^{a}{}_{c} \quad . \tag{3.3}$$

The function $\Delta$ (defined by equation (2.2)) satisfies the differential equation

$$\sigma^{;a}(\ln\Delta)_{;a} = 4 - \Box\sigma \tag{3.4}$$

Equations (3.3) and (3.4) together with the boundary condition $\Delta(x,x)=1$ imply

$$\Delta = (\gamma_1\gamma_2\gamma_3)^{\frac{1}{2}}/(\sin\gamma_1^{\frac{1}{2}}\sin\gamma_2^{\frac{1}{2}}\sin\gamma_3^{\frac{1}{2}}) \quad , \tag{3.5}$$

where $\gamma_i$ is the $i^{th}$ eigenvalue of the matrix $\Gamma_{bc}$ - we exclude the zero eigenvalue corresponding to the eigenvector $\sigma^{;c}$. (We note in passing that when the space-time is not symmetric, equations (3.3) and (3.5) still provide approximate solutions for x near to x'.)

The first two examples that we give are the Einstein static universe and the open Einstein universe. These symmetric space-times are locally characterised by the conditions

$$C_{abcd}=0 \qquad \text{and} \qquad R_{ab}=\tfrac{1}{3}R(g_{ab}+K_aK_b) \quad ,$$

where $K^a$ is a covariantly constant, unit magnitude, timelike vector field. The eigenvalues, $\gamma_i$, are given by

$$\gamma_1 = 0 \ , \ \gamma_2 = \gamma_3 = \tfrac{1}{6}R\{2\sigma + (K^a\sigma_{;a})^2\} \ . \tag{3.6}$$

The Einstein static universe has $R > 0$; the open Einstein universe has $R < 0$. In both cases $\Delta$ satisfies the equation

$$(\Box - \tfrac{1}{6}R)\Delta^{\frac{1}{2}} = 0 \ .$$

This enables the recursion relations (2.10) to be easily solved to yield

$$a_r = (r!)^{-1}(\tfrac{1}{6}-\xi)^r R^r \ ;$$

whence

$$\Lambda = \exp\{iw(\tfrac{1}{6}-\xi)R\} \ ,$$

and

$$G_L = \frac{\Delta^{\frac{1}{2}}}{16\pi\sigma}\lambda H_1^2(\lambda) \quad , \tag{3.7}$$

where $\quad \lambda^2 \equiv -2\{m^2+(\tfrac{1}{6}-\xi)R\}\sigma.$

When $m=0$, $\xi=\tfrac{1}{6}$ and $R=6/a^2$ this expression is clearly equal to the function $G_0$, the direct geodesic Green's function for the Einstein static universe.

For the open Einstein universe, $G_L$ is not singular other than for points joined by null geodesics and it is usually taken to be the global Green's function for the space-time. Of course, if we change the space-time topology from $R\times H^3$ , while keeping the same local curvature structure, $G_L$ may cease to be the appropriate global Green's function for the space-time but it would necessarily remain the appropriate local Green's function. The open Einstein universe (with its usual topology) clearly has $T_N^{ab}(u)=0$ when $|u\rangle$ is the global vacuum since this state is the same as the local vacuum. The renormalised stress tensor for this space-time in its natural vacuum state is also zero (Bunch (1978)).

Our third and final example is the de Sitter universe. The curvature tensor for this space is given by

$$R_{abcd} = \tfrac{1}{12} R(g_{ac}g_{bd} - g_{ad}g_{bc}) \quad ,$$

where $R > 0$. The eigenvalues, $\gamma_i$ , are all equal; they all have the value $R\sigma/6$, and so

$$\Delta = (R\sigma/6)^{3/2} \sin^{-3}(R\sigma/6)^{\frac{1}{2}}. \tag{3.8}$$

It is not easy to solve the recursion relations for the $a_n$'s in this case and instead we proceed as follows: To avoid unnecessary complications we set $m=0$ and $\xi=\frac{1}{6}$. We know that any purely geometrical Green's function for this space-time is a function of $\sigma$ alone (there are no other SO(1,4) invariant geometrical objects that are not constant). The general solution to the equation

$$(\Box - \tfrac{1}{6}R)G(\sigma) = -g^{-\frac{1}{2}}(x)\,\delta^4(x-x') \quad ,$$

is given by

$$G(\sigma) = \frac{i(R/6)}{16\pi^2}\left\{\frac{1}{1-\cos(R\sigma/6)^{\frac{1}{2}}+i\epsilon} + \frac{\alpha}{1+\cos(R\sigma/6)^{\frac{1}{2}}}\right\} \quad , \tag{3.9}$$

where $\alpha$ is a constant. $\alpha=0$ defines the global Green's function for the space (the presence of the second term is unacceptable as the function $1 + \cos(R\sigma/6)^{\frac{1}{2}}$ can be zero for points which cannot be joined by a null geodesic). The value of $\alpha$ that determines the local Green's function can be written in terms of coincidence limits of $a_n$'s as follows:

$$\alpha = -\tfrac{4}{3} + 2\sum_{r=2}^{\infty}(r-2)!\,(6/R)^r a_r(x,x,0) \quad , \tag{3.10}$$

where, as the notation suggests, the $a_n(x,x',\xi=0)$ satisfy the recursion relations (2.10) with $\xi=0$. We note that this is a special case of a more general result: For any space with a constant, non-zero Ricci scalar, the coincidence limit $w_0^L(x,x)$ for the local Green's function of the conformally invariant theory is given by the equation

$$w_0^L(x,x) = -\tfrac{1}{12}R + \tfrac{1}{2}\sum_{r=2}^{\infty}(r-2)!\,(6/R)^{r-1} a_r(x,x,0) \quad . \tag{3.11}$$

This equation is obtained from equation (2.11) by observing that the massless, conformally coupled Green's function can be obtained from equation (2.7) by setting $m^2 = R/6$ and $\xi = 0$. For de Sitter space $w_0^L(x,x)$ is related to $\alpha$ by the equation

$$w_0^L(x,x) = \tfrac{1}{72}R(3\alpha-2) \ .$$

We have not yet computed $\alpha$ beyond its representation provided by equation (3.10). One knows that, whatever its actual value, the normalised stress tensor for the global de Sitter vacuum will be zero since there are no non-zero, trace-free, de Sitter invariant tensors. The renormalised stress tensor for this state is a non-zero multiple of the metric tensor; in this case the conformal symmetry is broken by the renormalisation scheme.

## 4 CONCLUSION

The representation of $G_L$ as the integral (2.7) is not ideal. At present we have little idea of the convergence properties of the series (2.10) for general space-times. It may be that as our knowledge improves we shall have to refine this definition. (At the same time, it would be useful to make more precise the connection with the Feynman path integral. Unfortunately it is not enough to restrict the domain of integration to the space of paths that is the homotopy class of the direct geodesic. For the flat space $R^3 \times S^1$ example it is, in fact, sufficient; but for more complicated topologies and spaces the homotopy class of the direct geodesic will contain other geodesics. Somehow, one must weight the measure, in an invariant way, in favour of paths close to the direct geodesic. The WKB expansion essentially does just this: the deviation from the direct geodesic is supposed to be small in some sense.) These possible imperfections in the present definition of $G_L$ do not alter the general philosophy of normalisation, which we feel is of considerable importance. We hope that the formal definition that is given is sufficient to illustrate these ideas.

So far we have deliberately restricted our attention to free quantum field theories and have argued that these can be formulated without recourse to renormalisation theory even when space-time is curved. In our opinion, renormalisation theory in curved space-time, as in flat space-time, is properly reserved for the description of interacting field theories. As in flat space-time these renormalisations may break symmetries of the classical interacting field theory. For example, there will continue to be an axial current anomaly (Jackiw (1972)) when space-time is curved (which, of course, it is)

Finally we should like to observe that normalisation provides some insight into the nature of vacuum stress: The normalised stress tensor at x' for the global vacuum state of a given space-time is completely determined by a knowledge of the difference $G(x,x') - G_L(x,x')$ in a neighbourhood of x', where $G(x,x')$ is the global Feynman Green's function for the space-time. Suppose that the caustic surface corresponding to the point x' is close to x', then the function $G(x,x') - G_L(x,x')$ is rapidly varying in a neighbourhood of x' (in general, $G(x,x')$ will be regular for x on the caustic surface of x', whereas $G_L(x,x')$ will be singular there). We would, therefore, expect there to be a large vacuum stress-energy at x'. On the other hand, if x' is far from its caustic surface we would expect $G - G_L$ to be more slowly varying in a neighbourhood of x' and there to be a correspondingly lower stress-energy. We note that similar remarks may be applied to the stress near the surface of a perfect conductor. (In this case, $G - G_L$ will vary rapidly since G must vanish on the conductor whereas $G_L$ need not.) These comments also apply, if somewhat less directly, to the renormalised stress tensor: The point-separated subtraction terms for the renormalised stress tensor (Christensen (1976)) are obtained from the leading order WKB approximation to $G_L$.

REFERENCES

Adler, S.L., Lieberman, J. & Ng, Y.G. (1977). Regularisation of the stress-energy tensor for vector and scalar particles propagating in a general background metric. Ann.Phys.(N.Y.), 106, 279.

Brown, M.R., Ottewill, A.C. & Siklos, S.T.C. (1981). Comments on conformal Killing vector fields and quantum field theory. Phys.Rev.D, (to be published).

Bunch, T.S. (1978). Stress tensor of massless conformal quantum fields in hyperbolic universes. Phys.Rev.D, 18, 1844.

Christensen, S.M. (1976). Vacuum expectation value of the stress tensor in an arbitrary curved background: The covariant point-separation method. Phys.Rev.D, 14, 2490.

D'Eath, P.D. (1981). Perturbation methods in quantum gravity: The multiple-scattering expansion. Phys.Rev.D, 24, 811.

DeWitt, B.S. & Brehme, R.W. (1960). Radiation damping in a gravitational field. Ann.Phys.(N.Y.), 9, 220.

DeWitt, B.S. (1965). Dynamical Theory of Groups and Fields. New York: Gordon & Breach.

DeWitt-Morrette, C., Maheshwari, A. & Nelson, B. (1979). Path integration in non-relativistic quantum mechanics. Phys.Rep., 50C, 255.

Dowker, J.S. (1971). Quantum mechanics on group space. Ann.Phys.(N.Y.), 62, 361.

Dowker, J.S. & Critchley, R. (1976). Covariant Casimir calculations. J.Phys.A, 9, 535.

Duff, M.J. (1977). Observations on conformal anomalies. Nucl.Phys.B, 125, 334.

Grove, P.G. & Ottewill, A.C. (1981). Notes on "particle detectors". University of Oxford preprint.

Hadamard, J. (1923). Lectures on Cauchy's Problem in Linear Partial Differential Equations. New Haven: Yale University Press.

Hawking, S.W. (1979). The path integral approach to quantum gravity. In General Relativity - An Einstein Centenary Volume, ed. S.W. Hawking & W. Israel. Cambridge: Cambridge University Press.

Hawking, S.W. & Ellis, G.F.R. (1973). The Large Scale Structure of Space-Time. Cambridge: Cambridge University Press.

Jackiw, R. (1972). Field theoretic investigations in current algebra. In Lectures on Current Algebra and its Applications, eds. S.B. Treiman, R. Jackiw & D.J. Gross. Princeton: Princeton University Press.

Penrose, R. (1972). Techniques of Differential Topology in Relativity. Philadelphia: SIAM.

Unruh, W.G. (1976). Notes on black-hole evaporation. Phys.Rev.D, 14, 870.

Wald, R.M. (1977). The back reaction effect in particle creation in curved space-time. Commun.Math.Phys., 54, 1.

Wald, R.M. (1978). Trace anomaly of a conformally invariant quantum field in curved space-time. Phys.Rev.D, 17, 1477.

# INTRODUCTION TO QUANTUM REGGE CALCULUS.

Martin Rocek
California Institute of Technology, Pasadena, California 91125 USA

Ruth M. Williams
Girton College
and
Department of Applied Mathematics and Theoretical Physics,
University of Cambridge, Cambridge, England.

## *INTRODUCTION*

We have looked at some rudimentary aspects of an approach to the quantization of gravity using Regge's discrete description of general relativity (Regge 1961). First I'll review Regge Calculus, as this description is called, and then I'll tell you what we have been doing. I won't present any actual calculations--for those you should see Roček &Williams (1981, 1981a).

Regge Calculus describes Einstein's theory by using simplicial approximations of curved spacetimes. Its fundamental variables are a set of dynamical lengths, which are ***coordinate invariant*** geodesic lengths, and an incidence matrix that describes how these are connected. In $d$ dimensions, one joins together flat blocks, typically $d$-simplices, along flat faces, or ($d$-1)-simplices, so that the curvature sits on the hinges, or ($d$-2)-simplices which the faces share. I'll describe how this works explicitly in 2,3, and 4 dimensions; as I do so, I'll also describe the particular triangulations, or simplicial decompositions, of flat space that we have found useful.

The basic idea of Regge Calculus is easiest to visualize in two dimensions, where we approximate a curved surface by a skeleton of flat triangles joined along straight edges which meet at points. Imagine for example a pyramid as drawn in Fig. 1a; if we cut along one edge, we can flatten the pyramid to get the shape drawn in Fig. 1b. Notice that because the pyramid was curved, when we flatten it in this way, there is a gap or deficit; this deficit is a measure of the curvature which resided at the apex of the pyramid. Quantitatively, if we parallel transport a vector $\vec{v}$ along any path that encloses the apex, it is rotated by precisely the deficit angle $\varepsilon$ at the point; that is

$$\vec{v}\,' = R\,\vec{v}\,, \text{ where } R = \left(e^{\varepsilon U}\right)_{\mu\nu},\; U_{\mu\nu} = \begin{bmatrix} 0 & 1 \\ -1 & 0 \end{bmatrix},\; U_{\mu\nu}U_{\mu\nu} = 2. \tag{1}$$

It is easy to triangulate the plane, but the triangulation I'll describe here, which uses right triangles, has the virtue that it easily generalizes to higher dimensions. We simply draw a square lattice, and draw in all the diagonals in one direction as shown in Fig. 2a; this lattice can be generated by translating Fig 2b, which is just what you get when you take one corner of a square and connect it to the three other corners.

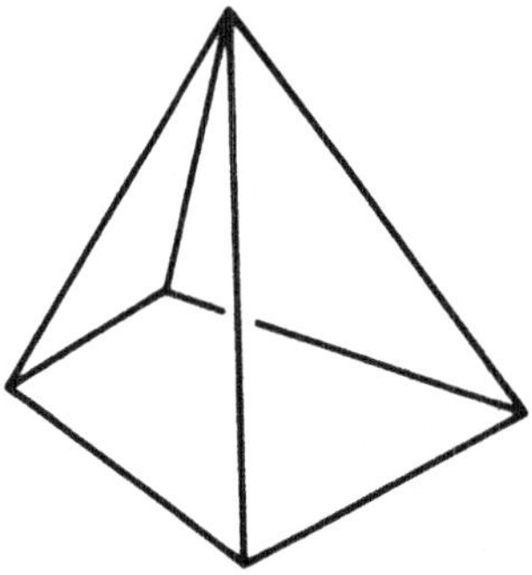

Figure 1 (a)

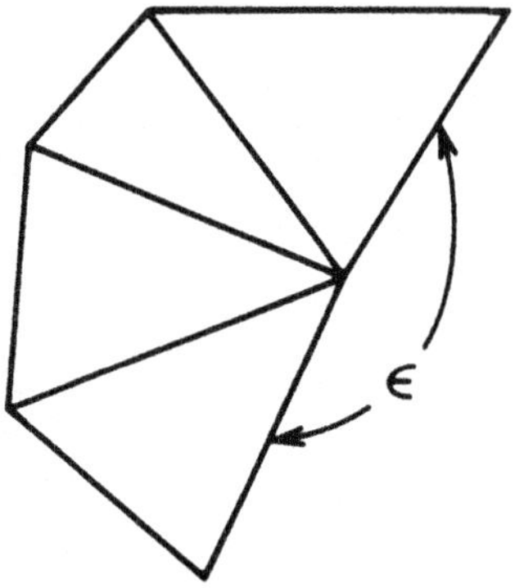


Figure 1 (b)

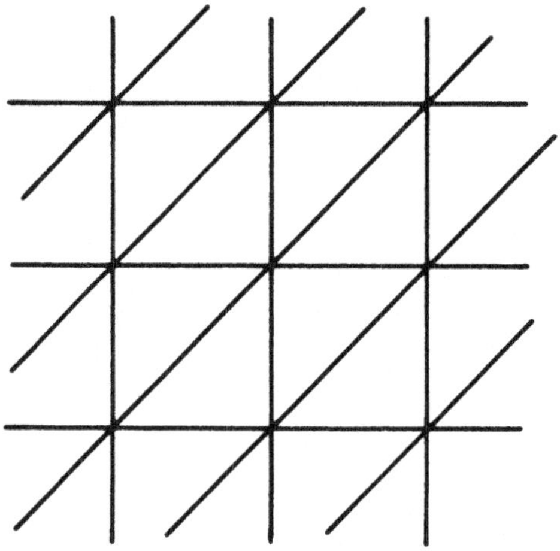

Figure 2 (a)

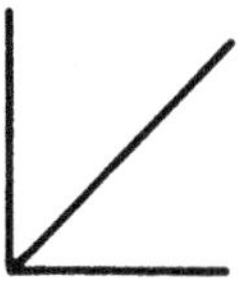

Figure 2 (b)

In three dimensions, we take flat tetrahedra and join them along flat triangles so that the curvature resides on the edges. It's difficult to visualize a curved three-dimensional space, but we can easily imagine the analog of Fig. 1a, that is, the flat figure we get once we cut along one of the faces; this is shown in Fig. 3. If one parallel transports a vector along any path enclosing an edge $l^\rho$ on which resides a deficit $\varepsilon$, one finds it rotates precisely as in the two-dimensional case, except that the rotation generator $U$ is now given by

$$U_{\mu\nu} = \frac{1}{l}\varepsilon_{\mu\nu\rho}l^{\rho} \tag{2}$$

where $l$ is the length of the edge and $\varepsilon_{\mu\nu\rho}$ is the alternating tensor in three dimensions. I won't try to draw our simplicial decomposition of flat three-dimensional space, but it is easy to understand: you get it by taking one corner of a cube and drawing vectors to the seven other corners, and then translating this figure in all three directions. The result is a cubic lattice where each cube is divided into six right tetrahedra (see Roček & Williams 1981).

Finally, in four dimensions, we take flat 4-simplices and join them along flat tetrahedra so that the curvature resides on the triangles. That's hard to draw, so I won't try; if one parallel transports a vector along any path enclosing a triangle, it rotates as before, but this time the rotation generator is, for two edges $l_a^\rho$, $l_b^\sigma$ that bound the triangle

$$U_{\mu\nu}^{(ab)} = \frac{1}{2A}\varepsilon_{\mu\nu\rho\sigma}l_a^\rho l_b^\sigma \tag{3}$$

where $A$ is the area of the triangle. By the way, it is easy to see how one can use Regge Calculus for classical relativity (Williams & Ellis 1980); since the simplices and their faces are flat, particles propagate in straight lines and their trajectories bend because of the way the blocks fit together; in particular, Williams & Ellis (1980) have shown that particles in a background which is a simplicial approximation to the Schwarzschild metric follow polygonal approximations to circular orbits.

The lattice we use in flat four dimensions is generated by translating a figure found by analogy with lower dimensions: take a hypercube, pick a corner, and connect all the 15 other corners to it. This gives a hypercubical lattice with each hypercube divided into 24 right 4-simplices.

*BIANCHI IDENTITIES*

Whereas the discrete approximations of Einstein's theory used in numerical relativity inevitably break the Bianchi identities, the curvature (deficit angles) in Regge Calculus satisfies Bianchi Identities. Regge (1961) found these identities, and has shown that they have a beautiful topological interpretation. It is simplest to

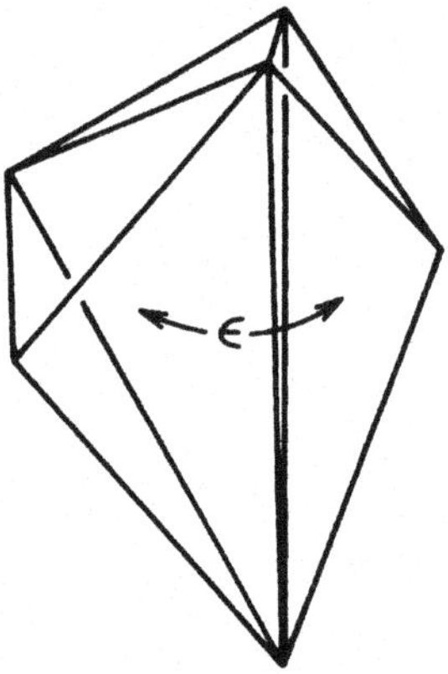


Figure 3

Figure 4

discuss these identities in three dimensions, but the generalization to higher dimensions is straightforward (in two dimensions, both in Regge Calculus and in ordinary general relativity, there are no Bianchi identities). Recall we found that if we parallel transport a vector along a path that encloses an edge, it rotates; if we parallel transport along a path that does not enclose an edge, nothing happens. Now consider the path shown in Fig. 4. It encloses, once, each of the edges shown; on the other hand, it is topologically trivial: it can be deformed without crossing any edges into a path that obviously encloses no edges. Consequently, it is clear that if we parallel transport a vector along the path shown, it will *not* rotate. This means that there is a relation among the deficits on the edges: the product of the rotation matrices on each edge is the identity matrix. This relation is the Bianchi identity, and Regge (1961) has shown that in the limit of small deficits it gives just the usual Bianchi identity of general relativity. The four-dimensional Bianchi identity states that the product of all the rotation matrices on all the triangles meeting along an edge is the identity matrix.

*THE ACTION*

One peculiarity of Regge Calculus that distinguishes it from, for example, lattice gauge theories, is that a Regge complex *is* a (singular) manifold; therefore, there is no need to invent an action: we simply evaluate the usual Einstein action for such a singular manifold.

Consider a manifold which is flat everywhere except on a triangle Δ, and assume that the curvature is constant on that triangle, that is

$$\sqrt{g}\; R = \varepsilon\, \delta^2(\Delta) \tag{4}$$

(it is not hard to show that if one parallel transports a vector around this triangle epsilon is just the deficit angle). Then

$$\int d^4x \sqrt{g}\; R = \int_\Delta d^2x\, \varepsilon = \varepsilon A \tag{5}$$

where A is the area of the triangle. If the curvature sits on many triangles, one has to sum (5), and we get Regge's (1961) result for the action

$$S = \sum_\Delta \varepsilon A = \frac{1}{2}\varepsilon^{\mu\nu}{}_{\rho\sigma} \sum_{(ab)} \varepsilon_{(ab)}\, U^{(ab)}_{\mu\nu}\, l^\rho_a\, l^\sigma_b \tag{6}$$

where in the second form, $(ab)$ labels the triangles Δ by two of their edges (see (3)).

The Regge difference equations follow from varying the action (6) with respect to independent fluctuations of the edge lengths; Regge (1961) has shown that, just as in the continuum theory, to first order one need not vary the curvature, that is:

$$\sum_\Delta \frac{\partial \varepsilon(l)}{\partial l}\delta l\, A(l) = 0 \;\Rightarrow\; \delta S = \sum_\Delta \varepsilon(l)\, \frac{\partial A(l)}{\partial l}\,\delta l. \tag{7}$$

*COMMENTS ON QUANTIZATION*

The title of this talk is perhaps somewhat misleading, as very little of what we actually describe has any direct relation to quantization. To quantize gravity using Regge Calculus one must first solve a large number of problems, some of which we discuss in the following sections. We have approached quantization from the point of view of path integrals, rather than canonical quantization; in this language, the basic problem is to define the generating functional

$$Z(J) = \int (dl)\, e^{-S(l) + \sum_{links} lJ} \tag{8}$$

There are many difficulties in defining such a path integral (see, however, Ponzano & Regge 1968, who propose a solution to essentially this problem in three dimensions; see also Hasslacher & Perry 1981). One obvious problem is to define a measure for the integration of the link lengths $l$; considerations of this led us to discuss the issue of gauge invariance in Regge Calculus, which we address in the final section of this talk. Another problem, which exists in the continuum theory and persists in Regge Calculus, is the unboundedness of the action; in the continuum theory, conformal transformations play a key role in understanding this problem (see Hawking 1979), and so we discuss conformal transformations for Regge Calculus. (Some rather speculative ideas on this subject are suggested in Roček & Williams 1981a). It does seem, however, that despite all these problems, defining at least a weak field perturbation theory is reasonably straightforward. We have actually found the free propagator for both the graviton and the Faddeev-Popov ghost fields in Regge Calculus, but as that is the main result of Roček & Williams (1981), we won't repeat the discussion here.

*CONFORMAL TRANSFORMATIONS*

We now turn to conformal transformations. The geometric meaning of a conformal transformation is clear: it corresponds to "stretching" the manifold independently at different points. How much one stretches at different points on the manifold is determined by a scalar function. This means that in Regge Calculus the stretching should be determined by a function that associates to each point a real number (Sorkin 1975). An obvious way to do this is to stretch each length by some function of the number at each of its endpoints; if we require that the product of two stretchings is again a stretching, that is, if we require that the conformal transformations form a group, we are led to an essentially unique function: some (arbitrary and irrelevant) power of the geometric mean. This is most easily shown by iterating successive infinitesmal transformations. Unfortunately, these conformal transformations are not satisfactory; geometry imposes certain restrictions on the edge lengths of a Regge skeleton: the triangle inequalities and their higher dimensional analogs.

However, a conformal transformation of the sort discussed above will not in general respect these restrictions, that is, it will take a skeleton that satisfies the inequalities into one that doesn't; if one tries to restrict the conformal transformations themselves, one loses the group property: it is possible to find two conformal transformations that preserve the inequalities but whose product does not.

The triangle inequalities are really just the statement that the volumes of all the simplices of various dimension are real; one might therefore hope that it would be possible to define conformal transformations by directly multiplying the volumes with a real number that is associated to each simplex. This would obviously preserve all the identities; unfortunately, we have been unable to find a way to make this idea work. The dynamical variables of the theory are after all the link lengths, and in the end the action of the conformal transformations must be expressible on them; there is no way we can see to express the idea of stretching volumes as a transformation of the link lengths. The basic problem is that each link is shared by many simplices, each of which want to stretch it by a different amount in order to preserve the triangle inequalities.

*PERTURBATION THEORY AROUND FLAT SPACE*

The main result of Roček & Williams (1981) showed that the weak field limit of Regge Calculus agreed with general relativity. We considered the free theory of small fluctuations about flat space, and showed that after a suitable change of variables one found the usual continuum result. I won't discuss any of the details here, but let me comment on one curious point. A hypercube has 16 vertices; therefore, if we consider all the lengths from one specific vertex to the others, we have 15 components per point. However, in four dimensions the metric is a symmetric tensor, and thus has only 10 components per point; we found that at the linearized level the theory took care of this discrepancy in a very neat way: the action was actually independent of one component (to second order in link length fluctuations about flat space) and dynamically constrained the four other extra components to vanish. Further, we found an interpretation of these extra components: they correspond to possible curvature within one hypercube, as opposed to curvature between different hypercubes.

Of the remaining 10 components, just as in the continuum theory, four are gauge transformations. In flat space it is clear what one means by gauge transformations: they are just variations of the link lengths induced by motions of the points that leave the space flat. In curved space, this idea is much more problematic.

*GAUGE TRANSFORMATIONS IN CURVED SPACE*

We must first decide what we mean by a gauge transformation; the notion we use meets the following requirements: in flat space, it reduces to the usual gauge transformations; in the continuum limit, it likewise reduces to the usual continuum transformations; in $d$ dimensions it has $d$ parameters per point; and it is an invariance, or at least, an approximate invariance of the action. We believe that *any* transformation which has these properties is a satisfactory candidate for a gauge transformation, and explicitly construct gauge transformations in the three-dimensional case. We also argue that there is at least an approximate gauge invariance in four dimensions, though it is likely that an exact invariance exists.

To construct gauge transformations in three dimensions, we need the concept of the *star* of a point. The star of a point in a three-dimensional skeleton is the set of points, links, and triangles contained in all the tetrahedra which share the point. Topologically, it looks like a triangulation of a two sphere with a point in its interior and links drawn from all the points on the surface to the interior point. The simplest possible star of a point C is shown in Fig. 5. In three dimensions, the star of a point has a unique embedding in flat four-dimensional space (but for a few exceptional cases, as was pointed out to us by Roger Penrose). This can be shown in two ways: by direct construction, and by an elegant counting argument. Both arguments are based on matching the number of given parameters, that is, the link lengths in the star to the number of parameters to be determined, that is, the coordinates of the points in flat four dimensions. If the star has $P$ points, not counting the center point, after we exclude possible rigid rotations of the star, there are $4P-6$ coordinates to determine. Now consider the simplest possible star (see Fig. 5): it has four points and $16-6=10$ links; therefore, it has a unique embedding. Now let's build up more complicated stars by adding points and links. Each time we add a point, we add four coordinates and four links; this is true if we add the point in the interior of a triangle, or on an already existing link (see Fig. 6). (The reason we have been unable to give a direct generalization of our construction to four dimensions is that in the case of a four-dimensional star, when one adds points on links or triangles, as opposed to the interior of a tetrahedra, one adds more links than coordinates). Thus we see that any 3-star has the right number of parameters to be embedded in flat 4-space. A quicker and more elegant argument can be given if one uses the Euler theorem. Recall that a 3-star is just a 2-sphere with links drawn to a point in the interior. If there are $P$ points on the 2-sphere, then there are $P$ links going to the center point. On the surface, the Euler theorem tells us that there is a relation between $T$, the number of triangles, $L$, the number of links, and $P$: $T-L+P=2$. On the other hand, $3T=2L$, and hence the number of links on the surface is $L=3P-6$, so when

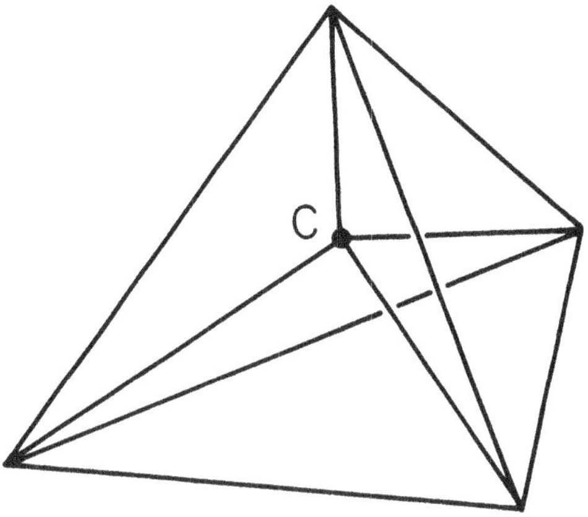


Figure 5

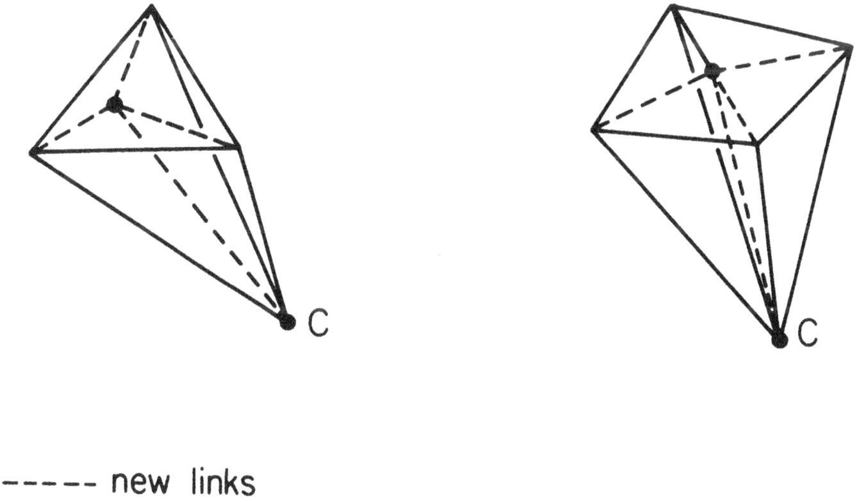


Figure 6

we add the links to the center point, we once again find that the number of links matches precisely the number of coordinates available in four dimensions.

We are now ready to define three-dimensional gauge transformations. To find the three parameter family of transformations at a point, we embed the point with its star in flat four dimensions. Now we consider motions of the point, not moving any of the other points in its star, that do not change the action; this defines a three parameter family of motions. Clearly these transformations meet all the requirements we made above.

There is another way to understand these transformations which has the virtue of emphasizing the relation of gauge invariance to the Bianchi identities and immediately generalizing to higher dimensions, but makes it hard to see that the invariance is exact. Because this discussion depends on the particular form of the action, which we have discussed only in four dimensions, we will present this form of gauge invariance only in four dimensions. In the continuum theory, gauge transformations take the form $\delta g_{\mu\nu} = \zeta_{\mu;\nu} + \zeta_{\nu;\mu}$, and invariance of the action is a consequence of the Bianchi identities (in three dimensions; in four dimensions, one needs only the contracted Bianchi identities). That is,

$$\delta S = \int d^4x \sqrt{g}\; G^{\mu\nu}\delta g_{\mu\nu} = -2\int d^4x \sqrt{g}\; G^{\mu\nu}{}_{;\nu}\zeta_\mu = 0 \iff G^{\mu\nu}{}_{;\nu} = 0. \tag{9}$$

As we saw above, in four dimensions the full Bianchi identities say that the product of all the rotation matrices associated with the deficits on the triangles sharing one edge is the identity matrix; the contracted Bianchi identities concern the rotation matrices associated with the deficits on the triangles sharing one point. The Bianchi identity naturally breaks into two parts: a symmetric part, and an antisymmetric part. If we expand the latter in powers of the deficits, we find that

$$\sum_b \varepsilon_{(ab)}\, U^{(ab)}_{\mu\nu} = o(\varepsilon^3) \tag{10}$$

where the sum is over all the triangles sharing the edge $a$ and labeled by $b$. The contracted identity takes the form

$$\sum_{(ab)} \varepsilon_{(ab)}\, U^{(ab)}_{\mu\nu}\, l^\rho_a = o(\varepsilon^3). \tag{11}$$

From (6), (7), and the normalization condition $U_{\mu\nu}U^{\mu\nu} = 2$ we find

$$\delta S = \varepsilon^{\mu\nu}{}_{\rho\sigma} \sum_{(ab)} \varepsilon_{(ab)}\, U^{(ab)}_{\mu\nu}\, l_a\, \delta l^\sigma_b. \tag{12}$$

Now if the star of a point were flat, it would be possible to choose a flat coordinate system that covered the point and its star; then moving the point would imply a set of $\delta l^\sigma_b$ which would be independent of the link under consideration. In general, we will have to change coordinate systems as we go from link to link, but it will be possible to choose these changes such that $\delta l_b^\sigma = \zeta^\sigma + \chi_b^\sigma$ where $\zeta$ is independent of $b$, and $\chi$ vanishes as the deficits $\varepsilon$ vanish. We don't know how fast $\chi$ can be made to vanish, but it seems plausible that $\chi = o(\varepsilon^2)$; then (11) and (12) imply $\delta S = o(\varepsilon^3)$. We note that in three dimensions an analogous argument holds; however, we know that in fact there is an *exact* invariance. It seems reasonable to conjecture that in four dimensions there is also an exact invariance whose precise relation to the approximate invariance found above is not clear.

Finally, it is interesting to ask, is all this needed? Perhaps we can ignore the gauge invariance, and quantize the system by blindly summing over *all* skeletons. We believe that this is *not* the case, and to study this problem we consider a simpler system with coordinate invariance. Mike Green and I have looked at a well known problem, the free particle, and have found that it is essential to fix the reparametrization invariance to get sensible results. It is well known that the Euclidean Klein-Gordon propagator can be written as a path integral

$$G(x,x') = \int_0^\infty dT \int (dx)\, e^{-\frac{1}{2}\int_0^T d\tau(\dot{x}^2+m^2)} \qquad (13)$$

It is possible to find the precise relation of this (see Roček & Williams 1981a) to a manifestly reparametrization invariant form, which, as it stands, does not make sense *because* of the infinite volume of the reparametrization group:

$$G(x,x') = \frac{1}{\text{awful}\,\infty} \int (dx)\, e^{-m\int_x^{x'} \sqrt{\dot{x}^2}} \qquad (14)$$

This derivation is a little tedious, and the interested reader is referred to Roček & Williams (1981a).

*ACKNOWLEDGEMENTS*

The number of people who have given useful comments and encouragement is too great to list here; we would particularly like to thank Stephen Hawking for many discussions and Chris Isham for his never failing enthusiasm in organizing this workshop. Finally we thank Claudio Rebbi for a very helpful comment on the subject of gauge invariance.

*REFERENCES*

Hasslacher, B. & Perry, M. J. (1981). Spin Networks are Simplicial Quantum Gravity Physics Letters 103B, 21-24.

Hawking, S. W. (1979). The path integral approach to Quantum Gravity. In General Relativity, eds. S. W. Hawking & W. Israel. Cambridge: Cambridge University Press.

Ponzano, G. & Regge, T. (1968). Semiclassical Limit of Racah Coefficients. In Spectroscopic and Group Theoretical Methods in Physics, eds. F. Bloch, S. G. Cohen, A. De-Shalit, S. Sambursky, & I. Talmi, pp. 1-58. New York, NY: John Wiley & Sons, Inc.

Regge, T. (1961). General Relativity without Coordinates. Nuovo Cimento 19, 558-571.

Roček, M. & Williams, R. M. (1981). Quantum Regge Calculus. Physics Letters 104B 31-37.

Roček, M. & Williams, R. M. (1981a). The Quantization of Regge Calculus. In preparation.

Sorkin, R. (1975). The electromagnetic field on a simplicial net. J. Math. Phys. 16, 2432-2440.

Williams, R. M. & Ellis, G. F. R. (1980). Regge Calculus and Observations: I. Formalism and Applications to Radial Motion and Circular Orbits. To appear in G.R.G.

# SPONTANEOUS SYMMETRY BREAKING IN CURVED SPACE-TIME

D.J. Toms
Imperial College, London, SW7 2BZ, England

In this talk I would like to discuss an approach which deals with some of the complications which arise when one attempts to study spontaneous symmetry breaking beyond the tree-graph level in situations where the effective potential may not be used. These situations include quantum field theory on general curved backgrounds or in flat space-times with non-trivial topologies. One example discussed briefly here is a report of some work which I have done in collaboration with L.H. Ford (Ford and Toms 1981).

## 1. INTRODUCTION

Recently there has been a great deal of work examining the role played by grand unified theories in the early universe. (See Ellis 1980;Linde 1979; Nanopoulos 1980). The usual approach is to take the effective potential (Coleman and Weinberg 1973; Jackiw 1974; Weinberg 1973) for the desired theory, include finite temperature corrections and then examine the consequences of any symmetry breaking. A problem with this is that the expressions for the effective potential used in these calculations are just those calculated in Minkowski space-time; however, they are used to deduce consequences of the field theory in curved space-time. One might expect that not only the curvature but also the space-time topology could be important, particularly in the early universe. What one really wants to be able to do is to study spontaneous symmetry breaking in curved space-time right from the start. Unfortunately there is an immediate snag if one attempts to follow the usual approach since the ground states of the theory will not in general be obtained from an effective potential.

In order to see this recall that the effective potential is defined by setting all fields in the effective action to constants:

$$V(\hat{\phi}) = \frac{1}{\mathrm{Vol}(M)} \quad \Gamma\left[\phi = \hat{\phi}\right] \quad . \tag{1}$$

Here Vol(M) denotes the volume of the space-time, and $\hat{\phi}$ denotes a constant field. The ground states of the theory are the minima of $V(\hat{\phi})$. Because it is necessary to set the fields to constants it does not make sense to talk about the effective potential in situations where constant non-zero fields are not allowed.

In curved space-time even at the classical level constant field solutions are not allowed in general. Consider a single scalar field which satisfies

$$\Box\phi + \xi R(x)\phi + U'(\phi) = 0 \tag{2}$$

where $R(x)$ is the scalar curvature, $U(\phi)$ is the classical potential, and $\xi$ is a dimensionless coupling constant which allows for a coupling of the scalar field to the background geometry. (We are assuming here that the background geometry is a classical fixed, but at this stage an arbitrary space-time. The scalar field is then treated as a test field (i.e. no back-reaction) on this classical background). Setting $\phi=\hat{\phi}$ in (2) where $\hat{\phi}$ is a constant non-zero field one finds that this requires

$$\xi R(x) = \text{constant}. \tag{3}$$

If this is to hold for a general $\xi$ then attention must be restricted to manifolds with a constant scalar curvature. An apparent way out of this is to take $\xi = 0$; however, when one considers radiative corrections to (2) there can arise an effective mass term which is not constant and which will prevent the existence of non-zero constant field solutions. (See section 3(B)). A number of authors (Domokos et al. 1975; Fleming et al. 1980; Grib et al. 1977, 1978a,b; Grib and Mostepanenko 1977) have considered the vacuum solutions to (2) in curved space-time and have found non-constant solutions in some cases.

This criticism will create problems for the broken-symmetric theory of gravity (Minkowski 1977; Zee 1979). In this approach the $\xi R\phi^2$ term in the scalar field Lagrangian gives rise to a term $\xi RV^2$ if the scalar field develops a constant vacuum expectation value V and the field is shifted in the usual manner. This term is then interpreted as the Hilbert-Einstein gravitational action where $G \propto V^{-2}$ gives the value of the Newtonian gravitational constant. If non-zero constant vacuum solutions are not allowed this will create trouble for such a scheme. Fleming (1980) has discussed a specific example where this occurs.

Another problem which may be circumvented by the observation that constant non-zero scalar fields may not be allowed is the problem of the cosmological constant which arises in theories with spontaneous symmetry breaking (Bludman and Ruderman 1977; Dreitlein 1974; Linde 1974; Veltman 1975). The problem here is that constant vacuum expectation values give rise to constant vacuum energy densities which is equivalent to having a cosmological constant about fifty orders of magnitude greater than the experimental upper bound. If constant scalar field solutions are not allowed then although non-zero vacuum energy densities may arise, they will not be constant and hence may not simply be interpreted in terms of a cosmological constant. Whether or not this still leads to trouble remains to be seen; however Davies and Unwin (1981) have recently suggested that if twisted scalar fields (Dowker and Banach 1978; Isham 1978) are responsible for symmetry breaking we may be living in a region of the universe where the vacuum energy density is small, although it could be large in other regions.

In addition to the effects of space-time curvature, there may be topological effects to be considered. It is known that quantum corrections in topologically non-trivial space-times can have an effect on vacuum stability (Banach 1981;Birrell 1981; Denardo and Spallucci 1980a,b,c; Ford 1980a,b; Gibbons 1978; Kennedy 1981; Shore 1980; Toms 1980b,c,d,1981) . This may appear more familiar if one recalls that finite temperature effects can restore symmetries (Dolan and Jackiw 1974; Weinberg 1974). Since in the functional integral for the partition function one sums over scalar fields which are periodic in Euclidean time with period $\beta$, one is really working in a flat space-time with the topology $S^1 x R^3$; that is, finite temperature effects may be regarded as having a topological origin since the Euclidean space-time one is working on is not simply connected.

If the space-time topology is non-trivial there exists the possibility of inequivalent field configurations (Banach and Dowker 1979; Dowker and Banach 1978; Isham 1978). In the case of real scalar fields which are cross-sections of real-line bundles whose base space is the space-time manifold M, the set of inequivalent real-line bundles is isomorphic to the cohomology group $H^1(M;Z_2) \approx \mathrm{Hom}(\pi_1(M),Z_2)$. (Isham 1978). If this cohomology group is non-trivial, then in addition to the usual or standard scalar field which is a cross-section of a product bundle, there will be additional twisted scalar fields which are cross-sections of non-product bundles. Since the cross-section of any non-product real-line bundle must vanish on at least one fibre (Milnor and Stasheff 1974) it is clear that the only constant twisted fields are those which vanish identically.

As a consequence of this, spontaneous symmetry breaking for twisted scalar fields is altered even at the classical level (Avis and Isham 1978). If one attempts to apply the effective potential method to study symmetry breaking beyond the tree-graph level, because the only constant twisted field is one which vanishes identically, all that one obtains is the energy density of the state $\phi=0$ (Toms 1980b). Symmetry breaking for twisted fields beyond the tree-graph level has been studied by Banach (1981), Ford (1980b), and Toms (1981).

Although only a single real scalar field has been considered, it is clear that the objections to using the effective potential discussed above hold also if there is more than one scalar field, and if they are coupled to gauge fields. In addition to these problems, Isham (1981a,b) has discussed how the breaking of a symmetry group G down to a subgroup H at the tree-level can be completely forbidden for topological reasons.

## 2. STABILITY AND THE EFFECTIVE ACTION

Although in general one cannot use the effective potential, one can still use the effective action as discussed in Toms (1981). Here, an alternate derivation of results in Ford and Toms (1981) is presented.

Consider the scalar field theory whose classical action is

$$I[\phi]=\int d^4x\ [-g(x)]^{\frac{1}{2}}\ \{\tfrac{1}{2}(\partial\phi)^2-\tfrac{1}{2}m^2\phi^2-\tfrac{1}{2}\xi R\phi^2-\tfrac{\lambda}{4!}\ \phi^4\} \tag{4}$$

Signature (+---) is used, and all quantities in (4) are bare. It was shown in Toms (1981) that the one-loop contribution to the effective action may be expressed as

$$\Gamma^{(1)}[\hat{\phi}] = -\frac{i}{2}\ \ln\mathrm{Det}\Delta_F+\frac{i}{2}\ \mathrm{Tr}\ln(1-\Delta_F\Phi) \tag{5}$$

$$= -\frac{i}{2}\ln\mathrm{Det}\Delta_F - \frac{i}{2}\sum_{n=1}^{\infty}\frac{1}{n}\mathrm{Tr}\left[(\Delta_F\Phi)^n\right]. \tag{6}$$

where $\hat{\phi}(x)$ is the background field which is taken to be arbitrary at this stage. $\Delta_F$ is the free Feynman propagator which satisfies

$$\left[\Box_x + m_R^2 + \xi_R R(x)\right]\Delta_F(x,x') = -\left[-g(x)\right]^{-\frac{1}{2}}\delta(x,x'). \tag{7}$$

(A subscript 'R' denotes a renormalized quantity). $\Phi$ is a biscalar distribution which may be taken to be

$$\Phi(x,x') = \left[-g(x')\right]^{-\frac{1}{2}}\,\delta(x',x)\left[\frac{\lambda}{2}\,\hat{\phi}^2(x)\right]. \tag{8}$$

$\Delta_F\Phi$ is shorthand for

$$\int d^4x'\left[-g(x')\right]^{\frac{1}{2}}\,\Delta_F(x,x')\Phi(x',x''). \tag{9}$$

Renormalization of the effective action proceeds as in flat space-time. (See Birrell (1981) and references therein). Assuming that we have now renormalized the one-loop effective action, the ground states will be the stable solutions to

$$\frac{\delta\Gamma[\hat{\phi}]}{\delta\hat{\phi}(x)} = 0 \tag{10}$$

where $\Gamma$ is the sum of the classical action (4) and the one-loop contribution in (5) or (6). Dropping terms of order $\lambda^2$ and higher, the wave equation including one-loop corrections (10) becomes

$$\Box\hat{\phi} + m_R^2\hat{\phi} + \xi_R R(x)\hat{\phi} + \frac{\lambda_R}{3!}\hat{\phi}^3 - i\,\frac{\lambda_R}{2}\,(\Delta_F(x,x))_f\hat{\phi} = 0 \tag{11}$$

where $(\Delta_F(x,x))_f$ represents the finite part of $\Delta_F(x,x)$ which remains after renormalization. It is observed to act like an effective mass term in (11). Noting that

$$(\Delta_F(x,x))_f = i\,\langle 0|\,\phi^2(x)\,|0\rangle_f \tag{12}$$

the result given in Ford & Toms (1981) is recovered.

Suppose that one has solved (11) for $\hat{\phi} = \hat{\phi}_V(x)$ and now one wishes to study the stability of this solution. (Note that $\hat{\phi}=0$ is always a solution). Write

$$\hat{\phi} = \hat{\phi}_V(x) + \psi(x) \tag{13}$$

and treat $\psi$ as a small perturbation. Substituting (13) in (11) and linearizing in $\psi$ one finds

$$\Box\psi + m_R^2\,\psi + \xi_R R(x)\psi + \tfrac{1}{2}\lambda_R\left[\phi_v^2(x) - i(\Delta_F(x,x))_f\right]\psi = 0. \qquad (14)$$

The solution for $\hat{\phi}_v$ will be stable provided that the solutions to (14) for the perturbations do not grow with time.

In static space-times it is sometimes very convenient to use Euclidean field theory. The condition for stability is then equivalent to the condition that the second functional derivative of the effective action have no negative eigenvalues (Toms 1981) .

There are then two main steps to perform here. The first involves calculating $(\Delta_F(x,x))_f$. This is very difficult and so far has only been done in simple examples. Even if one can calculate $(\Delta_F(x,x))_f$ the second step which involves solving (11) and the stability equation (14) will not be easy in general.

## 3. EXAMPLES

### (A) Twisted Scalar Field In $S^1 x R^3$

The simplest four-dimensional space-time which admits a twisted scalar field is a flat space-time where one of the spatial coordinates (say $x_1$) has been periodically identified with period L. In the case of a massless (at the tree-level) scalar field with a $\lambda\phi^4$ self-interaction one finds (Ford 1980b; Toms 1980a)

$$-\frac{i}{2}\,\lambda_R(\Delta_F(x,x))_f = -\,\frac{\lambda_R}{24L^2} \qquad (15)$$

which can be interpreted as a negative $(\text{mass})^2$. On the basis of one's experience with the effective potential it would appear that radiative corrections to the classical theory have made the tree-level vacuum state $\hat{\phi}_v=0$ unstable.

In order to see whether or not radiative corrections have really led to an instability of the tree-level vacuum state one requires the solution to (14). Because the twisted scalar field is antiperiodic in $x_1$ take

$$\psi(x)\alpha\; e^{-i\omega t}\; e^{-i(2n+1)\frac{\pi}{L}x_1}\; e^{-ik_2x_2-ik_3x_3} \qquad (16)$$

where $n=0, \pm 1, \pm 2, \ldots, -\infty < k_2, k_3 < \infty$. The lowest eigenfrequency is therefore given by

$$\omega_0^2 = \frac{\pi^2}{L^2}\left(1 - \frac{\lambda_R}{24\pi^2}\right) \tag{17}$$

which is seen to be positive provided that $\lambda_R \ll 1$ as one assumes for perturbation theory in $\lambda_R$ to make sense. Thus, even though radiative corrections have led to the generation of an imaginary mass - a situation one normally regards as a signature of vacuum instability - the twisted vacuum state $\hat{\phi}_v = 0$ remains stable.

If the twisted field is massive at the tree-level, in place of (15) one finds

$$-\frac{i}{2}\lambda_R\langle\Delta_F(x,x)\rangle_f = -\frac{1}{2}\lambda_R\int\frac{d^3k}{(2\pi)^3}(k^2+m_R^2)^{-\frac{1}{2}}\left[e^{L(k^2+m_R^2)^{\frac{1}{2}}}+1\right]^{-1} \tag{18}$$

where $m_R$ is the renormalized tree-level mass (Toms 1980a). In this case it may be seen that the lowest eigenfrequency of the perturbation in (14) is bounded by

$$\frac{\pi^2}{L^2} + m_R^2 - \frac{\lambda_R}{24L^2} < \omega_0^2 < \frac{\pi^2}{L^2} + m_R^2 \tag{19}$$

The vacuum state $\hat{\phi}_v = 0$ is even more stable for the massive field in the sense that the lowest eigenfrequency is larger than in the massless case.

Finally, consider a double-well tree-level potential

$$U(\phi) = \frac{\lambda}{4!}(\phi^2 - a^2)^2. \tag{20}$$

One normally regards $\phi = 0$ as unstable, and $\phi = \pm a$ as the stable ground states. However, if $\phi$ is a twisted field, $\phi = \pm a$ are not allowed solutions. For this potential at the tree-level the lowest eigenfrequency of the perturbation in (13) is found to be

$$\omega_0^2 = \frac{\pi^2}{L^2} - \frac{1}{6}\lambda_R a^2. \tag{21}$$

This quantity is negative, corresponding to an unstable perturbation which grows with time, provided that $L > L_c$ where

$$L_c = \frac{\pi}{a}\sqrt{\frac{6}{\lambda_R}} . \tag{22}$$

If $L < L_c$, then the perturbations do not grow with time and so $\hat{\phi}_v=0$ remains locally stable; that is, the twisted ground state can sit at the top of the $\phi=0$ hump of the potential (20) and remain locally stable. This behaviour was first discussed by Avis & Isham (1978). One-loop corrections do not significantly alter this picture except that the critical length at which $\hat{\phi}_v=0$ becomes unstable is decreased slightly from (22) to

$$(1 - \frac{\lambda_R}{24\pi^2})^{\frac{1}{2}} L_c \tag{23}$$

(B) Instabilities in an Expanding Universe

Consider massless $\lambda\phi^4$ theory in the spatially flat, topologically trivial, Robertson-Walker model whose scale factor is given by a power law:

$$ds^2 = dt^2 - a^2(t)(dx_1^2+dx_2^2+dx_3^2) \tag{24}$$

$$a(t) = \sigma t^c \tag{25}$$

where $\sigma$,c are constants. The scalar curvature is

$$R(t) = 6c\ (2c-1)t^{-2}. \tag{26}$$

If $0 \leq \xi < \frac{1}{6}$, then $\hat{\phi}_v=0$ is the locally stable tree-level vacuum state for all choices of c in (25).

In order to examine the stability at the one-loop level one requires an evaluation of (12). In the case $\xi=0$ this has been computed by Bunch & Davies (1978) using point-splitting regularization where it was found that

$$<0|\ \phi^2|\ 0>_f = - \frac{1}{96\pi^2}\ R\ \ell n\left(\frac{|R|}{\mu^2}\right) . \tag{27}$$

Here $\mu$ is an arbitrary constant (renormalization point) with units of mass which is not determined by the theory. The result for general $\xi$ may be found by using an argument based on the renormalization group and the invariance of (14) under a change of renormalization point. If we rescale $\mu \to \mu'$, then we must change $\xi \to \xi'$ by

$$\xi' = \xi + \frac{\lambda}{16\pi^2}(\xi - \tfrac{1}{6})\,\ell n\left(\frac{\mu'}{\mu}\right) \tag{28}$$

which is found (for $\xi \neq \frac{1}{6}$) by integrating the renormalization group equation for $\xi$. (See Ford & Toms (1981) for details). In order that (14) be invariant under this rescaling we must have

$$<0|\phi^2|0>_f = \frac{1}{16\pi^2}(\xi - \tfrac{1}{6})\,R\,\ell n\left(\frac{|R|}{\mu^2}\right) \tag{29}$$

An argument is presented in Ford & Toms (1981) to show that this term would be expected to be present in general. For a wide class of space-times, but certainly not all (see for example Candelas (1979)), it is found to be the only term present.

The exact solution to (14) is given in Ford & Toms (1981). The important thing is that there is now the possibility of unstable perturbations. This can be seen roughly by examining the sign of (29). Neglecting all constants, using (26) one has

$$<0|\phi^2|0>_f \sim \pm\,\ell n\,(\mu t) \tag{30}$$

where the plus sign holds if $c > \frac{1}{2}$ and the negative sign holds if $c < \frac{1}{2}$. If these one-loop corrections get too negative one might expect an instability to arise. If $c<\frac{1}{2}$, then for $t \lesssim \mu^{-1}$ the state $\hat{\phi}_v=0$ is expected to be stable, while for $t \gtrsim \mu^{-1}$ it is expected to be unstable. If $c>\frac{1}{2}$, then for $t \lesssim \mu^{-1}$ one expects $\hat{\phi}_v=0$ to be unstable, and if $t \gtrsim \mu^{-1}$ it is expected to be stable. These features are in qualitative agreement with results based on the exact solution.

The important feature here is that there is some critical time at which the perturbations either begin to grow, corresponding to $\hat{\phi}_v=0$ becoming unstable, or else become oscillatory, corresponding to $\hat{\phi}_v=0$ becoming stable. The discrete $Z_2$ symmetry $\phi \to -\phi$ can either be broken or restored as the universe expands. The critical time at which this occurs is invariant under the rescaling $\mu \to \mu'$, $\xi \to \xi'$. For large values of $|\ell n\mu t|$ the unstable perturbations behave like

$$t^{\frac{1-3c}{2}} \, |\ell n\mu t|^{-\frac{1}{4}} \exp \left\{\frac{2}{3} \, |B|^{\frac{1}{2}} \, |1-c|^{\frac{3}{2}} |\ell n\mu t|^{\frac{3}{2}}\right\} \tag{31}$$

where B is a constant. This growth is faster than any power of t, and hence faster than any power of the scale factor.

## 4. DISCUSSION AND CONCLUSIONS

We have seen how both the topology and curvature of a space-time may affect the stability of the vacuum state. There can be critical length scales or times beyond which symmetries may be broken or restored in certain cases. These features are certainly not present in Minkowski space-time and so would not show up in the usual types of early universe calculations.

Abbott (1981) has attempted to study the gravitational effects on symmetry breaking in an SU(5) model by adding an $R\phi^2$ term onto the standard flat space-time result for the effective potential. (See also Fujii (1981)). This has nothing to do with the effects discussed here as here the quantum corrections were computed in curved space-time, not in flat space-time. Just adding on an $R\phi^2$ term suffers from the criticisms of section 1 above.

Whether or not the effects discussed here will be important must await the application to more physically interesting models. It would also be interesting to include finite temperature effects, although how to do this in an expanding universe would appear to present difficulties. (For a recent discussion of the application of statistical ideas to the early universe see Dresden (1981)).

## ACKNOWLEDGEMENTS

I would like to thank L.H. Ford, C.J. Isham, and G. Kunstatter for useful discussions. I am grateful to the Natural Sciences and Engineering Research Council of Canada for financial support.

## REFERENCES

Abbott, L.F. (1981). Gravitational effects on the SU(5) breaking phase transition for a Coleman-Weinberg potential. CERN report TH.3018.

Avis, S.J. & Isham, C.J. (1978). Vacuum solutions for a twisted scalar field. Proc. Roy. Soc. (London) Ser. A, 363, p.581.

Banach, R. (1981). Effective potentials for twisted fields. J. Phys. A, 14, p.901.

Banach,R. & Dowker, J.S. (1979). Automorphic field theory: some mathematical issues. J. Phys. A, 12, p.2527.

Birrell, N.D. (1981). Interacting quantum field theory in curved spacetime. In Quantum Gravity II: A Second Oxford Symposium, ed. C.J.Isham, R. Penrose and D.W. Sciama. Oxford: Oxford University Press.

Bludman, S.A. & Ruderman, M.A. (1977). Induced cosmological constant expected above the phase transition restoring the broken symmetry. Phys. Rev. Lett., 38, p.255.

Bunch, T.S. & Davies, P.C.W. (1978). Non-conformal renormalized stress-tensors in Robertson-Walker spacetimes. J. Phys. A,11,p.1315.

Candelas, P. (1979). Vacuum polarization in Schwarzschild spacetime, Phys. Rev. D, 21, p.2185.

Coleman, S. & Weinberg, E. (1973). Radiative corrections as the origin of spontaneous symmetry breaking. Phys. Rev. D, 7, p.1888.

Davies, P.C.W. & Unwin, S.D. (1981). Why is the cosmological constant so small? Proc. Roy. Soc. (London) Ser.A., 377, p.147.

Denardo, G.& Spallucci, E. (1980a). Dynamical mass generation in $S^1xR^3$. Nucl. Phys. B, 169, p.514.

——————.(1980b). Symmetry restoration in conformally flat metrics. University of Trieste report.

——————.(1980c). Symmetry breaking and restoration in the Einstein universe. University of Trieste report.

Dolan, L. & Jackiw, R. (1974). Symmetry behaviour at finite temperature. Phys. Rev. D, 9,p.3320.

Domokos, G., Janson, M.M. & Kovesi-Domokos, S. (1975). Possible cosmological origin of spontaneous symmetry breaking. Nature (London), 257, p.203.

Dowker, J.S. & Banach, R. (1978). Quantum field theory on Clifford-Klein space-times. J. Phys. A, 11, p.2255.

Dreitlein, J. (1974). Broken symmetry and the cosmological constant. Phys. Rev. Lett. 33, p. 1243.

Dresden, M. (1981). Are statistical notions applicable in the early universe? Stony Brook report. ITP-SB-81-27.

Ellis, J. (1980). Grand Unified theories. CERN report TH.2942.

Fleming, H. (1980). On a broken-symmetric theory of gravity. Phys. Rev. D, 21, p.1690.

Fleming, H., Lyra, J.L., Prado, C.P.C. & Silveira, V.L.R, (1980). Spontaneous breakdown of symmetry in cosmological models. Lett. Nuovo Cimento, 27, p.261.

Ford, L.H. (1980a). Vacuum polarization in a non-simply connected space-time. Phys. Rev. D. 21, p.933.

——— (1980b). Instabilities in interacting quantum field theories in non-Minkowskian space-times. Phys. Rev. D, 22, p.3003.

Ford, L.H. & Toms, D.J. (1981). Dynamical Symmetry breaking due to radiative corrections in cosmology. Imperial College report.

Fujii, Y. (1981). Effects of space-time curvature on a cosmological first-order phase transition. Phys. Lett. B, 103, p.29.

Gibbons, G.W. (1978). Symmetry restoration in the early universe. J. Phys. A, 11, p.1341.

Grib, A.A. & Mostepanenko, V.M. (1977). Spontaneous breaking of gauge symmetry in a homogeneous isotropic universe of the open type. JETP Lett., 25, p.277.

Grib, A.A., Mostepanenko, V.M. & Frolov, V.M. (1977). Spontaneous breaking of gauge symmetry in a nonstationary isotropic metric. Theor. and Math. Phys., 33, p.869.

———.(1978a). Breaking of conformal symmetry and quantization in curved space-time. Theor. and Math. Phys., 37, p.347.

———.(1978b). Spontaneous breaking of CP symmetry in a non-stationary isotropic metric. Theor. and Math. Phys., 37, p.975.

Isham, C.J. (1978). Twisted quantum fields in a curved space-time. Proc. Roy. Soc. (London) Ser.A, 362, p.383.

———.(1981a). Spontaneous symmetry breaking and topological charge. Phys. Lett. B, 102, p.251.

———.(1981b). Space-time topology and spontaneous symmetry breaking. Imperial College report ICTP/80/81-22.

Jackiw, R. (1974). Functional evaluation of the effective potential. Phys. Rev. D, 9, p.1686.

Kennedy, G. (1981). Topological symmetry restoration. Phys. Rev.D, 23, p.2884.

Linde, A.D. (1974). Is the Lee constant a cosmological constant? JETP Lett., 19, p.183.

———. (1979). Phase transitions in gauge theories and cosmology. Rep. Prog. Phys., 42, p.389.

Milnor, J.W. & Stasheff, J.D. (1974). Characteristic Classes. Princeton, N.J.: Princeton University Press.

Minkowski, P. (1977). On the spontaneous origin of Newton's constant. Phys. Lett. B, 71, p.419.

Nanopoulos, D.V. (1980). Cosmological implications of grand unified theories. CERN report TH.2871.

Shore, G.M. (1980). Radiatively induced spontaneous symmetry breaking and phase transitions in curved space-time. Ann. Phys. (N.Y.), 128, p.376.

Toms, D.J. (1980a). Casimir effect and topological mass. Phys. Rev. D, 21, p.928.

———. (1980b). Symmetry breaking and mass generation by space-time topology. Phys. Rev. D, 21, p.2805.

———. (1980c). Interacting twisted and untwisted scalar fields in a non-Minkowskian space-time. Ann. Phys. (N.Y.), 129, p.334.

———. (1980d). Scalar electrodynamics in a non-simply connected space-time. Phys. Lett.A, 77, p.303.

———. (1981). Vacuum stability and symmetry breaking in non-Minkowskian space-times. Phys. Rev. D, to appear.

Veltman, M. (1975). Cosmology and the Higgs mass. Phys. Rev. Lett., 34, p.777.

Weinberg, S. (1973). Perturbative calculations of symmetry breaking. Phys. Rev. D, 7, p.2887.

———. (1974). Gauge and global symmetries at high temperatures. Phys. Rev. D, 9, p.3357.

Zee, A. (1979). Broken-symmetric theory of gravity. Phys. Rev. Lett., 42, p.417.

# SPONTANEOUS SYMMETRY BREAKING NEAR A BLACK HOLE†

M.S. Fawcett & B.F. Whiting,
Department of Applied Mathematics and Theoretical Physics,
Silver Street, Cambridge, CB3 9EW

## 1 INTRODUCTION

Black hole spacetimes provide a natural situation in which phase transitions are likely to be important. As a black hole emits thermal radiation it gets smaller and its temperature rises. At some point it will pass through the critical temperature of whatever theory describes the universe. Phase transitions at present under much discussion are the Weinberg/Salam transition at about 300 GeV, and the $SU_5$ GUTS model transition at $10^{15}$GeV. The masses at which a black hole would be radiating at these energies are respectively about $10^{11}$g and $10^{-2}$g. Both these are mini black holes.

Because of redshifting in a general curved spacetime, the temperature is a function of position. In a black hole spacetime the temperature is higher near the horizon than at large distances. This raises the possibility that one phase, the symmetric phase, could exist near the hole, while at large distances the field would be in the low temperature broken phase. At the critical temperature the mass of fluctuations goes to zero, so one could expect a sudden increase in the rate of emission which might be detectable as a burst of gamma rays (Hawking (1981)).

In this talk I will describe the calculation of $\langle\phi^2\rangle$ in a scalar theory, which determines the effective action at the one loop level. We compute values of $\langle\phi^2\rangle$ for a black hole in thermal equilibrium and for an incoming flux of thermal radiation at infinity, and take their difference to obtain results for a black hole radiating into empty space (see Hawking 1981 for discussion), since this more nearly approximates the situation for small black holes radiating at the present time. It turns out that $\langle\phi^2\rangle$ differs significantly from the flat space result only in a rather small region around the black hole. This suggests that $\langle\phi\rangle$ ,

† Delivered by M.S. Fawcett

the quantity that signals a phase transition, will not be significantly different near the horizon from at infinity. However, a thorough investigation remains to be completed.

In section 2 I will review the functional formalism relevant to spontaneous symmetry breaking. Section 3 will describe the regularization of the formal expression for $\langle\phi^2\rangle$ and section 4 will specialize this discussion to the Schwarzschild spacetime. In section 5 I will talk about the numerical methods needed to evaluate $\langle\phi^2\rangle$ . The evaluation of $\langle\phi^2\rangle$ for an incoming flux of purely thermal radiation follows in section 6. Finally, section 7 contains the results and a brief discussion.

## 2 SYMMETRY BREAKING AND RESTORATION

Spontaneous symmetry breaking refers to the situation where the vacuum of a theory does not respect a symmetry that the Lagrangian has. A simple example is a scalar field $\phi$ with a potential that is symmetric about $\phi = 0$, but for which 0 is not a minimum. Such a Lagrangian is invariant under $\phi \to -\phi$ but the expectation value of the field in the vacuum state is not zero. Only $\phi = 0$ would respect the symmetry under change of sign. To see whether a symmetry is broken we need some way of calculating $\langle\phi\rangle$ the expectation value of $\phi$. This can be done using the effective action $\Gamma[\phi]$ . The function $\phi$ which minimises $\Gamma$ is the expectation value of the field. The lowest approximation to $\Gamma$ is just $S[\phi]$ , the classical action. The quantum corrections to this are temperature dependent, introducing the possibility that the symmetry properties of the theory may also depend on temperature. Much work has been done on the restoration of spontaneously broken symmetries at high temperatures in flat space, notably Kirzhnits & Linde (1976), Weinberg (1974) and Dolan & Jackiw (1974). The problem can be slightly simplified in flat space because one can assume that the field which minimises $\Gamma$ will be independent of position, reducing what in general is a differential equation to an algebraic one. One uses instead of $\Gamma$ the effective potential $V(\phi)$, a function rather than a functional of $\phi$, which is $\Gamma$ with a constant argument and a factor of the volume of space-time removed. The above authors find that the broken symmetry is indeed restored above a critical temperature for simple scalar field theories, and scalar theories coupled to gauge fields. One can also sometimes make the assumption of the constancy of $\phi$ in a curved space-time, and some calculations have been done in these special cases. See e.g. Gibbons (1978).

In general, however, one must assume that $\langle\phi\rangle$ will be a function of position, and use the effective action. I shall briefly outline the definition of the effective action, and the perturbative method of evaluating it called the loop expansion. See e.g. Abers & Lee (1973).

The effective action $\Gamma$ is defined by the following expressions. In the path integral, the sum is over all the paths which are periodic in imaginary time with period $\beta = 1/T$, T being the temperature. This condition on the fields means we have thermal equilibrium at temperature T. We will therefore work entirely in Euclidean metrics, i.e. with signature ++++

$$e^{-W} = Z = \int [d\phi]\, e^{-S[\phi] - \int J\phi\sqrt{g}\, d^4x} \tag{2.1}$$

$$\Phi(x) \equiv \frac{1}{\sqrt{g}} \frac{\delta W}{\delta J(x)} = \langle\phi\rangle \tag{2.2}$$

then $$\Gamma[\Phi] = W[J] - \int J\Phi\sqrt{g}\, d^4x \tag{2.3}$$

where J is a functional of $\Phi$ through (2.2)

$$2.2\ \&\ 2.3 \Rightarrow \quad \frac{1}{\sqrt{g}} \frac{\delta\Gamma[\Phi]}{\delta\Phi(x)} = -J(x) \tag{2.4}$$

By (2.2) $\Phi = \langle\phi\rangle$ i.e. $\Phi$ is the expectation value of the field in the presence of a source, and is sometimes called the classical field although it is not a solution of the classical field equations. In fact it is a solution of the differential equation (2.4), which justifies calling $\Gamma$ the effective action.

There is a systematic way of evaluating $\Gamma$ in perturbation theory called the loop expansion. This consists of expanding the action in the functional integral about a field $\phi_0$ which satisfies the classical field

$$\frac{1}{\sqrt{g}} \frac{\delta S}{\delta\phi_0} = -J(x) \tag{2.5}$$

This means there is no linear term in the expansion. One then uses the quadratic part as a free Lagrangian and expands perturbatively in the remaining interaction terms.

The Lagrangian density we will be using is:

$$\mathcal{L} = \tfrac{1}{2}(\partial_\mu\phi)^2 - \tfrac{1}{2}m^2\phi^2 + \tfrac{g}{4}\phi^4 \tag{2.6}$$

The classical field equation is:

$$-(\Box + m^2)\phi + g\phi^3 = -J \tag{2.7}$$

Expanding L about a solution $\phi_0$ of (2.7) one obtains:

$$\begin{aligned}\mathcal{L}(\phi_0+\phi) + J(\phi_0+\phi) \\ = \mathcal{L}(\phi_0) + J\phi_0 + \tfrac{1}{2}(\partial_\mu\phi)^2 + \tfrac{1}{2}\phi^2(3g\phi_0^2 - m^2) \\ + g\phi_0\phi^3 + \tfrac{g}{4}\phi^4\end{aligned} \tag{2.8}$$

Thus

$$\Gamma[\Phi] = S[\phi_0] + \int J(\phi_0 - \phi)\sqrt{g}\,d^4x - \ln\int[d\phi]\exp\left[-\int d^4x\sqrt{g}\left\{\tfrac{1}{2}(\partial_\mu\phi)^2 + \tfrac{1}{2}\phi^2(3g\phi_0^2 - m^2) + g\phi_0\phi^3 + \tfrac{g}{4}\phi^4\right\}\right] \tag{2.9}$$

The one loop approximation is obtained by keeping only the quadratic term in the exponential of the path integral

$$\Gamma_1[\Phi] = S[\phi_0] + \int J(\phi_0 - \phi)\sqrt{g}\,d^4x - \ln\int[d\phi]\exp\left[-\int d^4x\sqrt{g}\left\{\tfrac{1}{2}(\partial_\mu\phi)^2 + \tfrac{1}{2}\phi^2(3g\phi_0^2 - m^2)\right\}\right] \tag{2.10}$$

The classical field $\Phi$ is to be calculated using (2.4). One finds that to $O(\hbar)$ $\Phi = \phi_0$, so we can write:

$$\Gamma_1[\Phi] \simeq S[\Phi] - \ln\int[d\phi]\exp\left[-\int d^4x\sqrt{g}\left\{\tfrac{1}{2}(\partial_\mu\phi)^2 + \tfrac{1}{2}\phi^2(3g\Phi^2 - m^2)\right\}\right] \tag{2.11}$$

So for a zero source, the classical field is a solution of

$$\frac{\delta\Gamma_1}{\delta\Phi} = \frac{\delta S}{\delta\Phi} + 3g\langle\phi^2\rangle\Phi = 0$$

i.e.

$$-\Box\Phi + \Phi\left(g\Phi^2 - m^2 + 3g\langle\phi^2\rangle\right) = 0 \tag{2.12}$$

We must also ensure that the solution to (2.12) is the minimum of $\Gamma$ so that it is stable, i.e. that,

$$\frac{\delta^2\Gamma}{\delta\Phi^2} \geqslant 0 \tag{2.13}$$

Equivalently we want the eigenvalues for perturbations of (2.12) to be non-negative. Perturbing the operator in (2.12) we obtain:

$$-\Box(\delta\phi) + \delta\phi\left(3g\Phi^2 - m^2 + 3g\langle\phi^2\rangle\right) + 3g\Phi\,\delta\langle\phi^2\rangle \tag{2.14}$$

We can ignore the last term as it is $O(g^2)$, $g \ll 1$. So we require that the lowest eigenvalue of the operator

$$A[\Phi] = -\Box + 3g\Phi^2 - m^2 + 3g\langle\phi^2\rangle$$

be positive for the given solution to (2.12). In flat space we can assume that $\Phi$ is constant, and we can immediately solve (2.12):

$$\Phi = 0 \qquad \text{or} \qquad \Phi = \pm\sqrt{\frac{m^2}{g} - 3\langle\phi^2\rangle} \tag{2.15}$$

The operator A whose eigenvalues must be positive for stability is:

$$A[0] = -\Box - m^2 + 3g\langle\phi^2\rangle \tag{2.16}$$

or

$$A\left[\pm\sqrt{\frac{m^2}{g} - 3\langle\phi^2\rangle}\right] = -\Box + 2(m^2 - 3g\langle\phi^2\rangle) \tag{2.17}$$

The spectrum of $-\Box$ is bounded below by zero, so the lowest eigenvalue of $A[0]$ is $-m^2 + 3g\langle\phi^2\rangle$. Thus the classical field is zero provided

$$-m^2 + 3g\langle\phi^2\rangle \geqslant 0 \tag{2.18}$$

The lowest eigenvalue of $A\left[\pm\sqrt{\frac{m^2}{g} - 3\langle\phi^2\rangle}\right]$ is $2(m^2 - 3g\langle\phi^2\rangle)$ and is positive provided:

$$-m^2 + 3g\langle\phi^2\rangle < 0 \tag{2.19}$$

Thus $\Phi = \pm\sqrt{\frac{m^2}{g} - 3\langle\phi^2\rangle}$ provided (2.19) holds. (All these statements are valid to one loop, of course).

The fluctuations $\langle\phi^2\rangle$ are a function of temperature, being zero at $T = 0$, and rising from that value. So at $T = 0$, (2.19) holds and $\Phi = \pm\frac{m}{\sqrt{g}}$ and the symmetry is broken. As T increases, there will come a point at which $\langle\phi^2\rangle = \frac{m^2}{3g}$. $\Phi$ will have decreased to zero, and above this point (2.18) holds and $\Phi = 0$ is the only stable solution. The symmetry under $\phi \to -\phi$ has been restored.

The calculation of $\langle\phi^2\rangle$ in flat space has been done by several people (see e.g. Linde 1979 for a review), the result for a massless free scalar field being:

$$\langle\phi^2\rangle = \frac{T^2}{12} \tag{2.20}$$

This means that the critical temperature $T_c$ is given, to one loop level by:

$$T_c^2 = \frac{4m^2}{g} \tag{2.21}$$

$T_c$ is large because g is small. In a general curved spacetime the calculation of $\langle\phi^2\rangle$, the solution of (2.12) and the checking of stability are more complicated.

The quantity $<\phi^2>$ in equation (2.12) should really be calculated using the path integral of (2.11) i.e. it depends on $\Phi$ which is a solution of (2.12). Thus (2.12) is not such a simple equation as it appears. This can be dealt with in the flat space analytic treatment where $\Phi$ is a constant. One calculates $<\phi^2>$ for an arbitrary mass, and then puts in the value of the temperature dependent mass that follows from (2.12). This is called the self consistent approximation. In the case of a curved spacetime where $\Phi$ is not constant, and calculations are being done numerically, one must implement the self consistent approximation iteratively - one calculates $<\phi^2>$ for a given constant mass, solves (2.12), puts the resultant $\Phi$ back into the expression for $<\phi^2>$ and proceeds. Since the calculation of $<\phi^2>$ is costly, we don't want to have to do this too many times.

## 3 THE CALCULATION OF $<\phi^2>$

The quantity $<\phi^2>$ needs a definition before it can be calculated. It is the regularized value of the Green's function as the two points are brought together. Thus if the operator in the wave equation for the scalar field is O, then a formal expression for $<\phi^2>$ is:

$$\langle\phi^2\rangle = \left[\sum_n \frac{\phi_n^2}{\lambda_n}\right]_{\text{regularized}} \tag{3.1}$$

where

$$O\phi_n = \lambda_n \phi_n \tag{3.2}$$

and $\phi_n$ is normalised by

$$\int d^4x \sqrt{g}\, \phi_n^2 = 1$$

if the space is compact.

If the space is compact, the eigenvalues $\lambda_n$ are discrete: if non-compact the spectrum is continuous and the sum becomes an integral. The sum or integral diverges and must be regularised. To see exactly how it diverges we can use the heat kernel or proper time method (de Witt 1979, Hawking, 1977). Let $K(x,x',t)$ be the heat kernel of the heat equation

$$\frac{\partial K}{\partial t} + O_x K = 0$$
$$K(x,x',0) = \delta(x,x') \tag{3.3}$$

Then K has a representation

$$K(x,x',t) = \sum_n \phi_n(x)\phi_n(x')e^{-\lambda_n t} \tag{3.4}$$

From (3.1),

$$\langle\phi^2\rangle = \left[\int_0^\infty K(x,x,t)\,dt\right]_{\text{regularised}} \tag{3.5}$$

The integral converges at the upper bound, but it is known (Gilkey 1975) that for small t, K has an asymptotic expansion:

$$K(x,x,t) \sim a_0(x)\,t^{-2} + a_1(x)\,t^{-1} + \dots \tag{3.6}$$

where for the operator

$$\begin{aligned} O &= -\Box + \mu^2(x) + \xi R \\ a_0 &= \frac{1}{16\pi^2} \\ a_1 &= \frac{1}{16\pi^2}\left(\left(\tfrac{1}{6} - \xi\right)R - \mu^2(x)\right) \end{aligned} \tag{3.7}$$

If we write (3.4) as

$$K(x,x,t) = \int_0^\infty \phi_\lambda^2(x)\,e^{-\lambda t}\,n(\lambda)\,d\lambda \tag{3.8}$$

where $n(\lambda)$ is the density of eigenvalues, we can show that

$$\phi_\lambda^2(x)\,n(\lambda) \sim a_0(x)\lambda + a_1(x) + \dots \tag{3.9}$$

This means that a regularised expression for $\langle\phi^2\rangle$ is:

$$\langle\phi^2\rangle = \int_0^\infty \left[\frac{1}{\lambda}\phi^2(\lambda,x)\,n(\lambda) - a_0 - \frac{a_1}{\lambda+c}\right] d\lambda \tag{3.10}$$

where c is an arbitrary real constant. This means our expression contains an arbitrary multiple of $a_1$. The coefficient of this term must be determined by experiment, or by imposing some physical condition. A reasonable condition that is sufficient to determine c is that in flat space at zero temperature $\langle\phi^2\rangle$ must vanish for any constant mass. If the mass is a function of position that becomes constant at infinity, then we would require that $\langle\phi^2\rangle$ be zero at infinity at zero temperature. In the next section I will show how to use this condition to derive an expression for $\langle\phi^2\rangle$ in the Schwarzschild spacetime.

## 4. $\langle\phi^2\rangle$ IN SCHWARZSCHILD SPACE-TIME

I will now derive in some detail the expression for $\langle\phi^2\rangle$ in a Schwarzschild space-time. The metric for Lorentzian signature Schwarzschild is:

$$ds^2 = -\left(1-\frac{2M}{r}\right)dt^2 + \left(1-\frac{2M}{r}\right)^{-1}dr^2 + r^2 d\Omega \tag{4.1}$$

The Euclidean metric is obtained by the substitution

$$\tau = it$$
$$ds^2 = \left(1-\frac{2M}{r}\right)d\tau^2 + \left(1-\frac{2M}{r}\right)^{-1}dr^2 + r^2 d\Omega \tag{4.2}$$

This metric can be made regular at the point $r = 2M$ by identifying the coordinate $\tau$ with period $8\pi M$. The singularity is then merely a coordinate singularity like that at the origin of polar coordinates.

Quantum field theory at finite temperature is performed using fields that are periodic in imaginary time with period $i\beta$, $\beta = 1/T$. Because of the identification of $\tau$, all fields on Euclidean Schwarzschild are periodic with period $8\pi M$. This means we are describing a black hole in thermal equilibrium with radiation at a temperature $T = 1/8\pi M$. To obtain results for a black hole radiating into empty space, one would have to subtract out the contribution of radiation from the heat bath at infinity. This contribution is found by scattering thermal radiation off the black hole - this calculation has been done by Bernard Whiting, which he discusses in section 6.

We wish to calculate $\langle\phi^2\rangle$ to one loop in the self consistent approximation. This approximation involves using a position dependent mass generated by $\langle\phi^2\rangle$ so we will have to iterate the procedure, each time putting in the mass calculated in the previous step. Therefore I will derive an expression for $\langle\phi^2\rangle$ that involves a mass that is an arbitrary function of the radial coordinate r.

We use (2.2)

$$\langle\phi^2\rangle = \left[\sum_n \frac{\phi_n^2}{\lambda_n}\right]_{reg}$$

with

$$\left(-\Box + \mu^2(x)\right)\phi_n = \lambda_n \phi_n \tag{4.3}$$

The operator $\Box$ on scalars in Euclidean Schwarzschild is:

$$\left(1-\frac{2M}{r}\right)^{-1}\frac{\partial^2}{\partial\tau^2} + \frac{1}{r^2}\frac{\partial}{\partial r}\left(r(r-2M)\frac{\partial}{\partial r}\right)$$
$$+\frac{1}{r^2}\frac{1}{\sin\theta}\frac{\partial}{\partial\theta}\left(\sin\theta\frac{\partial}{\partial\theta}\right) + \frac{1}{r^2}\frac{1}{\sin^2\theta}\frac{\partial^2}{\partial\psi^2} \tag{4.4}$$

Because $\mu^2$ is a function of r only, equation (4.3) is separable:

$$\phi = T_n(\tau)\,Y_{\ell m}(\theta,\psi)\,R_{n\ell p}(r) \tag{4.5}$$

where

$$\frac{d^2T}{d\tau^2} = -c_1 T_n \tag{4.6}$$

$$\frac{1}{\sin\theta}\frac{\partial}{\partial\theta}\left(\sin\theta\frac{\partial Y_{\ell m}}{\partial\theta}\right) + \frac{1}{\sin^2\theta}\frac{\partial^2 Y_{\ell m}}{\partial\psi^2} = c_2 Y_{\ell m} \tag{4.7}$$

$$\frac{1}{r^2}\frac{d}{dr}\left(r(r-2M)\frac{dR_{n\ell p}}{dr}\right) + \left(\lambda-\mu^2(r)+\frac{c_2}{r^2}-c_1\frac{r}{r-2M}\right)R_{n\ell p} = 0 \tag{4.8}$$

The separation constants $c_1$ and $c_2$ are determined by the conditions of periodicity in the coordinates.

The solutions to (4.6) are:

$$T_n = A\sin(\sqrt{c_1}\,\tau) + B\cos(\sqrt{c_1}\,\tau)$$

$\tau$ is identified with period $\beta = 8\pi M$, so we must have $c_1 = \left(\frac{2\pi}{\beta}\right)^2 n^2$ where n is a positive or negative integer or zero.

So we choose for the $T_n$ the functions:

$$T_0 = \frac{1}{\sqrt{\beta}}$$
$$T_n = \sqrt{\frac{2}{\beta}}\sin\left(\frac{2\pi n}{\beta}\tau\right) \qquad n>0$$
$$T_n = \sqrt{\frac{2}{\beta}}\cos\left(\frac{2\pi n}{\beta}\tau\right) \qquad n<0 \tag{4.9}$$

The normalization is chosen so that

$$\int_0^\beta T_n^2\,d\tau = 1 \tag{4.10}$$

Equation (4.7) for the angular functions is the usual equation for spherical harmonics, with solutions $Y_{\ell m}$. This means that $c_2 = -\ell(\ell+1)$. The normalization of the $Y_{\ell m}$ is such that

$$\int_{4\pi} |Y_{\ell m}(\theta,\psi)|^2 \, d\Omega = 1 \tag{4.11}$$

We can remove the explicit dependence of the radial equation on the mass of the black hole by defining a new independent variable $x = r/2M$, and introducing $\bar{\lambda} = (2M)^2\lambda$. We obtain:

$$\frac{d}{dx}\left(x(x-1)\frac{d}{dx}R_{n\ell p}\right) + \left(\bar{\lambda}x^2 - \bar{\mu}x^2 - \ell(\ell+1) - \frac{n^2}{4}\frac{x^3}{x-1}\right)R_{n\ell p} = 0 \tag{4.12}$$

We are interested in the range $1 < x < \infty$. The point $x = 1$ is a regular singular point, with the roots of the indicial equation $\pm n/2$. Our boundary condition at this point (the horizon) is that the solutions be regular, so near $x = 1$ we can write:

$$R_{n\ell p} = (x-1)^{n/2} \sum_{i=0}^{\infty} a_i (x-1)^i \tag{4.13}$$

Thus only $n = 0$ solutions contribute on the horizon. Candelas (1980) has made use of this fact to calculate $\langle\phi^2(2M)\rangle$ analytically, obtaining $T^2/3$.

If we impose boundary conditions at some radius r we obtain a discrete eigenvalue spectrum. The label p on $R_{n\ell p}$ refers to the radial eigenvalues for fixed n and $\ell$. The expression for $\langle\phi^2\rangle$ now becomes:

$$\langle\phi^2\rangle = \sum_{n=-\infty}^{\infty}\sum_{\ell=0}^{\infty}\sum_{m=-\ell}^{\ell}\sum_{p=0}^{\infty} \frac{1}{\lambda_n} T_n^2 Y_{\ell m}^2 R_{n\ell p}^2 \tag{4.14}$$

We can do the sum over m and combine the contributions from $\pm n$ to obtain:

$$\langle\phi^2\rangle = \frac{1}{4\pi\beta}\sum_{n=0}^{\infty} d_n \sum_{\ell=0}^{\infty}(2\ell+1)\sum_{p=0}^{\infty}\frac{1}{\lambda_{n\ell p}} R_{n\ell p}^2 \tag{4.15}$$

where $d_0 = 1$, $d_n = 2$ for $n > 0$.

Here the $R_{n\ell p}$ are normalized so that

$$\int_{2M}^{r_0} R_{n\ell p}^2 r^2 \, dr = 1$$

It is more convenient to normalize with

$$\int_1^{x_0} R_{n\ell p}^2 x^2 \, dx = 1 \tag{4.16}$$

With this normalization and putting $\bar{\lambda} = (2M)^2\lambda$, (4.15) becomes:

$$\langle\phi^2\rangle = \frac{1}{\beta^2}\sum_{n=0}^{\infty} d_n \sum_{\ell=0}^{\infty}(2\ell+1)\sum_{p=0}^{\infty}\frac{1}{\bar{\lambda}_{n\ell p}} R_{n\ell p}^2 \tag{4.17}$$

We are actually interested in the case in which the boundary is moved to infinity. Then the radial spectrum becomes continuous and the sum over p becomes an integral. We want this to be an integral over $\bar{\lambda}$, which can be done by looking at the asymptotic expression for the eigenvalues:

$$\bar{\lambda}_p - \frac{n^2}{4} - \bar{\mu}^2(x_0) \sim \frac{p^2\pi^2}{x_0^2}$$

$$\Rightarrow dp = \frac{x_0}{2\pi} \frac{d\bar{\lambda}}{\sqrt{\bar{\lambda} - \frac{n^2}{4} - \bar{\mu}^2(x_0)}} \tag{4.18}$$

It is assumed that $\mu^2 \to$ constant at infinity. Now we also change the normalization of the radial functions so that at large x,

$$R_{n\ell p} \sim \frac{1}{x} \sin\theta(x) \tag{4.19}$$

For large $X_0$ this differs from (4.16) by a factor $\sqrt{\frac{2}{x_0}}$. If we note that there are no normalizable solutions for $\bar{\lambda} < \bar{\mu}^2(\infty) + \frac{n^2}{4}$ the sum over p becomes:

$$\frac{1}{\pi} \int_{\bar{\mu}^2 + \frac{n^2}{4}}^{\infty} \frac{d\bar{\lambda}}{\bar{\lambda}} \frac{R^2_{n\ell}(\bar{\lambda},x)}{\sqrt{\bar{\lambda} - \bar{\mu}^2(\infty) - \frac{n^2}{4}}} \tag{4.20}$$

Putting this into (4.17) and changing the orders of the sums and the integral leads to

$$\begin{aligned} \langle\phi^2\rangle &= \int_{\bar{\mu}^2(\infty)}^{\infty} \frac{d\bar{\lambda}}{\bar{\lambda}} \frac{1}{\pi\beta^2} \sum_{n<2\sqrt{\bar{\lambda}-\bar{\mu}^2}} d_n \sum_{\ell=0}^{\infty} (2\ell+1) \frac{R^2_{n\ell}(\bar{\lambda},x)}{\sqrt{\bar{\lambda}-\bar{\mu}^2(\infty)-\frac{n^2}{4}}} \\ &= \int_0^{\infty} \frac{d\bar{\lambda}}{\bar{\lambda}+\bar{\mu}^2(\infty)} \frac{1}{\pi\beta^2} \sum_{n<2\sqrt{\bar{\lambda}}} d_n \sum_{\ell=0}^{\infty} (2\ell+1) \frac{R^2_{n\ell}(\bar{\lambda}+\bar{\mu}^2,x)}{\sqrt{\bar{\lambda}-\frac{n^2}{4}}} \end{aligned} \tag{4.21}$$

This divergent integral must now be regularized as discussed in the previous section. As was shown there, a finite answer can be obtained by subtracting from the integrand

$$\frac{4\pi^2}{\beta^2} a_0 + \frac{a_1}{\bar{\lambda} + \bar{c}} \tag{4.22}$$

where $\bar{c}$ is any constant. (The factor $4\pi/\beta^2$ is because we are integrating over $\bar{\lambda}$ not $\lambda$. The divergence is in fact independent of the temperature, as many people have pointed out.) To determine $\bar{c}$ we require that $\langle\phi^2\rangle = 0$ in flat space at T = 0 with a constant mass.

The regularized expression equivalent to (4.21) in flat space is:

$$\int_0^\infty \left( \frac{1}{\lambda+m^2} \frac{1}{\pi\beta} \sum_n d_n \sum_\ell \frac{1}{4\pi}(2\ell+1) \frac{R^2_{n\ell}(\lambda+m_3 r)}{\sqrt{\lambda - \frac{4\pi^2 n^2}{\beta^2}}} - a_0 - \frac{a_1}{\lambda+c} \right) d$$

where $c = \frac{4\pi^2}{\beta^2}\bar{c}$ (4.23)

The flat space radial eigenfunctions are simply spherical Bessel functions:

$$R_{n\ell} = \sqrt{\lambda - \frac{4\pi^2 n^2}{\beta^2}}\; j_\ell\left(\sqrt{\lambda - \frac{4\pi^2 n^2}{\beta^2}}\, r\right) \tag{4.24}$$

We can do the sum over $\ell$ to get:

$$\int_0^\infty \left( \frac{1}{\lambda+m^2} \frac{1}{\pi\beta} \frac{1}{4\pi} \sum_{n<\frac{\beta}{2\pi}\sqrt{\lambda}} d_n \sqrt{\lambda - \frac{4\pi^2 n^2}{\beta^2}} - a_0 - \frac{a_1}{\lambda+c} \right) d\lambda \tag{4.25}$$

Now we wish to take the limit $T \to 0$, or $\beta \to \infty$. We let

$$\frac{2\pi n}{\beta} \to y \,, \quad \sum_{n<\frac{\beta}{2\pi}\sqrt{\lambda}} d_n \frac{2\pi}{\beta} \longrightarrow 2\int_0^{\sqrt{\lambda}} dy \tag{4.26}$$

This gives

$$\int_0^\infty \left( \frac{1}{\lambda+m^2} \frac{1}{2\pi^2} \frac{2}{\pi} \int_0^{\sqrt{\lambda}} dy \sqrt{\lambda - y^2} - a_0 - \frac{a_1}{\lambda+c} \right) d\lambda$$

$$= \int_0^\infty \left( \frac{1}{16\pi^2} \frac{\lambda}{\lambda+m^2} - a_0 - \frac{a_1}{\lambda+c} \right) d\lambda \tag{4.27}$$

Now $a_0 = \frac{1}{16\pi^2}$, $a_1 = -\frac{m^2}{16\pi^2}$ so:

$$\langle \phi^2 \rangle_{T=0} = \frac{1}{16\pi^2} \int_0^\infty \frac{m^2(m^2-c)}{(\lambda+m^2)(\lambda+c)}\, d\lambda \tag{4.28}$$

This should be zero, so we choose $c = m^2$.

If the mass were a function of position that became constant a infinity, we would require that $\langle\phi^2\rangle = 0$ at infinity. Since $\langle\phi^2\rangle$ will approach the value appropriate to a constant mass as $x \to \infty$, the correct value of c to use is $c = m^2(\infty)$. Therefore the fully regularized version of (4.21) is:

$$\langle \phi^2 \rangle = \frac{1}{\beta^2} \int_0^\infty \left( \frac{1}{\pi} \frac{1}{\bar\lambda + \bar\mu^2(\infty)} \sum_n d_n \sum_\ell (2\ell+1) \frac{R^2(\bar\lambda + \bar\mu^2(\infty), x)}{\sqrt{\bar\lambda - \frac{n^2}{4}}} \right.$$

$$\left. -1 + \frac{\bar\mu^2(x)}{\bar\lambda + \bar\mu^2(\infty)} \right) d\bar\lambda \tag{4.29}$$

In the next section I shall discuss the numerical methods necessary to evaluate equation (4.29)

## 5. THE NUMERICAL EVALUATION OF $<\phi^2>$

The evaluation of (4.29) falls naturally into three stages:

a) the calculation of the normalized eigenfunctions

b) performing the sums over n and ℓ

c) performing the integral over λ

Each of these will be dealt with in turn.

### a) Calculation of the eigenfunctions

The radial eigenfunctions are solutions of equation (4.12) subject to the boundary condition of regularity at x = 1 and the normalization condition of going as $\frac{1}{x}$ times a function oscillating with unit amplitude at infinity. The numerical integration is easier if we transform to a new independent variable

$$x^* = x + \ln(x-1) \tag{5.1}$$

and a new dependent variable $y = xR$

Equation (4.12) then becomes:

$$\frac{d^2y}{dx^{*2}} = \left[\frac{n^2}{4} - \left(1-\frac{1}{x}\right)\left(\bar{\lambda} - \bar{\mu}^2(x) - \frac{1}{x^2}\ell(\ell+1) - \frac{1}{x^3}\right)\right]y \tag{5.2}$$

The range of X* is $-\infty$ to $\infty$ corresponding to $1 < x < \infty$. Equation (4.13) for the solution near x = 1 becomes:

$$y \sim C e^{qx^*}, \quad x^* \to -\infty \tag{5.3}$$

Equation (5.2) is of the form $y'' = f(x)y$. The first derivative is missing from this equation, and in fact we do not want to calculate it at every point. In this case an appropriate method to use is Numerov's method, which uses the iteration scheme:

$$\begin{aligned}&\left(1 - \frac{h^2}{12}f(x+h)\right)y(x+h)\\ &= \left(2 + \frac{5}{6}h^2 f(x)\right)y(x) - \left(1 - \frac{h^2}{12}f(x-h)\right)y(x-h)\\ &\qquad + \frac{h^6}{240}y^{vi}(\xi) \qquad\qquad x-h < \xi < x+h\end{aligned} \tag{5.4}$$

A WKB approximation reveals that $y^{vi}(\xi) \sim f^3(\xi)y$, so the error term is $\sim \frac{(h^2f)^3}{240}$. I choose a value of h every time I change the parameters ℓ and n, doing this in such a way that $h^2 f_{max}$ constant ($f_{max}$ being the maximum value of f for given n and ℓ). If $\bar{\mu}^2 = 0$ and $\bar{\lambda}$ is small compared to $\ell(\ell+1)$ the maximum occurs near $x = \frac{3}{2}$, and

$$f_{max} \sim \frac{n^2}{4} + \frac{4}{27}\ell(\ell+1) \tag{5.5}$$

The Numerov method requires two starting values. I calculate these by using the expansion (4.12) and converting to x* coordinates. I begin the integration at $x^* = -10$, which means $x - 1 \sim e^{-11}$. It is therefore only necessary to keep a couple of terms in the series in order to make the error in initial conditions small compared to the local truncation error.

The solutions must also be normalized so that they oscillate with unit amplitude at infinity. Because we can only integrate numerically out to some finite x, we must then match the solution on to some approximate solution. I use a WKB approximation, which will be good if $f(x^*)$ ($f(x^*)$ refers to everything within the square brackets in equation (5.2)) is not nearly zero and is not changing rapidly. , If $n \neq 0$, f always changes sign. We must integrate numerically at least past the zero of f in order to apply the WKB approximation.

I perform the normalization as follows. We have an equation of the form

$$y'' = f(x)\, y$$

We can write two independent solutions of this equation as:

$$y_1(x) = M(x)\sin\Theta(x)$$
$$y_2(x) = M(x)\cos\Theta(x) \tag{5.6}$$

where $M \underset{x\to\infty}{\longrightarrow} M_\infty$ , $\Theta' \underset{x\to\infty}{\longrightarrow} \Theta'_\infty = [-f(\infty)]^{\frac{1}{2}}$

To normalize, we need to know $M_\infty$ given y,(x) and $y_1(x)$. The Wronskian of $y_1$ and $y_2$ is:

$$y_1 y_2' - y_1' y_2 = -M^2\Theta' = \text{constant} = -M^2_\infty \Theta'_\infty \tag{5.7}$$

If we differentiate (5.6) and (5.7) we can then solve for $M^2_\infty$ in terms of $y_1, y_1^{\;1}$ and $\Theta^1$:

$$M^2_\infty = \frac{\Theta'(x)}{\Theta_\infty}\left[y^2 + \frac{1}{\Theta'^2}\left(y' + \frac{1}{2}\frac{\Theta''}{\Theta'}y\right)^2\right] \tag{5.8}$$

We have y(x) numerically, and can also calculate y′(x) numerically. We use the WKB approximation for Θ′, i.e.

$$\Theta'(x) \simeq [-f(x)]^{1/2}$$
$$\Theta''(x) \simeq -\frac{1}{2}\frac{f'}{\sqrt{-f}} \tag{5.9}$$

If we put these in (5.8) we obtain

$$M^2_\infty \simeq \sqrt{\frac{f(x)}{f(\infty)}}\left[y^2 - \frac{1}{f}\left(y' - \frac{1}{4}\frac{f'}{f}y\right)^2\right] \tag{5.10}$$

One can obtain an estimate for the error incurred in using this approximation by applying the standard analysis of the WKB approximation, which can be found in Olver (1974). I will not go through the details here, but the result is that for equation (5.2) the error is $O(x^{-3})$. In practice $x^* = 100$ gives sufficiently accurate results.

b) <u>The sums over n and $\ell$</u>

The next problem is performing the sums over n and $\ell$. The sum over n is not difficult - as I said earlier, if $n > 2\sqrt{\bar{\lambda} - \bar{\mu}^2(\infty)}$ the solutions are exponentially growing at infinity and are non-normalizable. So the sum over n is finite. The sum over $\ell$, however, is infinite, so we must cut it off at some finite $\ell$ and estimate or ignore the remainder.

Consider again the function f which multiplies y on the right hand side of equation (5.2). At any point $x^*$ where $f(x^*)>0$, the solution is exponentially growing. Hence when the solution is normalized, the value at $x^*$ will be greatly damped. However, if $f(x^*)<0$ there will be little change in amplitude between x and $\infty$, so the value of the solution at $x^*$ comes from values of $\ell$ that make $f(x^*)<0$. The value of $\ell$ at which f changes sign can easily be found from (5.2):

$$\ell(\ell+1) = x^2\left[\bar{\lambda} - \frac{1}{x^3} - \bar{\mu}^2(x) - \frac{x}{x-1}\frac{n^2}{4}\right] \tag{5.11}$$

This gives an effective upper cut-off to the sum over $\ell$. The decrease in contributions beyond this $\ell$ is at least exponential. The procedure I use is to sum over $\ell$ until f changes sign at the maximum $x^*$ I am interested in ($x^* = 10$), then continue summing until the contributions are sufficiently small. This usually means only one or two more $\ell$ values.

c) <u>The integral over $\lambda$</u>

It is difficult to get analytic information about the behaviour of the integrand of equation (4.29). Without some idea of what the integrand is like it is not possible to choose an appropriate numerical integration algorithm. As I will describe, we can get some information from the form of the integrand in flat space, but this is a little misleading as it turns out.

We expect that at large x the integrand will approach the flat space integrand, which for a zero mass is:

$$\frac{1}{\pi\lambda}\sum_{n<2\sqrt{\lambda}} d_n\sqrt{\lambda-\frac{n^2}{4}} \quad - 1 \tag{5.12}$$

As a function of $\sqrt{\lambda}$ this looks like figure 1.
The function is "periodic" with the minima every ½ in, $\sqrt{\lambda}$. There is a derivative singularity at each minimum which must be handled with care in a numerical integration. If one integrates between minima,

$$I_m = \int_{\frac{1}{4}m^2}^{\frac{1}{4}(m+1)^2}\left(\frac{1}{\pi\lambda}\sum_{n<2\sqrt{\lambda}} d_n\left(\lambda-\frac{n^2}{4}\right)^{\frac{1}{2}} - 1\right)d\lambda \tag{5.13}$$

one finds that $I_m \propto m^{-1.5} \propto \lambda^{-3/4}$ for large m. Using this one can extrapolate to get quite accurate values for the infinite integral.

The situation seems to be different in Schwarzschild. Figure 2 shows the integrand as a function of $\lambda$ for two values of x*, x* = - 10 and x* = 1.5. On the horizon the integrand is a smoothly oscillating decaying function with a period of 1. As one moves out from the horizon the period changes smoothly to reach ½ at about x = 7.5. There is no singularity in the derivative, but the slopes are getting steeper as x increases.

The smoothness makes the numerical integration easier - I am able to use a simple composite Newton-Cotes formula with equally spaced points. The major problem is that for large values of the integrand it is very expensive to evaluate (about 1 minute cpu time for each value of $\bar{\lambda}$ when $\bar{\lambda}$ is about 20, and the time is proportional to $\bar{\lambda}$). This means that it is necessary to extrapolate the integral on the basis of relatively few periods of the integrand. Nevertheless, this is what I have done and this leads to the results we present in section 7.

## 6) $\langle\phi^2\rangle$ FOR THERMAL RADIATION AT INFINITY (B.W.)

In this section we consider evaluating $\langle\phi^2\rangle$ for a thermal flux of scalar particles propagating in from infinity on a black hole (Schwarzschild) background space-time. In general, an expression for $\langle\phi^2\rangle$ depends on the boundary conditions assumed for the appropriate 'vacuum' state, and the quantity of interest here represents the difference between the Hartle-Hawking and Unruh 'vacua'. It can be written out explicitly from results of Candelas (1980), based on Green's function techniques, whence:

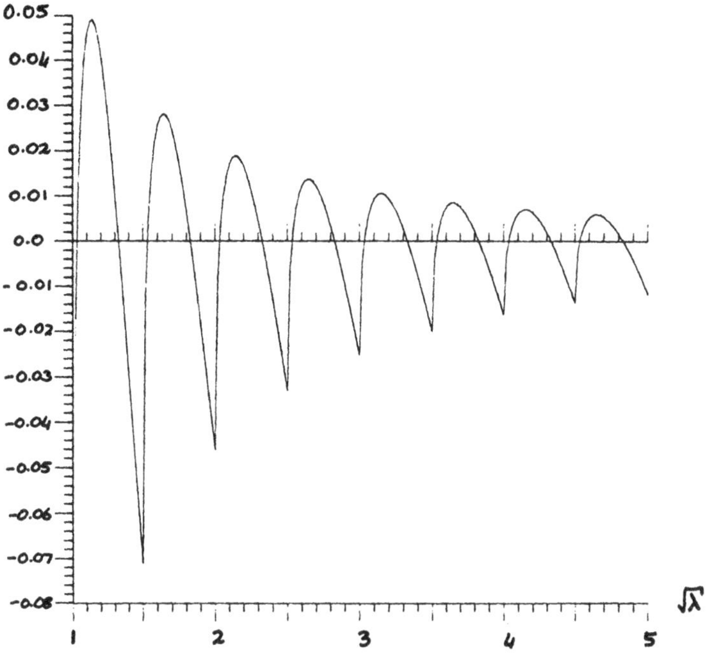


Figure 1. The integrand in flat space.

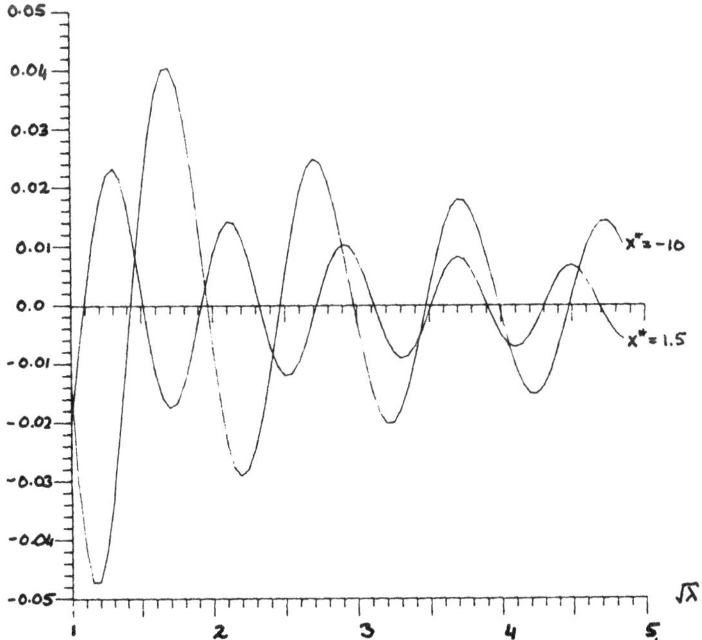


Figure 2. The integrand in Schwarschild at $x^* = -10$ and $x^* = 1.5$

$$\langle\phi^2\rangle_{IN} = \langle\phi^2\rangle_{HH} - \langle\phi^2\rangle_U$$
$$= \frac{1}{16\pi^2}\int_0^\infty \frac{d\omega}{\omega}.2.\left\{\exp(2\pi\omega/\kappa)-1\right\}^{-1}.\left[\sum_{l=0}^{\infty}(2l+1)|\overleftarrow{R}_l(\omega|r)|^2\right] \tag{6.1}$$

where $\kappa$ is the surface gravity of the black hole (=1/4M) and $R_\ell(\omega|r)$ represents a purely ingoing wave at the horizon (in a modified radial coordinate) with energy, $\omega$, and angular momentum, $\ell$, normalized to have unit inward flux (ie $1/r$) at infinity. The integral is well defined for small $\omega$ since the sum is $O(\omega^2)$ for all values of r, and there is an exponential damping factor for large $\omega$. As numerical calculation is required for the evaluation of the right-hand side of (6.1), we will be interested in an $\ell_{max}(\omega,r)$ and an $\omega_{max}(r)$ at which to truncate the sum and integral (respectively) without serious loss. Since numerical integration is also required in order to determine $R_\ell(\omega|r)$ it will be necessary to specify an $r_{min}$ and $r_{max}$ which limit the range of integration of the relevant (radial) wave equation:

$$\left\{\frac{1}{r^2}\frac{d}{dr}r^2\left(1-\frac{2M}{r}\right)\frac{d}{dr} + \frac{\omega^2}{\left(1-\frac{2M}{r}\right)} - \frac{l(l+1)}{r^2}\right\}R_l(\omega|r) = 0 \tag{6.2}$$

Before considering these restrictions further however, it will be useful to cast equation (6.2) into a more convenient form by non-dimensionalizing to $x = r/2M$ and $\tilde{\omega} = 2M\omega$ and by changing variables to $x^* = x + \ell n(x-1)$ and $R = \frac{r}{2M}\bar{R}$, to give

$$\left[\frac{d^2}{dx^{*2}} + \omega^2 - \left(1-\frac{1}{x}\right)\left\{\frac{l(l+1)}{x^2} + \frac{1}{x^3}\right\}\right]R = 0 \tag{6.3}$$

As x varies between 1 (horizon) and infinity, $x^*$ varies from $-\infty$ to $+\infty$. Since the effective potential goes to zero (sufficiently fast) at $x^* = \pm\infty$, R can be represented by inward and outward going waves in each of these regions as already referred to above.

Page (1977) has discussed a procedure for matching wave functions at large negative $x^*$ onto a boundary condition at $x^* = -\infty$, using a slightly modified radial coordinate (purely for analytical convenience). On the other hand, Candelas (1980) has shown that for large x, the sum in equation (6.1) is $4\omega^2$ so that for a finite error tolerance in $\langle\phi^2\rangle$ the integral in equation (6.1) can be readily truncated at some appropriate $\omega_{max}$ because of the exponential cut-off. For particular values of x & $\omega$, the contributions to equation (6.1) for $\ell \gtrsim \tilde{\omega}x = \omega r$ will become roughly exponentially small for increasing $\ell$, so that the loss due to truncation at some appropriate $\ell_{max}$ can also be kept within a given error bound. Finally, for

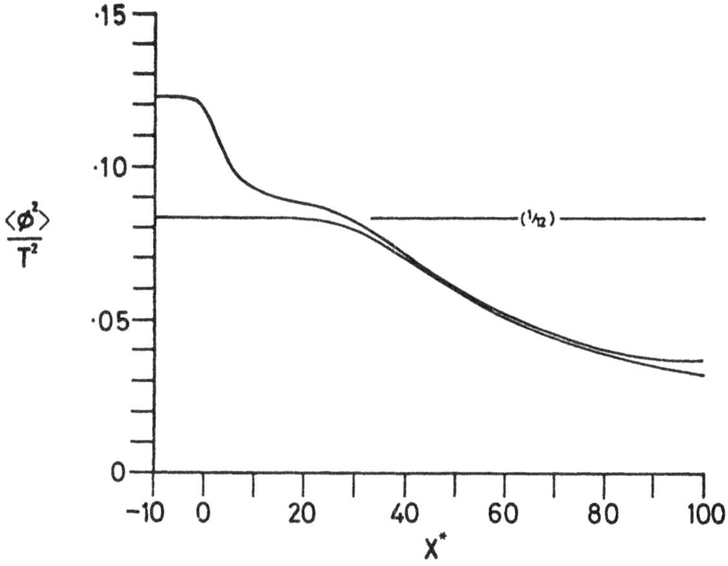


Figure 3. $\langle\phi^2\rangle/T^2$ for a flux of incoming thermal radiation in Schwarzschild (upper curve) and flat space (lower curve), with the $\ell$-sum truncated once errors at $x^* = 15$ are sufficiently small. The value at $\infty$ is also indicated (ie. 1/12).

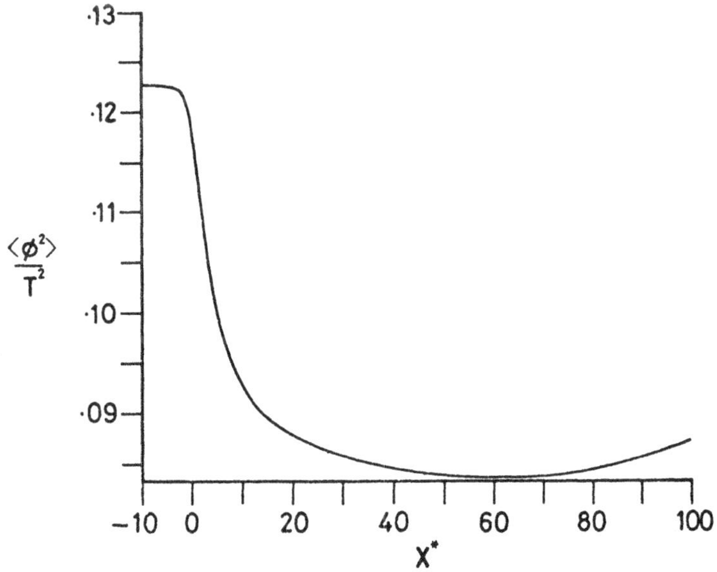


Figure 4. The 'corrected' $\langle\phi^2\rangle/T^2$ for incoming thermal radiation in Schwarzschild, shown as a departure from the value at $\infty$ .

large x, deviations from constant amplitude for the oscillations in R near $x = +\infty$ are $O(1/x^3)$ (eg, from WKB or similar arguments) and can actually be calculated in lowest order from several values of R, so that errors from terminating the numerical integration at some finite $x_{max}$ can also be kept within a specified bound.

For the problem at hand, in which values of $\langle\phi^2\rangle$ only very near the black hole are of much significance, satisfactory results were obtained with $\tilde{\omega}_{max} = 1$ and $x^*_{max} = 100$, both being constant for all values of $\ell$. Then $\ell_{max}$ was chosen so that the error from a similar calculation in a corresponding flat ($x^*$-) space was suitably small at some chosen large $x_c$ ($=15$) $\ll x^*_{max}$. The truncation error for the sum over $\ell$ was then given very well by the corresponding truncation error in the flat space calculation. Results for truncated $\ell$-sums in both the black hole and flat space cases are given in figure 3, and the black hole case 'corrected' for truncation is shown in figure 4, where it is seen that $\langle\phi^2\rangle$ differs from its flat space value only in a very small region near the black hole.

From the above discussion, it will be clear that this calculation (for zero mass) is orders of magnitude simpler than the Euclidean equilibrium calculation (section 5) which requires an extra $\lambda$ mass-like parameter right from the start but which then automatically gives results for a constant mass case also. However, for a position dependent mass, this advantage again disappears.

## 7) RESULTS AND DISCUSSION

The combined results are presented in figure 5. The upper line is $\langle\phi^2\rangle/T^2$ for a black hole in thermal equilibrium with (massless) radiation obtained as discussed in section 5. The lower line is $\langle\phi^2\rangle/T^2$ for a flux of purely thermal radiation at infinity, which is partly absorbed and partly scattered by the black hole: it has already been presented in figure 4. The difference between these, shown in figure 6, is $\langle\phi^2\rangle/T^2$ for a black hole radiating into empty space.

One can see that for the equilibrium case also, $\langle\phi^2\rangle$ differs from the flat space value of $T^2/12$ only in a small region about the hole. The value on the horizon for a massless field is that calculated by Candelas (1980). Because the region in which $\langle\phi^2\rangle$ is non-zero in figure 5 is so small, it seems unlikely that $\Phi$, the classical field and solution of equation (2.12), will differ significantly on the horizon from its value at infinity. This possibility has already been raised by Hawking (1981).

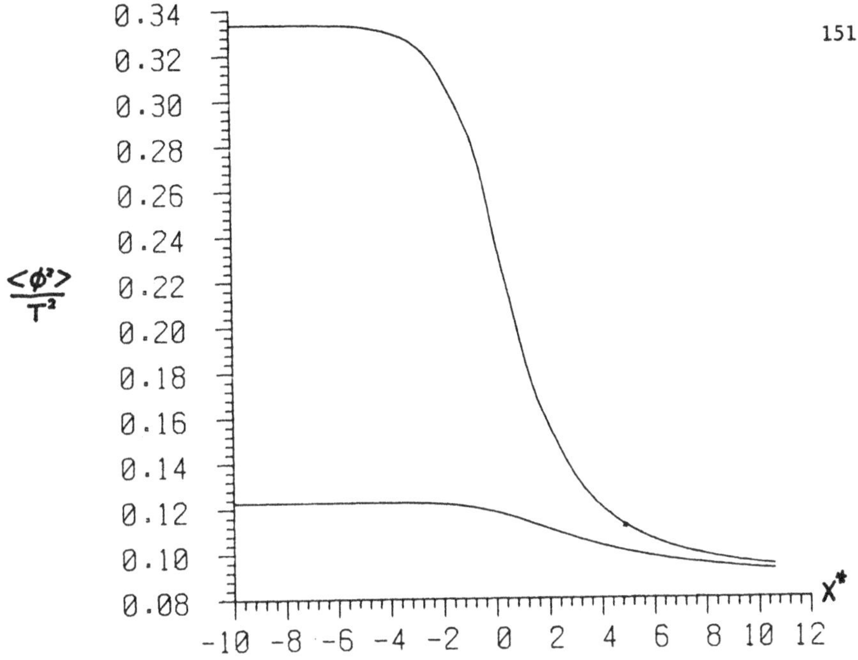


Figure 5. $<\phi^2>$ for thermal equilibrium and incoming radiation.

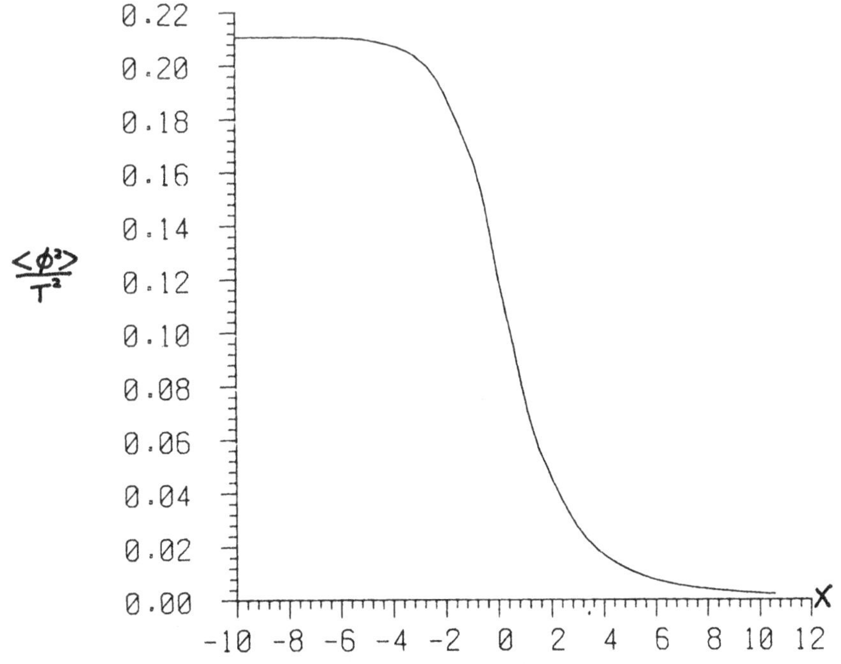


Figure 6. $<\phi^2>$ for outgoing radiation.

It would mean that the phase of the field was not a function of position, and that the symmetry would remain broken even near the hole, at least until the temperature became very high indeed.

The above calculations have not been done in the self consistent approximation. To do this we must solve (2.12) and put the resultant value of $\Phi$ into the expression for $\langle\phi^2\rangle$, i.e. use a position dependent mass $\mu^2(x) = 3g\Phi - m^2$. It is possible that there would be some reflection of the eigenfunctions off the boundary of this mass which would enhance $\langle\phi^2\rangle$ near the horizon. This may happen to such an extent that $\Phi$ will become small on the horizon and symmetry restoration may occur. This calculation will be completed in the near future.

Another way of increasing $\langle\phi^2\rangle$ is to couple the scalar field to a gauge field. The gauge field contributes a term to $\langle\phi^2\rangle$ which, for flat space, has been derived by Dolan & Jackiw (1974). One of us (MSF) is looking at the feasibility of doing this calculation numerically for Schwarzschild. It is significantly more difficult because one must deal with coupled equations, but may nevertheless be worth doing.

The numerical results obtained for a black hole in thermal equilibrium allow one to put forward a rather interesting conjecture (B.W.) From Candelas (1980) we can immediately write:

$$\langle\phi^2\rangle_{HH} = \frac{1}{16\pi^2}\int_0^\infty \frac{d\omega}{\omega}\left[\coth\left(\frac{\pi\omega}{\kappa}\right).\sum_{l=0}^{\infty}(2l+1)\left\{|\overrightarrow{R}_l(\omega|r)|^2 + |\overleftarrow{R}_l(\omega|r)|^2\right\} - \frac{4\omega^2}{(1-\frac{2M}{r})}\right] - \frac{M^2/(48\pi^2)}{r^4(1-\frac{2M}{r})} \quad (7.1)$$

($\overleftarrow{R}$ is defined in an analogous way as for $\overrightarrow{R}$) in which the integral over $\omega$ would naturally become a (problematical) sum over imaginary integer values of $\omega/\kappa$ on the Euclidean section. By a slight rearrangement, equation (7.1) can also be written as the following:

$$\langle\phi^2\rangle_{HH} = \frac{1}{16\pi^2}\int_0^\infty \frac{d\omega}{\omega}.\coth\left(\frac{\pi\omega}{\kappa}\right).\left[\sum_{l=0}^{\infty}(2l+1)\left\{|\overrightarrow{R}_l(\omega|r)|^2 + |\overleftarrow{R}_l(\omega|r)|^2\right\} - \frac{4\omega^2}{(1-\frac{2M}{r})}\right] + \frac{T^2}{12}\frac{\left(1-\left(\frac{2M}{r}\right)^4\right)}{(1-\frac{2M}{r})} \quad (7.2)$$

where T is the Hawking temperature of the black hole ($T = \kappa/2\pi = (8\pi M)^{-1}$ ). The last term in equation (7.2) is always finite (!), while for convergence of the integral we require the square bracket to be $O(\omega^2)$ as $\omega \to o$ (this is usual), and to fall off at least as some negative power of $\omega$ as $\omega \to \infty$ for all values of r.

Candelas (1980) has shown that, as a function of r, the quantity in the square brackets in equation (7.2) is not divergent as $r \to 2M$ because of the contribution from $\sum_{l=0}^{\infty}(2l+1).|\overrightarrow{R}_l(\omega|r)|^2$. However, in the limit $r \to 2M$, the last term in equation (7.2) gives precisely the contribution ($T^2/3$) he is already mentioned as having obtained for $\langle\phi^2\rangle_{HH}$ on the horizon, so we can make the even stronger statement that the integral in equation (7.2) is exactly zero for $r = 2M$. It is also known from results in the same paper that, as a function of r, the square bracket goes to zero as $r \to \infty$ consistent with the requirement that we obtain the flat space result $T^2/12$) there, as again given exactly by the last term in equation (7.2), and showing that the integral term must also be zero for $r = \infty$. What seems to be a rather remarkable result, and one which, so far, can be made only with the numerical data at hand, is that, over the range of r for which the numerical data are considered to be reliable (ie, calculational errors are small), they give a result virtually indistinguishable from the contribution of only the last term in equation (7.2), suggesting that, for Schwarzschild, the integral term may be exactly zero for all values of r.

Page (1981) has made the observation that the last term in equation (7.2) is given exactly by $1/12 \left\{ T^2_{loc} - T^2_{acc} \right\}$ where $T_{loc}$ is the locally red-shifted temperature of the thermal radiation and $T_{acc}$ is the temperature of the "acceleration radiation" corresponding to the local value of acceleration. He has also shown (Page, 1981) that the above conjecture holds exactly, for example, in the multi-instanton metrics (Page, 1979) and a number of other exactly solvable cases, though it obviously depends on the boundary conditions involved. Further numerical accuracy is currently being sought to test the validity of this conjecture: meanwhile an analytic determination of its applicability here would be most welcome.

ACKNOWLEDGEMENTS

We would like to thank S.W. Hawking for suggesting this problem, and for help and advice. Discussions with G.W. Gibbons and particularly with D.N. Page at the London Conference have been very fruitful and we are most grateful for them. One of us (B.W.) has been supported by the S.R.C., which allowed part of this work to be undertaken.

## REFERENCES


Abers E. & Lee B. (1973) Gauge Theories Phys.Rep 9C, 1-141

Candelas, P. (1980) Vacuum Polarization in Schwarzschild Space-time. Phys.Rev. D 21, 2185-2202.

de Witt (1979) Quantum Gravity: the new synthesis. In General Relativity: An Einstein Centenary Survey, eds. S.W. Hawking & W. Israel pp. 680-745 Cambridge University Press.

Dolan & Jackiw (1974) Symmetry Behaviour at High Temperature. Phys.Rev.D12, 3320-3341.

Gibbons, G.W. (1978) Symmetry Restoration in the Early Universe. J.Phys.A 11, 1341-1345.

Gilkey, P.B. (1975) The Spectral Geometry of a Riemannian Manifold. J.Diff.Geom. 10, 601-618.

Hawking, S.W. (1977) Zeta Function Regularization of Path Integrals in Curved Space-time. Comm. Maths.Phys. 55, 133-148.

Hawking, S.W. (1981) Interacting Fields Around a Black Hole. Comm.Math.Phys. 80, 421-442.

Kirzhnits & Linde (1976) Symmetry Behaviour in Gauge Theories. Ann.Phys. 101, 195-238.

Linde, A.D. (1979) Phase Transitions in Gauge Theories and Cosmology. Rep.Prog.Phys. 42 389-437.

Olver, F.W.J. (1974) Asymptotics and Special Functions. Academic Press.

Page, D.N. (1977) Particle Emission Rates From a Black Hole III. Charged Leptons From a Non-rotating Hole. Phys.Rev. 16D, 2402-2411.

Page, D.N. (1979) Green's Functions for Gravitational Multi-Instantons. Phys.Lett. 85B, 369-372.

Page, D.N. (1981) Discussions and calculations during this London Conference. See also Thermal Stress Tensors in Static Einstein Spaces. Phys. Rev. D25, 1499-1509 (1982).

Weinberg, S. (1974) Gauge and Global Symmetries at High Temperature. Phys.Rev.D12, 3357 - 3378.

# YANG-MILLS VACUA IN A GENERAL THREE-SPACE

G. Kunstatter
Department of Physics
University of Toronto
Toronto, Ontario
Canada M5S 1A7

## 1 INTRODUCTION

The purpose of the talk is to illustrate with a few simple examples the effect that the non-trivial topology of three-space might have on the vacuum structure of a canonically quantized Yang-Mills theory. The present discussion is based on a more detailed analysis done recently by Isham and Kunstatter (1981 a,b).

We shall consider the canonical quantization of Yang Mills theories with gauge group, G, on an arbitrary compact connected three space, $\Sigma$. Non-trivial spatial topology can occur in quantum field theory due to boundary conditions(such as periodicity) imposed on the fields in flat space-time, the presence of a classical background gravitational field which affects the large scale structure of three-space, or due to small scale quantum fluctuations at the Planck length. Moreover, Poincaré's conjecture states that all compact three manifolds except those homeomorphic to the three-sphere are multiply connected. As we shall see, the presence of a non-trivial fundamental group, $\pi_1(\Sigma)$, of three space enlarges considerably the space of zero-energy classical Yang-Mills fields which can exist on $\Sigma$. Consequently, the quantum vacuum structure is quite different from the n-vacua which occur in the standard (Jackiw and Rebbi, 1976; Callan _et al_ 1976) one point compactification of Euclidean three-space.

Since we do not have a full quantum Yang-Mills theory, the usual approach is to examine the classical vacua (zero-energy solutions) in the hope that the quantum vacuum can in some sense be built up perturbatively about them. Consider the Yang-Mills Hamiltonian density in the temporal gauge ($A_0$=0):

$$H(x,t) = \tfrac{1}{2}[\frac{\partial}{\partial t} A_i^a(x,t)]^2 + \tfrac{1}{4}(F_{ij}^a)^2 , \qquad (1.1)$$

where $F^a_{ij} = A^a_{i,j} - A^a_{j,i} + gc^{abc}A^b_iA^c_j$ .

The zero-energy configurations of this Hamiltonian obey

$$\frac{\partial}{\partial t} A^a_i(x,t) = 0, \tag{1.2}$$

$$F^a_{ij}(x,t) = 0 . \tag{1.3}$$

We are interested in static potentials with zero spatial curvature. The space of zero-energy solutions is then simply the space of flat connections in principal G-bundles over $\Sigma$, which we shall henceforth denote by $F$.

We wish to classify the naive (degenerate) quantum vacuum states. These will be "peaked" around the classical vacua and eventually yield a unique quantum ground state via tunnelling. For simplicity we shall therefore only consider as distinct those classical vacua which are separated by a potential barrier. In other words, we define two zero energy classical configurations to be <u>physically equivalent</u> for the purpose of quantization if there exists a continuous family of zero-energy solutions which interpolates between them. In the path integral approach this implies that there exist Euclidean interpolating solutions with arbitrarily small action. This definition yields the result that the space of "quantum vacua" (i.e. equivalence classes of zero-energy solutions) is simply:

$$\{|0\rangle\} = \pi_0(F) , \tag{1.4}$$

where $\pi_0(F)$ denotes the zeroth homotopy set of $F$. This classification is particularly useful for the topological analysis we shall present.

When $\Sigma$ is simply connected ($\pi_1(\Sigma) = 0$), $F_{ij} = 0$ implies that

$$A_i(x) = \Omega(x)\ \partial_i\Omega^{-1}(x) , \tag{1.5}$$

where $\Omega:\Sigma\to G$ is a gauge function. The space of zero energy potentials is therefore in bijective correspondence with the space of (differentiable) maps from $\Sigma$ into G and we have that

$$\{|0\rangle\} = \pi_0(F) = \pi_0(G^\Sigma) = [\Sigma,G] \tag{1.6}$$

where $[\Sigma,G]$ denotes the group of homotopy classes of gauge functions and we have used the standard notation $Y^X$ to denote the function space of maps from X into Y given the compact open topology. All maps will be taken as base point preserving unless otherwise stated. Moreover $\pi_1(\Sigma)=0$ implies that $[\Sigma,G]=Z$ and there exists a countable infinity of degenerate n-vacua, in complete analogy with the standard SU2 theory on $S^3$.

When $\Sigma$ is multiply connected, elements of $[\Sigma,G]$ are labelled by members of the cohomology groups $H^1(\Sigma, \pi_1(G)) = \mathrm{Hom}(\pi_1(\Sigma), \pi_1(G))$ and $H^3(\Sigma, \pi_3(G)) = Z$. Thus one expects the naive Yang-Mills vacua to be classified by a "primary winding number" (element of $H^1$) and a "secondary winding number" (element of $H^3$). In the following it will be shown that such a classification is incorrect because it ignores zero energy classical solutions which are not pure gauge, but are nonetheless well-defined on a multiply-connected three space.

## 2 FLAT CONNECTIONS ON A MULTIPLY-CONNECTED THREE-SPACE

When $\pi_1(\Sigma) \neq 0$, $F_{ij} = 0$ no longer implies Eq. (1.5) globally. The space of zero-energy solutions is larger than the space of gauge functions due to the existence of:

(1) non-trivial principle G-bundles on $\Sigma$ which admit flat connections.

(2) flat connections in trivial G-bundles with non-trivial (discrete) holonomy group.

A homotopically non-trivial loop in $\Sigma$ might lift horizontally to a curve in the bundle space whose endpoints lie on different elements of the fibre, G. If this is the case, the connection is not pure gauge and induces a homomorphism, $h:\pi_1(\Sigma)\to G$ from the equivalence classes of loops in $\Sigma$ into the gauge group.

Conversely, given any h in the space of homomorphisms from $\pi_1(\Sigma)$ into G (which we will henceforth denote by $\mathrm{Hom}(\pi_1(\Sigma), G)$), one can construct a flat connection in a principal G-bundle over $\Sigma$ using the following standard technique (Milnor 1957; Kamber and Tondeur 1968):

Consider the universal covering space, $\hat{\Sigma}$, of $\Sigma$. This is a principle fibre bundle over $\Sigma$ with (discrete) group $\pi_1(\Sigma)$ and natural action denoted by $y.\gamma$, $y\in\hat{\Sigma}$, $\gamma\in\pi_1(\Sigma)$. Given any $h\in\mathrm{Hom}(\pi_1(\Sigma),G)$ define the right action of $\pi_1(\Sigma)$ on G by $g.\gamma = h(\gamma^{-1})g$. Using this action, construct in the usual way (Kobayashi and Nomizu 1969) a fibre bundle $\mathcal{E}$ over $\Sigma$ with fibre G and structure group $\pi_1(\Sigma)$ associated with $\hat{\Sigma}$. That is, $\xi = \hat{\Sigma}x_h G$ is the set of points $(y,g)$ in $\hat{\Sigma}xG$ modulo the equivalence relation $\{(y.\gamma, h(\gamma^{-1})g) \equiv (y,g)\}$. Now consider $\xi$ as the bundle space for a principle G-bundle, with action $[y,g]_h g' = [y,gg']_h$ and define a flat connection by taking the canonical lift

$$x_t \mapsto [y_t, g]_h \in \xi \qquad (2.1)$$

where $x_t$ denotes a curve in $\Sigma$ and $y_t$ denotes its canonical lift in $\hat{\Sigma}$. Of course h does not determine a connection in $\xi$ uniquely since a bundle automorphism of $\xi$ (gauge transformation) yields a new connection induced by the same homomorphism, with the same holonomy group (up to inner conjugation).

We shall now illustrate the above construction with a simple example. Consider a U1 Yang-Mills theory over $S^1$. $\pi_1(S^1) = Z$ and $\hat{\Sigma}$ is just the real line. Hom(Z,U1) is of the form:

$$\mathrm{Hom}(Z,U1) = \{h_\lambda : n \mapsto e^{i2\pi n\lambda},\ \lambda \in [0,1]\} \qquad (2.2)$$

For a particular $h_\lambda$, $\hat{\Sigma}x_h\, G = Rx_{h_\lambda}\, U1$ is obtained by taking the cylinder RxU1 and identifying

$$[y,e^{i\theta}]_{h_\lambda} \equiv [y+n, e^{-i2\pi n\lambda} e^{i\theta}]_{h_\lambda},\quad \forall n \in Z\ . \qquad (2.3)$$

$\xi$ is therefore a torus twisted along its axis, as shown in Figure 1, where the points marked with an x are to be identified. A loop which wraps around $S^1$ n-times lifts via Eq.(2.1) to a straight line in the twisted torus (see Fig.1), so that the endpoints of the lift are displaced with respect to each other on the fibre by $e^{-i2\pi n\lambda}$. The holonomy group in this case is the image of $h_\lambda(\pi_1(\Sigma))$ in U1, namely Z when $\lambda$ is irrational and $Z_s$ when $\lambda = r/s$, where r,s are integers with no common dividers.

In this example $\xi(S^1,U1)$ is trivial as a U1 bundle, but this will not be true in general, as shown in the following section.

## 3 NON-TRIVIAL FLAT BUNDLES

If homomorphisms exist between $\pi_1(\Sigma)$ and G which induce non-trivial G-bundles, then the space of zero energy configurations, $F$, splits into disconnected components. In this case, there exist distinct quantum vacua labelled by the topological charges which classify the inequivalent G-bundles. Moreover, these topological charges will be conserved because smooth transitions (in either real or imaginary time) between connections on different bundles cannot occur.

We shall illustrate the effects of non-trivial flat bundles with the example of a U1 bundle over $RP_3$, which is the space obtained by identifying antipodal points on a three-dimensional sphere. It is of particular physical interest because it provides an alternative compactification of Euclidean space via weaker boundary conditions than usually used (Sciuto 1979). The set of all U1 bundles over $RP_3$ is classified by $H^2(RP_3, \pi_1(U1)) = Z_2 = (e,a)$ (Avis and Isham 1979). There are consequently two such bundles, $\xi_e$ and $\xi_a$ labelled by the identity element, e, of $Z_2$ and the generator, a, of $Z_2$, respectively. Naturally, $\xi_e$ is trivial and admits a canonical flat connection. We wish to know whether $\xi_a$ admits flat connections as well.

The fundamental group of $RP_3$ is $Z_2$, and there are exactly two homomorphisms from $Z_2$ into U1, namely the trivial homomorphism, $h_0$, and a non-trivial one, $h_1$, which maps the generator of $Z_2$ into minus the identity element in U1. $h_0$ induces $\xi_e$, while $h_1$ induces $\xi_a$, although the proof of the latter is not obvious. In this example, there are thus two distinct vacuum sectors labelled by the elements of $Z_2$. Since U1 is abelian, the group of bundle automorphisms in both sectors is just $U1^{RP_3}$, but $[RP_3, U1]$ =0 so the bundle sectors do not divide further into "n-vacua".

The zero-energy configurations in the non-trivial sector have been discussed in a somewhat different context by Asorey and Boya (1979), who point out that they describe Yang-Mills potentials in a non-trivial U1 bundle but with no magnetic monopoles. The charge of a magnetic monopole is associated with the first real Chern class of the bundle which takes its values in $H^2(\Sigma,R)$. The important thing to note is that the magnetic charges are not sufficient to label the inequivalent U1 bundles on $\Sigma$. What is required is an element of $[\Sigma, BU1] = H^2(\Sigma,Z)$, (BU1 denotes the universal base space for U1 bundles, see Avis and Isham (1979)) in order to completely classify such bundles. In particular, torsional elements of $H^2(\Sigma,Z)$ label inequivalent flat U1 bundles, but will correspond to the zero element in $H^2(\Sigma,R)=H^2(\Sigma,Z)\otimes R$ (Eguchi et al. 1980).

Further examples of non-trivial flat bundles are provided by $SUN/Z_N$ gauge theories on $S^1xS^1xS^1$. These have been discussed in some detail by 't Hooft (1980) in the context of quark confinement. The characteristic classes which label such bundles take their values in

$$H^2(S^1xS^1xS^1,\ \pi_1(SUN/Z_N)) = Z_N \oplus Z_N \oplus Z_N \qquad (3.1)$$

There are consequently $N^3$ inequivalent bundles. A recent analysis by Ambjorn and Flybjerg (1980) points out that the generalized "magnetic charge" defined by 't Hooft labels the inequivalent $SUN/Z_N$ bundles. Moreover, they show that field configurations exist with arbitrary "magnetic charge" and zero energy. In other words, the bundles corresponding to the elements in Eq.(3.1) admit flat connections. This example is of physical interest because of the possible relation between these generalized magnetic charges and confinement ('t Hooft 1980).

## 4 FLAT CONNECTIONS ON A TRIVIAL BUNDLE

### 4a General Discussion

We now restrict ourselves to the special case where $\xi = \hat{\Sigma}x_hG$ is trivial as a G-bundle. When $\xi$ is trivial any flat connection can be represented locally by (Isham and Kunstatter 1981a,b)

$$A_i(x) = D(y)\partial_i D(y)^{-1} \qquad (4.1)$$

where $x \in \Sigma$, $y \in \hat{\Sigma}$ and $D(y)$ is a function from $\hat{\Sigma}$ into G which obeys

$$D(y\cdot\gamma) = D(y)h(\gamma) \qquad (4.2)$$

for some homomorphism in $Hom(\pi_1(\Sigma)\ G)$. Note that $A_i(x)$ is single-valued on $\Sigma$, even though the generalized gauge function $D(y)$ is multi-valued. Moreover, it was shown by Isham and Kunstatter(1981b) that the space $F_t$ of flat connections in a trivial G-bundle over $\Sigma$ is in bijective correspondence with the space, $\mathcal{D}$, of all (differentiable) functions in $G^{\hat{\Sigma}}$ which satisfy Eq.(4.2).

The group of gauge functions $\Omega(x) \in G^{\Sigma}$ has the following natural action on the space $\mathcal{D}$:

$$(\Omega\cdot D)(y) := \Omega(p(y))D(y) \qquad (4.3)$$

where p(y) denotes the projection from $\hat{\Sigma}$ onto $\Sigma$. This leads to the important result (proven in Isham and Kunstatter 1981b) that $\mathcal{D}$ is the bundle space for a locally trivial principal fibre bundle with fibre $G^{\Sigma}$ over the space $\mathcal{R}$, where $\mathcal{R} = \mathrm{Hom}_t(\pi_1(\Sigma),G)$ is the space of homomorphisms from $\pi_1(\Sigma)$ into G which induces the trivial bundle. It is now possible to give a complete (although somewhat formal) classification of the quantum vacuum sectors given by

$$\pi_0(F_t) = \pi_0(\mathcal{D}) \tag{4.4}$$

This classification is achieved by using the information on the zeroth homotopy set $\pi_0(\mathcal{D})$ provided by the following homotopy exact sequence (in an exact sequence the kernel of every map in the sequence is equal to the image of the preceding map ) of the fibre bundle,

$$\rightarrow \pi_1(\mathcal{R},h) \xrightarrow{\partial} [\Sigma,G] \xrightarrow{i_*} \pi_0(\mathcal{D},D_h) \xrightarrow{q_*} \pi_0(\mathcal{R},h) \rightarrow 0 \ . \tag{4.5}$$

Here, h represents some arbitrarily chosen base point in $\mathcal{R}$ and $D_h$ the corresponding base point in $\mathcal{D}$. We have also used the standard result that $\pi_0(G^{\Sigma}) = [\Sigma,G]$. By exactness of the sequence in (4.5), it is clear that for a given basepoint, h, the components $\pi_0(\mathcal{D},D_h)$ which map onto the connected component of $\mathcal{R}$ containing h are given by the image of $i_*$. But this is just

$$[\Sigma,G]/\ker i_* = [\Sigma,G]/\partial\pi_1(\mathcal{R},h) \tag{4.6}$$

where $\partial\pi_1(\mathcal{R},h)$ denotes the image of the map $\partial:\pi_1(\mathcal{R},h) \rightarrow [\Sigma,G]$. By choosing basepoints in different components of $\mathcal{R}$ we have that $\pi_0(\mathcal{D})$ is completely classified by

$$\{|0>\} = \pi_0(\mathcal{D}) = \{|c_h;\ell>\} \tag{4.7}$$

where $c_h \in \pi_0(\mathcal{R})$ is a connected component of $\mathcal{R}$ and $\ell \in [\Sigma,G]/\partial\pi_1(\mathcal{R},h)$.

The important thing to note about Eq.(4.7) is the fact that the multiple-connectedness of the spatial manifold can have two effects on the Yang-Mills vacuum. Firstly, when $\pi_0(\mathcal{R}) \neq 0$, the number of vacuum sectors is greater than expected by looking at just the homotopy classes of gauge functions $[\Sigma,G]$. We shall call the extra sectors corresponding to non-trivial components of $R$ "holonomy sectors". Secondly, when a particular component of $R$ is not simply connected, the set of vacua in that holonomy sector may be smaller than $[\Sigma,G]$ due to the presence of the non-trivial little group $\partial\pi_1(R,h)$. We shall now give simple examples of both these phenomena.

4b Increase of Vacua

Consider an SU2 Yang-Mills gauge theory over $RP_3$. In contrast to the U1 example in section 3, all SU2 bundles on which such a theory can be constructed are classified by $H^2(RP_3, \pi_1(SU2)) = 0$, since SU2 is simply connected. Thus all bundles are trivial. However, there still exist two homomorphisms from $\pi_1(RP_3) = Z_2$ into SU2, both of which factor through the $Z_2$ centre of SU2. In particular

$$R = \{h_0, h_1\} \qquad (4.8)$$

where $h_0$ denotes the trivial homomorphism and $h_1$ maps the generator of $Z_2$ into minus the identity of SU2. $R$ therefore consists of two disconnected points $h_0$ and $h_1$, and $D$ splits into two disconnected bundles over these points.

It now remains to calculate $[\Sigma,G]/\partial\pi_1(R,h_0)$ and $[\Sigma,G]/\partial\pi_1(R,h_1)$ Since a single point is simply connected, both $\partial\pi_1(R,h_0)$ and $\partial\pi_1(R,h_1)$ consist only of the identity. Moreover $[\Sigma,G] = H^3(RP_3,Z) = Z$ so that

$$\{|0>\} = \{|h_0;n>, \ |h_1;m> \ , \ n,m\in Z\} \qquad (4.9)$$

In other words, there exist an integer's worth of gauge sectors in each of the two holonomy sectors. We have therefore shown that the presence of flat connections with holonomy group $Im(h_1) = Z_2$ causes a doubling of vacuum sectors.

This example is of particular interest because we can explicitly construct local representatives of the zero-energy configurations in each vacuum sector. The covering space of $RP_3$ is $S^3$, and SU2 is topologically also a three sphere. If we give $\hat{\Sigma} = S^3$ the coordinates $y_\mu$, $\mu=0,1,2,3$ such that $|y|^2 = 1$, then the action of $\pi_1(\Sigma) = Z_2$ on $\hat{\Sigma}$ takes $y_\mu$ into $-y_\mu$.

Now define

$$D^n_{h_o}(y) = (y_o + i\ \vec{y}\cdot\vec{\sigma})^{2n} \quad , \quad n \in Z \tag{4.10}$$

$$D^n_{h_1}(y) = (y_o + i\ \vec{y}\cdot\vec{\sigma})^{2n+1} \quad , \quad n \in Z \tag{4.11}$$

where $\vec{y} = (y_1, y_2, y_3)$ and $\vec{\sigma} = (\sigma_1, \sigma_2, \sigma_3)$ are the Pauli spin matrices. It can easily be seen that

$$D^n_{h_o}(-y) = D^n_{h_o}(y) \tag{4.12}$$

$$D^n_{h_1}(-y) = -D^n_{h_1}(y) \tag{4.13}$$

so that $D^n_{h_o}$ and $D^n_{h_1}$ represent elements of the function space $\mathcal{D}$ in the trivial and non-trivial holonomy sectors respectively. The corresponding connections are:

$${}^{(0)}A^n_i(x) = D^n_{h_o}(y)\ \partial_i\ D^n_{h_o}(y)^{-1} \tag{4.14}$$

$${}^{(1)}A^n_i(x) = D^n_{h_1}(y)\ \partial_i\ D^n_{h_1}(y)^{-1}\ . \tag{4.15}$$

If we define the quantity (Ademollo et al. 1978)

$$\psi[A] = \frac{1}{24\pi^2} \int_{RP_3} d^3x\ \epsilon_{ijk} Tr(A_i A_j A_k) \tag{4.16}$$

in the usual way, then it can be verified that $\psi[{}^{(0)}A^n] = n$ and $\psi[{}^{(1)}A^n] = (2n+1)/2$ measure the number of times the image of the "gauge function", D(y) wraps around the topological three-sphere SU2. When $\Sigma = S^3$ this is precisely the topological winding number of the configuration. In this sense, the vacuum configurations of our example in the trivial holonomy sector are said to have integral "winding number", while those in the non-trivial holonomy sector have half-integral "winding number". These latter configurations have been discussed by Mayer and Viswanathan (1979) but not in the context of flat connections with holonomy group $Z_2$.

## 4c Decrease of Vacua

We wish to know when the number of vacua in a particular holonomy sector is less than the number of homotopy classes of gauge functions. In other words, which elements of $[\Sigma,G] = H^1(\Sigma,\pi_1(G)) \oplus H^3(\Sigma,\pi_3(G))$ lie in the image of the map $\partial:\pi_1(R,h) \to [\Sigma,G]$? Although this is a difficult question to answer in general, significant results can be obtained in the trivial holonomy sector (the component of $R$ which contains the trivial homomorphism). In this case, an equivalent question is: when can two homotopically inequivalent pure gauge configurations $A_i^{(0)}=\Omega_0(x)\partial_i\Omega_0(x)^{-1}$ and $A_i^{(1)} = \Omega_1(x)\partial_i\Omega_1(x)^{-1}$ be joined by a continuous path $A_i^{(\lambda)}(x)=D_\lambda(y)\partial_i D_\lambda(y)^{-1}$? Thus we are looking for a continuous family of functions $D_\lambda(y)$ in $\mathcal{D}$ which interpolate between $D_0(y) = \Omega_0(p(y))$ and $D_1(y) = \Omega_1(p(y))$.

In the trivial holonomy sector the map $\partial$ can be completely characterized by known maps, and the necessary and sufficient conditions for the existence of interpolating functions $D_\lambda(y)$ have been determined for certain special cases. The details are given in Isham and Kunstatter (1981). Here I shall just give some results:

A. $\pi_1(R,h) = 0$ unless $\pi_1(\Sigma)$ contains elements of infinite order. Thus for $\pi_1(\Sigma)$ finite, the entire group $[\Sigma,G]$ is needed to classify the gauge sectors, as in the SU2 example over $RP_3$.

B. If the abelian part of $\pi_1(\Sigma)$ contains at least one factor of Z (i.e. the first Betti number of the three-manifold is non-zero), then the primary winding number can always be "killed off" by the map $\partial$, so that only the secondary winding number remains in the classification of the vacua.

C. If $\pi_1(\Sigma)$ is abelian and infinite, the secondary winding number always survives.

It would be of great interest to obtain results for more general three-manifolds. In particular, since the secondary winding number plays a vital role in the standard discussions of $\theta$-vacua on $S^3$, it is important to discover whether it is possible to interpolate between configurations with different secondary winding numbers.

As a trivial but useful example of how the map $\partial$ causes the collapse of the "n-vacua", consider again the U1 Yang-Mills theory on $S^1$.

The following pure gauge configurations

$$A^{(0)} = 0 = 1 \frac{\partial}{\partial\theta}(-1) \tag{4.17}$$

$$A^{(1)} = -i2\pi m = e^{i2\pi m\theta}\frac{\partial}{\partial\theta} e^{-i2\pi m\theta} \tag{4.18}$$

cannot be joined by a continuous path of configurations which are also pure gauge, because $\Omega_0(\theta)$ and $\Omega_1(\theta)$ are homotopically inequivalent maps from $S^1$ into U1. However, the configurations

$$A^{(\lambda)} = \lambda A^{(1)} + (1-\lambda)A^{(0)} \quad , \lambda \in [0,1] \tag{4.19}$$

do interpolate between $A^{(0)}$ and $A^{(1)}$ with zero energy, because U1 is abelian. Thus, according to the definition in Section 1, $A^{(0)}$ and $A^{(1)}$ belong in the same vacuum sector.

In order to make contact with the analysis of the preceding sections note that $A^{(\lambda)}$ may be rewritten as:

$$A^{(\lambda)} = D_\lambda(y)\frac{\partial}{\partial y} D_\lambda(y)^{-1} \tag{4.20}$$

where $D_\lambda(y) = e^{i2\pi m\lambda y}$, and $y \in R$, the covering space of U1. Moreover

$$D_\lambda(y+n) = D_\lambda(y)h_\lambda(n) \quad , \quad n \in \pi_1(S^1) = Z \tag{4.21}$$

where $h_\lambda(n) = e^{i2\pi mn\lambda}$ is a homomorphism from Z into U1, and y+n denotes the natural action of $\pi_1(S^1)$ on R. Thus $A^{(\lambda)}$ is a flat U1 connection on $S^1$ with holonomy group $h_\lambda(Z)$, and $\{D_\lambda(y)\}$ is the set of interpolating functions we are looking for.

According to Eq.(2.2) the space of homomorphisms from Z into U1 is parameterized by a real parameter $\lambda$, which is only defined modulo 1 ($\lambda$ = 0,1 yield the same homomorphism) $\mathcal{R}$ can therefore be given the topology of $S^1$. In addition, the homotopy classes of gauge functions are given by

$$[S^1,U1] = [S^1,S^1] = \pi_1(S^1) = Z \tag{4.22}$$

so that the fibre $G^\Sigma$ of $\mathcal{D}$ in this example has a countable infinity of connected components. The fibre bundle $\mathcal{D}$ is illustrated in fig.2, and it can now be clearly seen how the topology of the space of homomorphisms, $R$, and the topology of the function space $G^\Sigma$ "conspire" to collapse the vacuum sectors: $\pi_0(R) = 0$, $\pi_1(R,h_0) = Z$ and $\pi_0(G^\Sigma) = Z$. The map $\partial$ takes the homotopy class, m, of a closed loop $\{\rho_\lambda : \lambda \mapsto h_\lambda | h_0 = h_1 = 1\}$ which wraps around $R = S^1$ m times, and maps it into the homotopy class of the endpoint of its lift $\{D_\lambda : \lambda \mapsto D_\lambda(y) | D_\lambda(y+n) = D_\lambda(y)h_\lambda(n) \; \forall \lambda \in [0,1],\ n \in Z\}$. Since we are considering pointed maps, $D_0(y)$ is the trivial map, and $D_1(y)$ will obviously lie in the m-th component of the fibre over $h_0$. Moreover, any component in the fibre can be reached in this way so the map $\partial$ is onto and $[\Sigma,G]/\partial\pi_1(R,h_0) = 0$. There is consequently only one vacuum sector and the bundle space $\mathcal{D}$ is connected as shown in fig.2.

## 5 CONCLUSIONS

We have seen that non-trivial spatial topology can have striking effects on the canonical Yang-Mills vacuum. The number of vacuum sectors increases when the spatial manifold admits non-trivial flat bundles or when the space of homomorphisms which induces the trivial bundle is disconnected. The number of vacuum sectors can decrease due to the presence of continuous families of flat connections with non-trivial holonomy group. There still remains considerable work before the analysis outlined above is complete. In particular the classification of vacua in non-trivial bundles and in the non-trivial holonomy sectors of trivial bundles must be obtained. In addition, it would be interesting to examine the quantum corrections to the classical theories in these non-trivial sectors, in order to discover whether there exist physical quantities which allow us to distinguish between them, as for example in the theory of twisted scalar fields (Isham 1978). These and other aspects of the effects of spatial topology on quantum field theory are currently under investigation.

## ACKNOWLEDGEMENTS

I would like to thank C.J. Isham for many helpful discussions and a critical reading of the manuscript. I am also grateful for the hospitality shown me at Imperial College, London, where the work described above was carried out.

## REFERENCES

Ademollo, M., Napolitano, T. and Sciuto, S. (1978). Nucl. Phys., B134, 477.

Ambjorn, J. and Flybjerg H. (1980). Phys. Letts., 97B, 241.

Asorey, M. and Boya, J.J. (1979). J. Math. Phys., 20, 2379.

Avis, S.J. and Isham, C.J. (1979). In Recent Developments in General Relativity, eds. S. Deser and M. Levy, New York: Plenum Press.

Callan, C.G., Dashen, R.F. and Gross, D.J. (1976). Phys. Letts., 63B, 334.

Eguchi, T., Gilkey, P.B. and Hanson, A.J. (1980). Phys. Rep., 66, 215.

't Hooft, G. (1980). In Acta Physica Austriaca, Suppl. XXII, Springer Verlag.

Isham, C.J. (1978a). Proc. Roy. Soc. (London), 362A, 383.

Isham, C.J. (1978b). In Essays in Honour of Wolfgang Yourgau, ed. A. van der Merwe, New York: Plenum Press.

Isham, C.J. and Kunstatter, G. (1981a). Phys. Letts., 102B, 417.

Isham, C.J. and Kunstatter, G. (1981b). To be published in J. Math. Phys.

Jackiw, R. and Rebbi, C. (1976). Phys. Letts. 37, 179.

Kamber, F. and Tondeur, P. (1968). Lectures in Mathematical Physics, vol. 67, Springer Verlag.

Kobayashi, S. and Nomizu, K. (1969). Foundations of Differential Geometry vol. I, New York: Interscience.

Mayer, D.H. and Viswanathan, K.S. (1979). Rep. Math. Phys. (G.B.), 16, 281.

Milnor, J.W. (1957). Comm. Math. Helv., 32, 215.

Sciuto, S. (1979). Phys. Reports, 49, 181.

Figure 1: The wavy line denotes the canonical horizontal lift (Eq.2.1) in the flat U1 bundle over S' defined by Eq.(2.3). Points marked by x are identified.

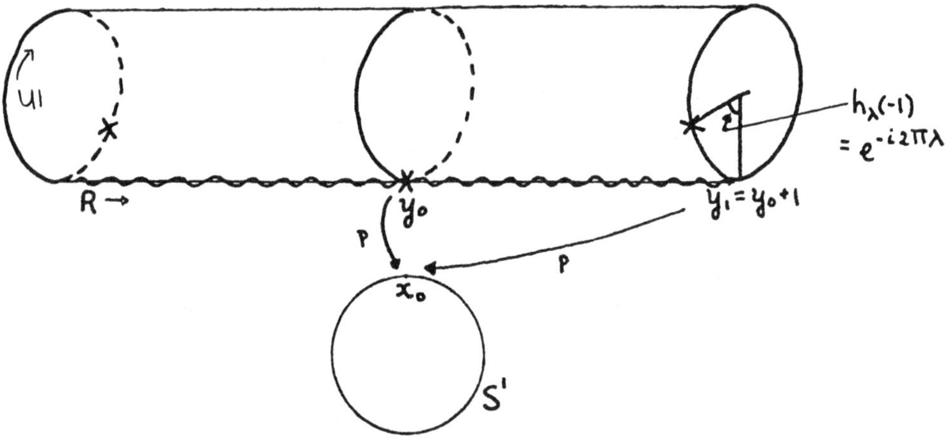


Figure 2: $\mathcal{D}$ represents the space of flat U1 connections on S'. It forms a fibre bundle over $\mathcal{R}$=Hom(Z,U1). Note that $\mathcal{D}$ is connected even though the (infinite dimensional) fibre, $G^\Sigma$, is disconnected.

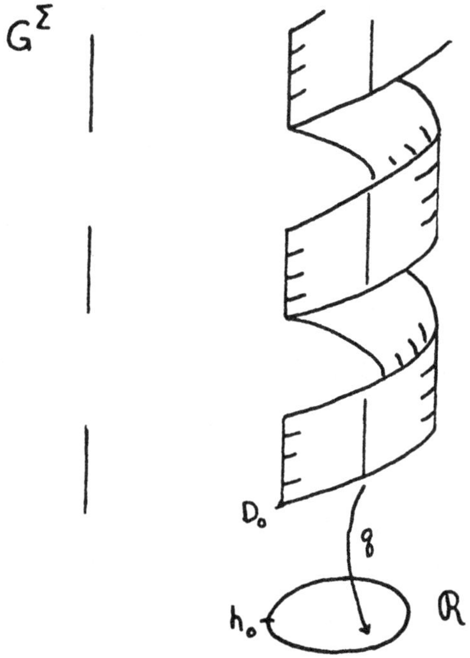

# FERMION FRACTIONIZATION IN PHYSICS

R. Jackiw
Center for Theoretical Physics
Massachusetts Institute of Technology
Cambridge, Massachusetts 02139

## I INTRODUCTION

I shall discuss a novel quantum mechanical/topological phenomenon that has become a focus for contemporary research in theoretical and experimental physics as well as in mathematics. The effect is as easy to describe as it is puzzling to comprehend: when a fermion is introduced to a soliton, the fermion can split up — it can fractionize — and its quantum numbers are shared by the resulting states. In this way, fractional charge can emerge in a theory where all basic constituents carry integral charges.

This unexpected behavior was first noted in purely mathematical, but otherwise unmotivated, studies of soliton-fermion interactions.[1] The current interest arises from the circumstance that a one-dimensional fermion-soliton system can be physically realized in condensed matter situations, as a description of an electron [fermion] in the presence of a domain wall [soliton]. In polyacetylene, for example, the peculiar experimental effects that have been observed[2] are explained by the peculiarities of fermions in the presence of solitons.[3] What makes the theory especially beautiful is that its predictions depend little on details of any particular model. Rather, very general topological and quantum mechanical considerations suffice to establish the occurrence of charge fractionization. But of course the practical relevance of the result depends on the specific physical setting in which it is encountered.

There are many topics to discuss: the original calculations in relativistic quantum field theory and the subsequent ones in condensed matter field theory which established fermionic solitons with charge ½, per degree of freedom;[1,3,4] the physical situation of polyacetylene and similar quasi-one dimensional organic polymers, which provide a theoretical and experimental laboratory for soliton physics;[2,3] the recently posited generalizations which give rise to other fractional charges in condensed matter systems, as well as in intriguing relativistic field theories;[5,6,7] and finally the mathematical concepts which relate this phenomenon to the non-trivial topological structures that are present.

All this cannot be covered here; therefore I shall select for discussion only the relativistic field theory calculations, with an emphasis on the topological properties which give rise to charge ½ and, in a generalization, to arbitrary fractions. However, since the polyacetylene story is so important for establishing the physical actuality of the phenomenon, I shall review it briefly and pictorially.

## II THE POLYACETYLENE STORY

Polyacetylene is a material consisting of parallel chains of carbon atoms, with electrons moving primarily along the chains, while hopping between chains is strongly suppressed. Consequently, the system is effectively one-dimensional, and can be represented as in Fig. 1a. The distance between carbon atoms is about 1Å.

If the atoms are considered to be completely stationary, i.e. rigidly attached to their equilibrium lattice sites, electron hopping along the chain is a structureless phenomenon, described by a probability amplitude, which to be sure depends on the distance between the atoms, but since that quantity never changes neither does the hopping amplitude.

Fig. 1(a) The rigid lattice of polyacetylene; the carbon atoms are 1 Å apart
(b),(c) The effect of Peierls' instability is to shift the carbon atoms .04 Å to the right (A) or to the left (B), thus giving rise to a double degeneracy.

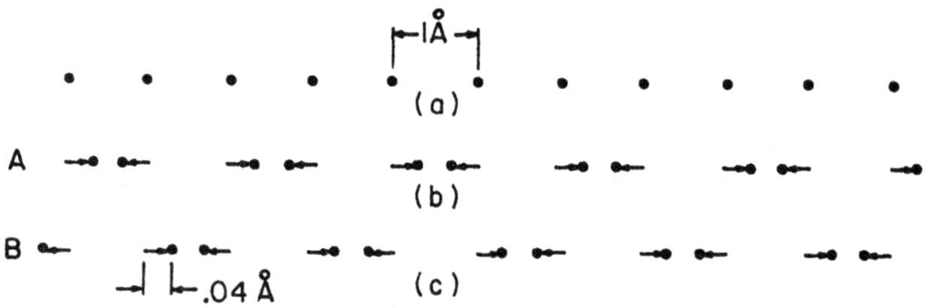

However, the atoms can be displaced from their rigid lattice positions, for a variety of reasons, like zero-point motion, thermal excitation, etc. It might be thought that these effects merely give rise to small oscillations about the rigid-lattice sites, and produce only a slight fuzzing of the undistorted-lattice situation.

In fact this is not correct; something more dramatic takes place. Rather than oscillating about the rigid-lattice site, the atoms first shift a distance of about .04 Å and then proceed to oscillate around the new, slightly distorted location. That this should happen was predicted by Peierls, and is called the Peierls instability.[8] Due to reflection symmetry, there is no difference between a shift to the right or a shift to the left; the material chooses one or the other, thus breaking spontaneously the symmetry, and giving rise to doubly degenerate vacua, called A and B, as is illustrated in Figs. 1b and 1c. The chemical bonding patterns are illustrated in Figs. 2a and 2b, where the double bond connects atoms that are closer together, and the single bond those that are further apart than in the [unphysical] rigid lattice.

If the displacement is described by a field $\phi$ which depends on the position x along the lattice, the so-called phonon field, then Peierls' instability, as well as detailed dynamical calculations[3] indicate that the energy density $V(\phi)$, as a function of constant $\phi$, has the double-well shape, depicted in Fig. 3a. The symmetric point $\phi=0$ is unstable; the system in its ground state must choose one of the two equivalent ground states $\phi = \phi_0 = \pm.04$Å. In the ground states, the phonon field has uniform

Fig. 2(a),(b) Pattern of chemical bonds in vacua A and B.
(c) Two solitons inserted into vacuum B.

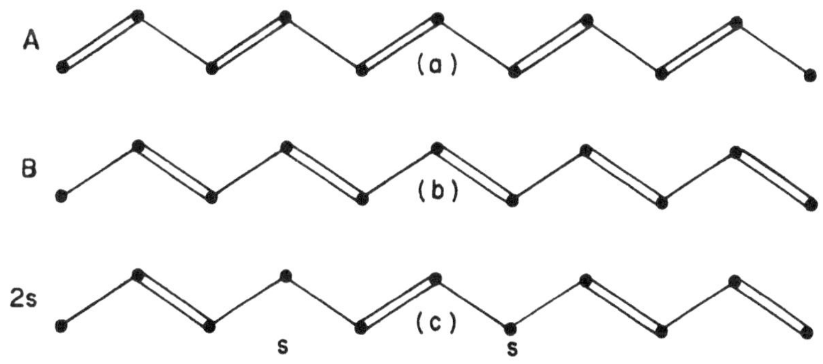

values, independent of x; as is shown in Fig. 3b.

By now it is widely appreciated that whenever the ground state is degenerate there frequently exist additional stable states of the system, for which the phonon field is non-constant. Rather, as a function of x, it interpolates, when x passes from negative to positive infinity, between the allowed ground states. These are the famous solitons, or kinks. For polyacetylene they are also depicted in Fig. 3b, and they correspond to domain walls which separate regions with vacuum A from those with vacuum B [solitons], and vice versa [anti-solitons].

Fig. 3(a) Energy density $V(\phi)$, as a function of a constant phonon field $\phi$. The symmetric stationary point, $\phi=0$, is unstable. Stable vacua are at $\phi=+|\phi_0|$, (A) and $\phi=-|\phi_0|$, (B).

(b) Phonon fields corresponding to stable states. The two constant fields, $\pm|\phi_0|$, correspond to the two vacua (A and B). The two kink fields, $\pm\phi_s$, interpolate between the vacua and represent domain walls.

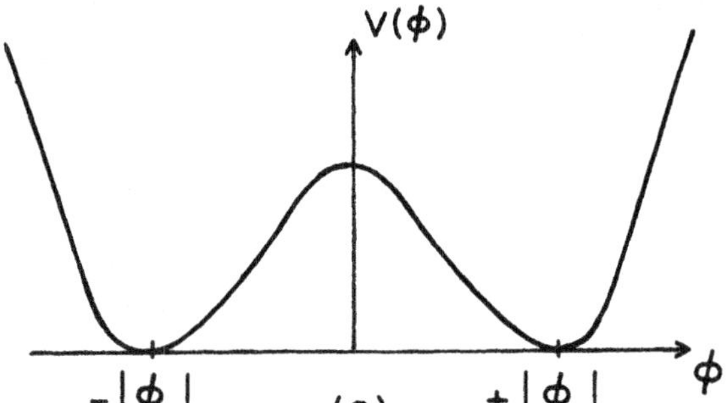

(a)

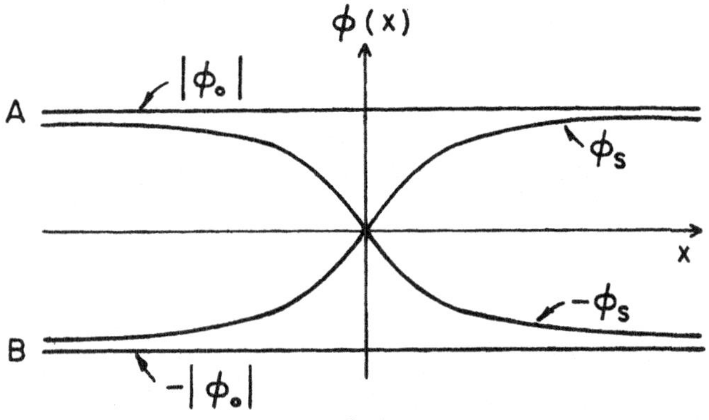

(b)

Consider now a polyacetylene sample in the B vacuum, but with two solitons along the chain, as depicted in Fig. 2c. Let us count the number of links in the sample without solitons (Fig. 2b) and compare with the number of links where two solitons are present (Fig. 2c). It suffices to examine the two chains only in the region where they differ, i.e. between the two solitons. Vacuum B exhibits 5 links, while the addition of two solitons decreases the number of links to 4. The two soliton state exhibits a deficit of one link. If now we imagine separating the two solitons a great distance, so that they act independently of one another, then each soliton carries a deficit of half a link, and the quantum numbers of the link, for example the charge, are split between the two states. This is the essence of fermion fractionization.

It should be emphasized that we are not here describing the familiar situation of an electron moving around a two-center molecule, spending "half" the time with one nucleus and "half" with the other. Then one might say that the electron is split in half, on the average; however fluctuations in any quantity are large since the one-nucleus state is not an eigenstate of any physical observable. In our soliton example, the fractionization is without fluctuations; in the limit of infinite separation one achieves an eigenstate with fractional eigenvalues.

We must however remember that the link in fact corresponds to two states: an electron with spin up and another with spin down. This doubling obscures the dramatic charge ½ effect, since everything must be multiplied by 2 to account for the two states. So in polyacetylene, a soliton carries a charge deficit of one unit of electric charge. Nevertheless charge fractionization leaves a spur: the soliton state has net charge, but no net spin, since all the electron spins are paired. If an additional electron is inserted into the sample, the charge deficit is extinguished, and one obtains a neutral state, but now there is a net spin. These spin-charge assignments [charged - without spin, neutral - with spin] are unexpected, but in fact have been observed,[2] and provide experimental verification for the soliton picture of polyacetylene.

Moreover, it has been suggested that in another one-dimensional system, TTF-TCNQ under pressure, the fundamental fractionization should be 1/3 per state, and the doubling due to spin would no longer obscure the charge fractions, which should be experimentally observable.[5] Other fractions also appear to be physically attainable.[6]

## III CHARGE ½ FRACTIONS IN RELATIVISTIC QUANTUM FIELD THEORY

I shall now provide a dynamical calculation which shows how charge ½ arises in a relativistic field theory describing fermions in interaction with solitons. Although a fully consistent, second quantized analysis has been developed for this problem,[9] it suffices here to discuss a simpler, approximate, first quantized approach, where the soliton is described by an external, non-dynamical c-number field. The fermion dynamics are governed by a Dirac Hamiltonian, $H(\phi)$, which also depends on a background field $\phi$, with which the fermions interact. In the vacuum sector, $\phi$ takes on a constant value $\phi_0$, appropriate to the vacuum. When a soliton is present, $\phi$ becomes the appropriate, static soliton profile $\phi_s$. We need not be any more specific; in particular the discussion is not limited to one spatial dimension. Thus in one dimension the soliton can be the kink of Fig. 3b; in two, it is an Abrikosov [Nielsen-Olesen] vortex;[10] in three, the 't Hooft-Polyakov monopole.[11] We need not even insist on the explicit soliton profiles that solve the corresponding non-linear field equations; all that we require is that the topology [i.e. the large distance behavior] of the profiles be non-trivial. The one-dimensional model is relevant for polyacetylene; the Dirac Hamiltonian arises not because the electrons are relativistic; but rather it emerges in a certain well-formulated approximation [which ignores electron spin][12] to the microscopic theory.[3]

To analyze the system we need the eignmodes, both in the vacuum and soliton sectors.

$$H(\phi_o)\psi_E^o = E^o\psi_E^o \tag{3.1}$$

$$H(\phi_s)\psi_E^s = E^s\psi_E^s \tag{3.2}$$

As is familiar, there will be in general negative energy solutions and positive energy solutions. [For polyacetylene, the negative energy solutions correspond to the states in the valence band; the positive energy ones, to the conduction band.] In the ground state, all the negative energy levels are filled, and the ground state charge is the integral over all space of the charge density $\rho(x)$; which in turn is constructed from all the negative energy solutions.

$$\rho(x) = \int_{-\infty}^{0} dE\, \rho_E(x) \tag{3.3}$$

$$\rho_E(x) = \psi_E^*(x)\psi_E(x)$$

Of course integrating (3.3) over x will produce an infinity; to renormalize we measure all charges relative to the ground state in the vacuum sector. Thus the soliton charge is

$$Q = \int dx \int_{-\infty}^{0} dE \,\{\rho_E^s(x) - \rho_E^o(x)\} \tag{3.4}$$

Eq. (3.4) may be completely evaluated without explicitly specifying the soliton profile, nor actually solving for the negative energy modes, provided H possesses a further property. We assume that there exists a conjugation symmetry which takes positive energy solutions of (3.1) and (3.2) into negative energy solutions. That is, we assume that there exists a unitary matrix M, such that

$$M\psi_E = \psi_{-E} \tag{3.5}$$

An immediate consequence, crucial to the rest of the argument, is that the charge density at E is an even function of E.

$$\rho_E(x) = \rho_{-E}(x) \tag{3.6}$$

[Clearly, one may allow a phase factor, or even complex conjugation, in the right-hand side of (3.5), without affecting (3.6)]. Charge conjugation is an example of the required symmetry; it holds in the polyacetylene theory.

Whenever one solves a conjugation symmetric Dirac equation, with a topologically interesting background field, like a soliton, there always are, in addition to the positive and negative energy solutions related to each other by conjugation, self-conjugate, normalizable zero-energy solutions. That this is indeed true can be seen by explicit calculation, and will be presently demonstrated when we exemplify the general discussion. However, the occurrence of the zero mode is also predicted by very general mathematical theorems about differential equations. These so-called "index theorems" count the zero eigenvalues, and insure that the number is non-vanishing whenever the topology of the background is non-trivial.[13] We shall assume that there is just one zero mode, described by a normalizable wave function $\psi_0$. This minimal situation can be achieved by taking a minimal, non-trivial background topology.[1]

To evaluate (3.4), we first recall that the wave functions are complete both in the soliton sector and in the vacuum sector.

$$\int_{-\infty}^{\infty} dE \ \psi_E^*(x)\psi_E(y) = \delta(x-y) \tag{3.7}$$

As a consequence, it follows that

$$\int_{-\infty}^{\infty} dE \ [\rho_E^s(x) - \rho_E^o(x)] = 0 \tag{3.8}$$

In the above completeness integral over all energies, we record separately the negative energy contributions, the positive energy contributions, and for the soliton, the zero-energy contribution. Since the positive energy charge density is equal to the negative energy one, by virtue of (3.6), we conclude that (3.8) may be equivalently written as an integral over negative E.

$$\int_{-\infty}^{0} dE [2\rho_E^s(x) + \psi_o^*(x) \ \psi_o(x) - 2\rho_E^o(x)] = 0 \tag{3.9}$$

After a rearrangement of terms, Eq. (3.9) shows that

$$Q = -\tfrac{1}{2} \int dx \ \psi_o^*(x) \ \psi_o(x) = -\tfrac{1}{2} \tag{3.10}$$

This is the final result: the soliton's charge is $-\frac{1}{2}$; a fact that follows from completeness [Eq. (3.7)] and conjugation symmetry [Eq. (3.6)]. It is seen in (3.10) that the zero-energy mode is essential to the conclusion. The existence of the zerc modes in the conjugation symmetric case is assured by the non-trivial topology of the background field. The result is otherwise completely general, and holds in any number of dimensions.

## IV ARBITRARY FRACTIONAL CHARGE IN RELATIVISTIC QUANTUM FIELD THEORY

Since charge ½ follows from completeness of the wave functions in the presence of a conjugation symmetry, generalizing to arbitrary fractions requires abandoning the conjugation symmetry. The original suggestion for a generalization came from condensed matter physics,[5,6] and now has been realized in relativistic quantum field theory.[7] However, as we shall see, the field theoretical version is not yet completely satisfactory: the fractionization is arbitrary, parameterized by the arbitrary amount of conjugation symmetry breaking; there is no understanding how this arbitrariness is fixed. [In the condensed matter applications[5,6], physical considerations determine all parameters, leading to definite fractions.]

I shall present a simple one-dimensional model, for which it will be possible to show that the soliton's charge is a transcendental function of the parameters. The calculation will demonstrate that the effect again depends little on the details, and requires almost no explicit computation; though of course more is needed than in the very general charge ½ examples. It is important to know that the generalization is not solely a one-dimensional effect; similar results have been established in three dimensions.[7]

Consider the following Dirac Hamiltonian $H(\phi)$, depending on a background field $\phi$.

$$H(\phi) = \sigma^2 p + \sigma^1 \phi + \sigma^3 \varepsilon \tag{4.1}$$

$$p = \frac{1}{i}\frac{d}{dx}$$

Since we are in one spatial dimension, the Dirac algebra is realized by 2x2 matrices; we take them to be the Pauli matrices: Dirac's $\alpha$ matrix is $\sigma^2$, $\beta$ is $\sigma^1$, and $\sigma^3$ corresponds to a pseudoscalar coupling. When $\varepsilon=0$, there exists a conjugation symmetry since $\sigma^3$ anti-commutes with $H|_{\varepsilon=0}$. However, with non-vanishing $\varepsilon$, which we take to be positive, the conjugation symmetry is lost.

To evaluate the charge (3.4), we need the negative energy modes, both in the vacuum sector $\phi=\phi_0$, and in the soliton sector $\phi=\phi_s$. All we need to know about the soliton profile is that it interpolates between the vacua.

$$\phi_s \underset{x\to\pm\infty}{=} \pm|\phi_0| \tag{4.2}$$

The vacuum sector is trivial: the wave functions are plane waves and the spectrum is continuous, beginning at $\pm\sqrt{\phi_0^2+\varepsilon^2}$ ; $E^0(k) = \pm\sqrt{k^2+\phi_0^2+\varepsilon^2}$ .

In the soliton sector, let us note first the existence of a discrete bound state at $E^s=\varepsilon$. The wave function is proportional to

$$\begin{pmatrix} e^{-\int^x dx' \phi_s(x')} \\ 0 \end{pmatrix}$$

This is normalizable precisely because $\phi_s$ tends to opposite limits at opposite ends of the real line. Observe that in the conjugation symmetric limit, $\varepsilon=0$, this is the zero-energy bound state mentioned before. It is present here as well, but shifted from the symmetric value. Its occurrence will play an important role in our analysis.

To proceed, we need more explicit information about the eigenmodes. Upon writing $\psi_E$ as $\binom{u}{v}$, and working out the Pauli matrices, we see that v is determined by u.

$$v = \frac{1}{E+\varepsilon}(\partial_x+\phi)u \tag{4.3}$$

The function u satisfies a Schrödinger-like equation

$$(-\partial_x^2 + \phi^2 - \phi')u = (E^2-\varepsilon^2)u \tag{4.4}$$

$$\phi' = \frac{d}{dx}\phi$$

with a "potential" $\phi^2-\phi'$, which is the constant $\phi_0^2$ in the vacuum sector, and tends to $\phi_0^2$ at infinite x in the soliton sector.

The soliton sector Schrödinger equation has a bound state, $u^s=e^{-\int^x dx' \phi_s(x')}$; this is just the Dirac state at $E=\varepsilon$, previously identified. We shall assume that the soliton profile is sufficiently weak, so that it supports no other normalizable solutions to (4.4) in the soliton sector. [It is obvious how to modify the analysis when additional bound states are present.] The remaining eigenmodes of (4.4) lie in the continuum, which begins at $E^2-\varepsilon^2=\phi_0^2$.

It is now straightforward to construct the negative energy solutions of the Dirac equation, from the continuum solutions $u_k$ of the Schrödinger equation.

$$\psi_k = \begin{pmatrix} \sqrt{\frac{E+\varepsilon}{2E}}\, u_k \\ \frac{-1}{\sqrt{2E(E+\varepsilon)}}\, (\partial_x + \phi) u_k \end{pmatrix}$$

$$H(\phi)\psi_k = E\psi_k$$

$$E = -\sqrt{k^2+\phi_o^2+\varepsilon^2} \tag{4.5}$$

Proper normalization of $\psi_k$ is assured, when $u_k$ is properly normalized.

The charge density, at a given [negative] energy E, is given by $\psi_k^*(x)\psi_k(x)$, which according to (4.5) is

$$\begin{aligned} \rho_k(x) &= \frac{E+\varepsilon}{2E} |u_k(x)|^2 + \frac{1}{2E(E+\varepsilon)} |(\partial_x+\phi)u_k|^2 \\ &= |u_k(x)|^2 + \frac{1}{4E(E+\varepsilon)} \partial_x^2 |u_k(x)|^2 \\ &\quad + \frac{1}{2E(E+\varepsilon)} \partial_x \left[|u_k(x)|^2 \phi\right] \end{aligned} \tag{4.6}$$

The second equality follows from the first by virtue of the Schrödinger equation. The charge is the integral of the above over all x and k in the soliton sector, minus a similar integral in the vacuum sector. However in the vacuum sector $|u_k|^2$ is a constant, as is $\phi$, so the last two terms in (4.6) vanish. This leaves

$$\begin{aligned} Q = &\int_{-\infty}^{\infty} dx \int_{-\infty}^{\infty} \frac{dk}{(2\pi)} \left[|u_k^s(x)|^2 - |u_k^o(x)|^2\right] \\ &+ \int_{-\infty}^{\infty} \frac{dk}{(2\pi)} \frac{1}{4E(E+\varepsilon)} \{\partial_x |u_k^s(x)|^2 + 2|u_k^s(x)|\phi_s(x)\}\Big|_{x=-\infty}^{x=\infty} \end{aligned} \tag{4.7}$$

The double integral on the right hand side can be evaluated by completeness. The $u_k^o$ represent all the Schrödinger modes in the vacuum sector, while the $u_k^s$ are one short of a complete set in the soliton sector, since the normalizable bound state is not included among them. [It corresponds to positive energy Dirac mode; the negative energy modes are constructed from the continuum Schrödinger solutions.] Hence the first term on the right-hand side of Eq. (4.7) contributes -1 to Q. To evaluate the remaining terms, we need to know the Schrödinger modes, in the presence of the soliton, but only at $x=\pm\infty$. These may be expressed in terms of transmission and reflection coefficients.

$$\begin{aligned} u_k(x) &\xrightarrow[x\to]{} Te^{ikx} \\ u_k(x) &\xrightarrow[x\to]{} e^{ikx}+Re^{-ikx} \end{aligned} \tag{4.8}$$

We thus get

$$Q = -1 + \int_{-\infty}^{\infty} \frac{dk}{(2\pi)} \frac{|\phi_o|}{2E(E+\varepsilon)} \{|T|^2 + (|R|^2 + 1)\} \tag{4.9}$$

where oscillatory terms have been dropped and the plus sign between the term from $x=+\infty$ and the terms from $x=-\infty$ arises because of the sign reversal in $\phi_s(x)$. Unitarity

$$|T|^2 + |R|^2 = 1 \tag{4.10}$$

permits the final evaluation.

$$Q = -\frac{1}{\pi} \tan^{-1}\left|\frac{\phi_o}{\varepsilon}\right| \tag{4.11}$$

In the conjugation symmetric limit, $\varepsilon\to 0$, and the previous result, $Q = -\frac{1}{2}$, is regained.

## V DISCUSSION

The results here presented demonstrate a novel and fascinating quantum mechanical phenomenon, which was previously unsuspected. The emergence of fractional quantum numbers is related to other examples of odd quanta in topologically non-trivial settings, like the occurrence of spin ½ in a bosonic theory,[14] or the presence of hidden variables [vacuum angles] in the description of physical states.[15]

Charge fractionization possesses an elegant mathematical description. Observe that the charge (3.4) may also be written as an integral of a charge density.

$$Q = \int_{-\infty}^{\infty} dx\ \rho(x) \qquad (5.1a)$$

$$\rho(x) = -\frac{1}{2\pi}\frac{d}{dx}\tan^{-1}\left(\frac{\phi_s(x)}{\varepsilon}\right) \qquad (5.1b)$$

It has been suggested that for weakly varying background fields the above formula holds even when $\varepsilon$ depends on x.[7] It has been further suggested that for weakly time-varying background fields, (5.1b ) is the time-component of a conserved topological current.[7]

$$j^\mu = -\frac{1}{2\pi}\varepsilon^{\mu\nu}\partial_\nu \tan^{-1}\left(\frac{\phi_s}{\varepsilon}\right)$$

$$= \frac{1}{2\pi}\varepsilon^{\mu\nu}\varepsilon_{ab}\hat{\phi}_a\partial_\nu\hat{\phi}_b$$

$$\hat{\phi}_1 = \phi_s/\sqrt{\phi_s^2+\varepsilon^2}$$

$$\hat{\phi}_2 = \varepsilon/\sqrt{\phi_s^2+\varepsilon^2} \qquad (5.2)$$

I have already mentioned that similar effects are found in other dimensions. For example, in three spatial dimensions, where the monopole is the topological soliton, the analogue to (5.2) is[7]

$$j^\mu = \frac{1}{12\pi^2}\varepsilon^{\mu\alpha\beta\gamma}\varepsilon_{abcd}\{\hat{\phi}_a(\mathcal{D}_\alpha\hat{\phi})_b(\mathcal{D}_\beta\hat{\phi})_c(\mathcal{D}_\gamma\hat{\phi})_d + \frac{3}{4}eF_{\alpha\beta,ab}\hat{\phi}_c(\mathcal{D}_\gamma\hat{\phi})_d\} \qquad (5.3a)$$

Here the SU(2) monopole is considered within a SU(2)xSU(2) gauge theory. Taking the fourth component of the scalar field to be $\varepsilon$, and the fourth components of the gauge field to be zero, while setting the remaining components to the monopole profile, one finds the same result as in one dimension .

$$Q = -\frac{1}{\pi}\tan^{-1}\left|\frac{\phi_0}{\varepsilon}\right| \qquad (5.3b)$$

There is another mathematical connection worth describing. In the formula for the charge (3.4), assume for a moment that the states are normalizable, and the negative energy integral is replaced by a sum. It is then clear that Q measures the difference between the number of negative eigenmodes of the Dirac equation in the presence of the soliton, relative to the situation without the soliton. A more continuous description is achieved by parametrizing the background field by a parameter $\alpha$ such that $\phi$ varies continuously from $\phi_o$ to $\phi_s$, as $\alpha$ passes from $-\infty$ to $\infty$; for example,

$$\phi(x,\alpha) = [\theta(-\alpha)\phi_o + \theta(\alpha)\phi_s(x)] \tanh |\alpha| \qquad (5.4)$$

The energy eigenvalues E of the Dirac equation will now depend on $\alpha$, $E(\alpha)$, and in the course of $\alpha$'s variation some number will flow across E=0. The charge is a measure of this spectral flow, which is topologically determined.[16] Of course if the system were finite, so that the levels are truly discrete and the wavefunctions normalizable, that number would be an integer; however, on the infinite space, with continuum wavefunctions, a non-integral value is possible.

What is the physical import of fermion fractionization? It is clearly relevant to condensed matter phenomena, especially in one dimension. Moreover, particle physicists are intrigued by the quark-like charge assignments, which emerge dynamically, rather than being postulated a priori However, it is not at all apparent how one might replace the successful hadron phenomenology based on quarks, by one based on fractionally charged solitons. Nevertheless, let me record one more amusing formula. In the models considered, the mass of the fermion is $M_F=\sqrt{\phi_o^2+\varepsilon^2}$. We may eliminate $\varepsilon$ in favor of the charge, with the help of (4.11), which is also true in three dimensions, see (5.3b). There results a mass formula,

$$M_F = \frac{|\phi_o|}{|\sin \pi Q|} \qquad (5.5)$$

which has the intriguing property that it takes the same value, both for $Q=\frac{1}{3}$ and $Q=\frac{2}{3}$.

In gravity theory it should be possible to establish similar effects. One needs to develop the concept of a static, soliton-like background geometry. Then the solutions of the Dirac equation on the solitonic space should have zero modes, which would be interpreted as states with fractional fermionic quantum numbers. However, none of this has been worked out as yet.

It is very satisfying to contemplate a physical idea, with range of application from chemistry, to condensed matter physics, to quantum field theory and gravity theory, and finally to mathematics and topology. This is certainly striking evidence for the unity of all these disciplines, and provides another example of the power of mathematics to uncover unexpected physical phenomena.

ACKNOWLEDGEMENT

In addition to the Nuffield Workshop, held in London, England, during August, 1981, this material was presented at the ICTP Field Theory Workshop, Trieste, Italy, July, 1981; and at the European Physical Society Conference in Lisbon, Portugal, July, 1981. I thank the organizers of all three events for giving me the opportunity to lecture. The research was supported in part through funds provided by the U.S. DEPARTMENT OF ENERGY (DOE) under contract DE-AC02-76ER03069.

REFERENCES

1. R. Jackiw and C. Rebbi, Phys. Rev. D13, 3398 (1976).
2. For a review of the experiments, see A. Heeger, Comments Solid State Phys. 10, 53 (1981).
3. W. P. Su, J. R. Schrieffer and A. Heeger, Phys. Rev. Lett. 42, 1698 (1979) and Phys. Rev. B22, 2099 (1980); see also M. Rice, Phys. Lett. 71A, 152 (1979).
4. R. Jackiw and J. R. Schrieffer, Nucl. Phys. B190 [FS3], 253 (1981).
5. W. P. Su and J. R. Schrieffer, Phys. Rev. Lett. 46, 738 (1981).
6. M. Rice and E. Mele, Xerox preprint, April 1981; J. Gammel and J. Kumhansl, Cornell Preprint, #4519, July, 1981.
7. J. Goldstone and F. Wilczek, Phys. Rev. Lett. 47, 968 (1981).
8. R. Peierls, Quantum Theory of Solids, Clarendon Press, Oxford (1955).
9. The approach to fermionic problems, based on J. Goldstone and R. Jackiw, Phys. Rev. D11, 1486 (1975), is given in Ref. 1. For a summary, see R. Jackiw, Rev. Mod. Phys., 49, 681 (1977).
10. A. Abrikosov, Zh. Eksp. Teor. Fiz. 32, 1442 (1957) [JETP 5, 1174 (1957)]; H. Nielsen and P. Olesen, Nucl. Phys. B61, 45 (1973).
11. G. 't Hooft, Nucl. Phys. B79, 276 (1974); A. Polyakov, Zh. Eksp. Teor. Fiz. Pis'ma Red. 20, 430 (1974) [JETP Lett. 20, 194 (1974)].
12. H. Takayama, Y. Lin-Liu, and K. Maki, Phys. Rev. B21, 2388 (1980).
13. For a summary of the index theorems, see for example T. Eguchi, P. Gilkey and A. Hanson, Phys. Reports 66, 213 (1980).
14. R. Jackiw and C. Rebbi, Phys. Rev. Lett. 36, 1116 (1976); P. Hasenfratz and G. 't Hooft, Phys. Rev. Lett. 36, 1119 (1976); J. Friedman and R. Sorkin, Phys. Rev. Lett. 44, 1100 (1980).
15. S. Coleman, R. Jackiw and L. Susskind, Ann. Phys. (NY) 93, 267 (1975); R. Jackiw and C. Rebbi, Phys. Rev. Lett. 37, 172 (1976); C. Callan, R. Dashen and D. Gross, Phys. Lett. B63, 334 (1976); C. Isham, J. Math. Phys. (in press) and this volume.
16. M. F. Atiyah, V. K. Patodi and I. M. Singer, Math. Proc. Camb. Phil. Soc. 79, 71 (1976).

# PART II

# SUPERGRAVITY

# THE NEW MINIMAL FORMULATION OF N = 1 SUPERGRAVITY AND ITS TENSOR CALCULUS

M.F. Sohnius *)

The Blackett Laboratory, Imperial College, London SW7 2BZ

P.C. West

Department of Mathematics, King's College, London WC2 R2LS

## INTRODUCTION

Non-extended supergravity is the generally relativistic field theory of a graviton with spin 2 and a massless spin-$\frac{3}{2}$ Majorana particle, the gravitino. It therefore describes 2 (bosonic) + 2 (fermionic) degrees of freedom for physical states. In terms of fields, the graviton is described by the vierbein $e_\mu{}^a$ with 16 real field components and the gravitino by a Majorana vector-spinor $\psi_{\mu\alpha}$ with also 16 real components. The gauge-invariances of the theory are four general coordinate transformations and six local Lorentz rotations, both bosonic, and the four components of an N=1 local supersymmetry transformation which is fermionic in nature. This leaves 6+12 off-shell field components for $e_\mu{}^a$ and $\psi_{\mu\alpha}$. Since supersymmetry requires matching numbers of bosonic and fermionic degrees of freedom, we see that an off-shell formulation must have 6+0, modulo 4+4, auxiliary fields. A version with no fermionic and six bosonic ones is called a "minimal" off-shell supergravity.

The first off-shell supergravity was found by Breitenlohner (1977a) and has 14+8 auxiliary fields. The first minimal one was found independently by Stelle & West (1978a) and by Ferrara & van Nieuwenhuizen (1978a). This "old minimal" set contained as auxiliary fields a scalar $M$, a pseudoscalar $N$ and an axialvector $b_a$. Recently (Sohnius & West 1981b), we have proposed an alternative minimal set which seems to become known as the "new minimal" set. This contains as auxiliary fields an axialvector $A_\mu$ and an antisymmetric tensor $a_{\mu\nu}$, both with gauge-freedoms

$$A_\mu \rightarrow A_\mu + \partial_\mu \alpha \qquad ; \qquad a_{\mu\nu} \rightarrow a_{\mu\nu} + \partial_\mu \beta_\nu - \partial_\nu \beta_\mu$$

*) address as of 1 Oct 1981: CERN, Geneva

which reduce the total number of degrees of freedom to six, as required. A linearized version of this formulation seems to have first been given by Akulov, Volkov & Soroka (1977). No one, however, properly took into account the chiral gauge-invariance, and thus all attempts to complete the theory to its full non-linear version were bound to fail. We thank W. Siegel for pointing out to us the history of this problem.

The present article falls into two parts, roughly according to our seminars at the Workshop. First we repeat and extend the derivation in our previous paper which was based on a study of possible supersymmetry current multiplets. As suggested by Ogievetsky & Sokatchev (1977) already as early as 1976, we then study the linearized supergravity to which such a current multiplet acts as source, just as the energy-momentum tensor is the source for the gravitational field. The linearized supergravity multiplet is derived as the contragradient of the supercurrent multiplet, and by comparison of the linearized transformation laws with the general geometric formulas of supergravity, the torsion constraints can be read off which define the particular off-shell non-linear supergravity.

We then proceed, after a short discussion of the possible implications of the existence of this new formulation, to first review the rigid tensor calculus and then generalize it to our version of supergravity. We introduce the general, the chiral, the gauge, the linear and the kinetic supermultiplets and show how they are related to each other. We derive multiplication rules and, finally, action formulas for the general multiplet and the chiral multiplet of chiral weight 2.

We then use the action formulas to derive the supergravity action itself and the couplings of the Wess-Zumino model and the Maxwell-Einstein model (super-QED). We discuss the restrictions imposed by the presence of the chiral transformations in the superalgebra, and we show that these are strong enough to rule out a particular form of cosmological constant, but not strong enough to rule out all possible higher-loop counterterms. Finally, we show that a model with a Fayet-Iliopoulos mechanism can be coupled without problems.

## MINIMAL POINCARE SUPERCURRENT MULTIPLETS

The only supercurrent multiplet which is irreducible is that for <u>conformal supersymmetry</u> (Ferrara & Zumino 1975). It has 8 + 8 field components and consists of a conserved axialvector $C_\mu$, a conserved Majorana vector-spinor $J_{\mu\alpha}$ and a conserved symmetric tensor $\Theta_{\mu\nu}$,

$$0 = \partial^\mu C_\mu = \partial^\mu J_\mu = \partial^\mu \Theta_{\mu\nu} \quad , \tag{1}$$

with tracelessness conditions for $J_{\mu\alpha}$ and $\Theta_{\mu\nu}$,

$$0 = \gamma^\mu J_\mu = \eta^{\mu\nu}\Theta_{\mu\nu} \ . \tag{2}$$

The transformation laws under supersymmetry transformations with infinitesimal Majorana spinor parameter $\zeta$ are (conventions: $\eta_{00} = \varepsilon_{0123} = +1$; $\gamma_5{}^2 = -1$; $\sigma_{\mu\nu} = \frac{i}{2}[\gamma_\mu,\gamma_\nu]$ ):

$$\begin{aligned}
\delta C_\mu &= \bar{\zeta}\gamma_5 J_\mu \\
\delta J_\mu &= i\,\gamma^\nu\left(2\Theta_{\mu\nu} - \gamma_5\,\partial_\nu C_\mu - \tfrac{1}{2}\,\varepsilon_{\mu\nu\kappa\lambda}\,\partial^\kappa C^\lambda\right)\zeta \\
\delta\Theta_{\mu\nu} &= -\tfrac{i}{4}\,\bar{\zeta}\,\left(\sigma_{\mu\lambda}\partial^\lambda J_\nu + \sigma_{\nu\lambda}\partial^\lambda J_\mu\right)
\end{aligned} \tag{3}$$

and the algebra closes on the multiplet as

$$[\delta_1,\delta_2] = 2i\;\bar{\zeta}_1\gamma^\mu\zeta_2\;\partial_\mu \quad . \tag{4}$$

The tracelessness conditions (2) result in conservation laws for the moments $ix\cdot\gamma J_\mu$, $x^\nu\Theta_{\mu\nu}$ and $2x_\mu x^\lambda\Theta_{\nu\lambda} - x^2\Theta_{\mu\nu}$, i.e. for the currents of conformal supersymmetry transformations, dilatations and special conformal transformations. Since Poincaré supersymmetry has none of these invariances, condition (2) must be violated for a Poincaré supercurrent multiplet.

The fields of a <u>reducible multiplet</u> of supersymmetry generally fall into two classes which we denote by $\mathcal{A}$ and $\mathcal{B}$. The class $\mathcal{B}$ forms a submultiplet, while the transformation laws for $\mathcal{A}$ involve the fields of $\mathcal{B}$ as well as those of $\mathcal{A}$. We can write this symbolically as

$$\begin{aligned}
\delta\mathcal{A} &= \mathcal{A} + \mathcal{B} \\
\delta\mathcal{B} &= \mathcal{B} \quad .
\end{aligned} \tag{5}$$

A reducible multiplet can thus be constrained by setting $\mathcal{B} = 0$.

The <u>Poincaré supercurrent multiplet</u> of a theory which has a conformal limit must be of this form $(\mathcal{A},\mathcal{B})$, with $\mathcal{A}$ the above conformal supercurrent multiplet, and with $\mathcal{B}$ containing the traces $\gamma\cdot J$ and $\Theta_\mu{}^\mu$ together with other fields. $\mathcal{B}$ is then called the "trace-multiplet" or "multiplet of anomalies".

If the Poincaré supercurrent multiplet is to be minimal, the trace-multiplet must be as small as possible, i.e. have 4 + 4 field components. Since it should also contain a spinor of dimension $\frac{7}{2}$ and a scalar of dimension 4, namely $\gamma\cdot J$ and $\Theta_\mu{}^\mu$, there are only two candidates for a minimal $\mathcal{B}$ , the chiral multiplet

$$\delta A = \bar{\zeta}\varphi \quad ; \quad \delta B = \bar{\zeta}\gamma_5\varphi$$

$$\delta\varphi = -(F + \gamma_5 G)\zeta - i\,\partial\!\!\!/\,(A + \gamma_5 B)\zeta \tag{6}$$

$$\delta F = i\,\bar{\zeta}\,\partial\!\!\!/\,\varphi \quad ; \quad \delta G = i\,\bar{\zeta}\gamma_5\partial\!\!\!/\,\varphi$$

and a parity-flipped version of the gauge multiplet

$$\delta v_\mu = i\,\bar{\zeta}\,\gamma_\mu\,\lambda$$

$$\delta\lambda = -i\,\sigma^{\mu\nu}\zeta\,\partial_\mu v_\nu - \gamma_5\zeta\,D \tag{7}$$

$$\delta D = i\,\bar{\zeta}\gamma_5\partial\!\!\!/\,\lambda \quad ,$$

on which the algebra (4) is represented only modulo a field-dependent gauge transformation

$$v_\mu \rightarrow v_\mu + \partial_\mu v \quad . \tag{8}$$

Other trace-multiplets are reducible and non-minimal and will eventually lead to non-minimal versions of supergravity. Breitenlohner's fields (1977a), e.g., are related to a spinor trace-multiplet $\boldsymbol{\chi}$ with a constraint $\bar{D}\sigma_{\mu\nu}\boldsymbol{\chi} = 0$.

There is a general procedure for finding a Poincaré supercurrent multiplet. One first writes down the most general multiplet into which a non-conserved axialvector $C_\mu$ can vary: this is a general real multiplet with an additional vector index ( $C_\mu$ ; $\chi_\mu$ ; $M_\mu$ , $N_\mu$ , $v_{\mu\nu}$ ; $\lambda_\mu$ ; $D_\mu$ ):

$$\delta C_\mu = \bar{\zeta}\gamma_5\,\chi_\mu \quad ; \quad \delta\chi_\mu = \ldots . \; ; \quad \ldots \tag{9}$$

(see eqs. (42) for full details). This multiplet, of course, contains no conserved quantities whatsoever, neither $\chi_\mu$ nor $\lambda_\mu$ are $\gamma$-traceless, $v_{\mu\nu}$ is neither symmetric nor traceless. This large multiplet with 32 + 32 field components contains a 24 + 24 component submultiplet, consisting of

$$i\gamma\cdot\chi\ ;\ M_\mu\ ,\ N_\mu\ ,\ v_{[\mu\nu]} - \tfrac{1}{2}\,\varepsilon_{\mu\nu\kappa\lambda}\,\partial^\kappa C^\lambda\ ,\ \partial\cdot C\ ,\ v_\mu{}^\mu\ ;$$

$$\lambda_\mu - i\,\partial\!\!\!/\chi_\mu\ ,\ \partial\cdot\chi\ ;\ \partial^\nu v_{(\mu\nu)}\ ,\ D_\mu + \Box\, C_\mu\ .$$

If this submultiplet is set to zero, we are left with the conformal supercurrent multiplet (3), with $J_\mu \equiv \chi_\mu$ and $\Theta_{\mu\nu} \equiv -\tfrac{1}{2}\, v_{(\mu\nu)}$.

As a next step, we identify $i\gamma\cdot\chi$ with a spinor out of our chosen trace-multiplet. Let us do this for the axial gauge-multiplet (7):

$$-\,i\gamma\cdot\chi = \gamma_5\lambda\ . \tag{10a}$$

The supersymmetry variations of this condition lead to the following set of constraints

$$0 = M_\mu = N_\nu = \lambda_\mu - i\,\partial\!\!\!/\chi_\mu = D_\mu + \Box\, C_\mu\ ; \tag{10b}$$

$$v_{[\mu\nu]} - \tfrac{1}{2}\,\varepsilon_{\mu\nu\kappa\lambda}\,\partial^\kappa C^\lambda = -\tfrac{1}{2}\,\varepsilon_{\mu\nu\kappa\lambda}\,\partial^\kappa v^\lambda\ ;\quad v_\mu{}^\mu = D\ ; \tag{10c}$$

$$0 = \partial\cdot C = \partial\cdot\chi = \partial^\nu v_{(\mu\nu)}\ , \tag{10d}$$

and the Poincaré multiplet has become, with $J_\mu \equiv \chi_\mu$ and $\Theta_{\mu\nu} \equiv -\tfrac{1}{2}\, v_{(\mu\nu)}$ which are conserved due to (10d):

$$\begin{aligned}
\delta C_\mu &= \bar\zeta\,\gamma_5\, J_\mu \\
\delta v_\mu &= \bar\zeta\,\gamma_5\gamma_\mu\gamma\cdot J \\
\delta J_\mu &= i\,\gamma^\nu\big(2\Theta_{\mu\nu} - \gamma_5\,\partial_\nu C_\mu - \tfrac{1}{2}\,\varepsilon_{\mu\nu\kappa\lambda}\,\partial^\kappa(C^\lambda - v^\lambda)\big)\zeta \\
\delta\Theta_{\mu\nu} &= -\tfrac{1}{4}\,\bar\zeta\big(\sigma_{\mu\lambda}\partial^\lambda J_\nu + \sigma_{\nu\lambda}\partial^\lambda J_\mu\big)\ .
\end{aligned} \tag{11}$$

This is our current multiplet of Sohnius & West (1981b). In the earlier paper we had replaced $v_\mu$ by the dual of its curl, $t_{\mu\nu} \equiv \tfrac{1}{4}\,\varepsilon_{\mu\nu\kappa\lambda}\,\partial^\kappa v^\lambda$, so that the algebra closed without gauge transformations (8). Instead, we had then the conservation law for $t_{\mu\nu}$.

If we had chosen the chiral multiplet (5) as trace multiplet, i.e.

$$- i\gamma \cdot \chi = \varphi \;, \tag{12a}$$

we would have found a set of constraints different from (10), namely

$$\begin{aligned} 0 &= M_\mu - \partial_\mu A = N_\mu - \partial_\mu B = \lambda_\mu - i\, \partial\!\!\!/ \chi_\mu + i\, \partial_\mu \gamma \cdot \chi \\ &= D_\mu + \Box C_\mu - \partial_\mu \partial \cdot C \end{aligned} \tag{12b}$$

$$0 = v_{[\mu\nu]} - \tfrac{1}{2}\, \varepsilon_{\mu\nu\kappa\lambda}\, \partial^\kappa C^\lambda = v_\mu{}^\mu - F = \partial \cdot C - G \tag{12c}$$

$$0 = \partial \cdot \chi - \partial\!\!\!/ \gamma \cdot \chi = \partial^\nu v_{(\mu\nu)} - \partial_\mu\, v_\nu{}^\nu \;, \tag{12d}$$

which leads to the Poincaré supercurrent multiplet of Ferrara & Zumino (1975) with currents $J_\mu \equiv \chi_\mu - \gamma_\mu \gamma \cdot \chi$ and $\Theta_{\mu\nu} \equiv -\tfrac{1}{2}\left( v_{(\mu\nu)} - \eta_{\mu\nu}\, v_\lambda{}^\lambda \right)$, which are conserved due to eqs. (12d), and without a conserved axialvector, $\partial \cdot C \neq 0$.

## LINEARIZED SUPERGRAVITY

Since the currents should be the sources for the fields of supergravity, we now consider the multiplet which is contragradient to (11). It will contain fields $h_{\mu\nu}$, $\psi_\mu$, $A_\mu$ and $V_\mu$ with transformation laws such that

$$\mathcal{L} = -\tfrac{1}{2}\, h^{\mu\nu} \Theta_{\mu\nu} + \bar{\psi}^\mu J_\mu + (A^\mu + \tfrac{3}{2}\, V^\mu) C_\mu + \tfrac{1}{2}\, V^\mu v_\mu \tag{13}$$

transforms as a total derivative. For every one of the conservation laws (1) there will be a gauge-freedom in the contragradient multiplet,

$$\begin{aligned} \delta_{\text{gauge}}\, h_{\mu\nu} &= \partial_\mu \xi_\nu + \partial_\nu \xi_\mu \\ \delta_{\text{gauge}}\, \psi_\mu &= \partial_\mu \varepsilon \\ \delta_{\text{gauge}}\, A_\mu &= \partial_\mu \alpha \;, \end{aligned} \tag{14}$$

and for the gauge-freedom $v_\mu \rightarrow v_\mu + \partial_\mu v$ there will be a conservation law

$$0 = \partial^\mu V_\mu \;. \tag{15}$$

The required invariance of $\mathcal{L}$ determines the supersymmetry transformations up to terms of the form (14):

$$\delta h_{\mu\nu} = -2i\,\bar{\zeta}\,(\gamma_\mu\psi_\nu + \gamma_\nu\psi_\mu) \tag{16a}$$

$$\delta\psi_\mu = \frac{i}{4}\sigma^{\kappa\lambda}\zeta\,\partial_\kappa\,h_{\lambda\mu} - \gamma_5\zeta\,(A_\mu + V_\mu) - \frac{i}{2}\sigma_{\mu\nu}\gamma_5\zeta\,V^\nu \tag{16b}$$

$$\delta A_\mu = \bar{\zeta}\gamma_\mu\sigma_{\kappa\lambda}\gamma_5\partial^\kappa\psi^\lambda \tag{16c}$$

$$\delta V_\mu = -i\,\varepsilon_{\mu\nu\kappa\lambda}\,\bar{\zeta}\gamma^\nu\partial^\kappa\psi^\lambda \tag{16d}$$

The algebra closes on the fields only modulo field-dependent gauge transformations of the form (14).

The constraint (15) can be solved in terms of an antisymmetric tensor

$$V^\mu = \tfrac{1}{4}\,\varepsilon^{\mu\nu\kappa\lambda}\,\partial_\nu a_{\kappa\lambda} \tag{17}$$

This solution introduces a further gauge-freedom

$$\delta_{\text{gauge}}\,a_{\mu\nu} = \partial_\mu\beta_\nu - \partial_\nu\beta_\mu \tag{18}$$

and the transformations of $a_{\mu\nu}$,

$$\delta a_{\mu\nu} = 2i\,\bar{\zeta}\,(\gamma_\mu\psi_\nu - \gamma_\nu\psi_\mu) \tag{19}$$

represent the algebra only modulo field-dependent gauge transformations of this form.

## NON-LINEAR SUPERGRAVITY

The general geometric formulation of supergravity in superspace or as a gauge theory with field-dependent algebra has been extensively treated elsewhere (Akulov et.al. 1977; Breitenlohner 1977a; Wess & Zumino 1977; Wess 1978; Breitenlohner & Sohnius 1980; Sohnius 1981). The transformation laws for super-vielbein $E_M{}^A$ and structure-group connections $\Omega_M$ are in general given by

$$\delta E_M{}^A = \mathcal{D}_M\,\zeta^A + E_M{}^C\,\zeta^B\,T_{BC}{}^A$$
$$\delta\Omega_M = E_M{}^C\,\zeta^B\,R_{BC}$$

but here it suffices to give the supersymmetry transformations of the vierbein $e_\mu{}^a$, the Rarita-Schwinger field $\psi_{\mu\alpha}$ and the chiral gauge field $A_\mu$:

$$\delta_S e_\mu{}^a = e_\mu{}^b \bar{\zeta}^\alpha T_{\alpha b}{}^a + \bar{\psi}_\mu{}^\beta \bar{\zeta}^\alpha T_{\alpha\beta}{}^a \tag{20a}$$

$$\delta_S \psi_\mu{}^\alpha = \mathcal{D}_\mu \bar{\zeta}^\alpha + e_\mu{}^a \bar{\zeta}^\beta T_{\beta a}{}^\alpha + \bar{\psi}_\mu{}^\gamma \bar{\zeta}^\beta T_{\beta\gamma}{}^\alpha \tag{20b}$$

$$\delta_S A_\mu = e_\mu{}^a \bar{\zeta}^\alpha f_{\alpha a} + \bar{\psi}_\mu{}^\beta \bar{\zeta}^\alpha f_{\alpha\beta} \tag{20c}$$

$\mathcal{D}_\mu \bar{\zeta}$ is a short-hand notation for

$$\mathcal{D}_\mu \bar{\zeta} \equiv \partial_\mu \bar{\zeta} + \frac{i}{4} \omega_\mu{}^{ab} \bar{\zeta}\sigma_{ab} - A_\mu \bar{\zeta}\gamma_5 \tag{21}$$

Indices $\mu,\nu,\ldots$ refer to $x$-space vectors, indices $a$, $b$, ... to vectors under local Lorentz transformations, and indices $\alpha$, $\beta$, ... to spinors under such rotations. The *only* quantities where we allow implicit conversion of a local into a "world" vector by means of $e_\mu{}^a$ or its inverse $e_a{}^\mu$, are $\gamma$-matrices and $\psi_\mu$. In particular, $\varepsilon^{abcd}$ is a numerically invariant scalar while $\varepsilon^{\mu\nu\kappa\lambda}$ is a numerically invariant density.

The "torsions" $T$ and the "chiral field strengths" $f$ in eqs. (20) are constrained by Bianchi-identities (Wess & Zumino 1977).

In the linearized limit we have

$$e_\mu{}^a \simeq \delta_\mu{}^a + \tfrac{1}{2} h_\mu{}^a \quad ; \quad g_{\mu\nu} \equiv e_\mu{}^a e_\nu{}^b \eta_{ab} \simeq \eta_{\mu\nu} + h_{\mu\nu} \tag{22}$$

and by direct comparison of eq. (20a) with (16a) we read off the constraints

$$T_{\alpha\beta}{}^a = 2i\ (\gamma^a C)_{\alpha\beta} \quad ; \quad T_{\alpha a}{}^b = 0 \tag{23}$$

C is the charge-conjugation matrix, $C^{-1}\gamma_a C = -\gamma_a{}^T$, and should not be confused with a field. The "conventional" constraint

$$T_{ab}{}^c = 0 \tag{24}$$

now expresses the Lorentz connection $\omega_\mu{}^{ab}$ in terms of $e_\mu{}^a$ and $\psi_\mu$,

$$\omega_{\mu ab} = \tfrac{1}{2} \left(e_a{}^\nu \partial_\mu e_{\nu b} - e_b{}^\nu \partial_\mu e_{\nu a}\right) - \tfrac{1}{2} \left(e_a{}^\kappa e_b{}^\lambda - e_a{}^\lambda e_b{}^\kappa\right)\partial_\kappa g_{\lambda\mu}$$
$$-i \left( \bar{\psi}_\mu \gamma_a \psi_b - \bar{\psi}_\mu \gamma_b \psi_a + \bar{\psi}_a \gamma_\mu \psi_b \right) , \tag{25}$$

so that in the linearized limit $\omega_{\mu ab}$ is given by

$$\omega_{\mu ab} \simeq -\tfrac{1}{2} \left( \partial_a h_{b\mu} - \partial_b h_{a\mu} \right) \tag{26}$$

We now directly compare the linearized limit of (21b) with (16b) and find the further constraints

$$T_{\alpha\beta}{}^{\gamma} = 0 \tag{27a}$$

$$T_{a\alpha}{}^{\beta} = (\gamma_5)_\alpha{}^\beta V_a - \frac{i}{2}(\sigma_{ab}\gamma_5)_\alpha{}^\beta V^b \tag{27b}$$

Before we proceed to check the Bianchi-identities for consistency of the constraints, we define the curls

$$T_{\mu\nu}{}^{\alpha} \equiv \mathcal{D}_\mu \bar{\psi}_\nu{}^\alpha - \mathcal{D}_\nu \bar{\psi}_\mu{}^\alpha \tag{28a}$$

$$R_{\mu\nu}{}^{ab} \equiv \partial_\mu \omega_\nu{}^{ab} + \omega_\mu{}^{ac}\,\omega_{\nu c}{}^{b} - (\mu \leftrightarrow \nu) \tag{28b}$$

$$f_{\mu\nu} \equiv \partial_\mu A_\nu - \partial_\nu A_\mu \quad , \tag{28c}$$

with $\mathcal{D}_\mu \bar{\psi}_\nu$ defined in complete analogy with eq. (21), and their super-covariant equivalents

$$T_{ab} \equiv e_a{}^\mu e_b{}^\nu T_{\mu\nu} - \bar{\psi}_a{}^\alpha T_{\alpha b} + \bar{\psi}_b{}^\beta T_{\beta a} + \bar{\psi}_a{}^\alpha \psi_b{}^\beta T_{\beta\alpha} \tag{28d}$$

(similar for $f$ and $R$). We also define the supercovariant derivative

$$\mathcal{D}_a \equiv e_a{}^\mu \left( \partial_\mu - \delta_S(\psi_\mu) - \delta_C(A_\mu) - \delta_L(\omega_\mu{}^{ab})\right) \quad , \tag{29}$$

where $\delta_S$, $\delta_C$ and $\delta_L$ are supersymmetry transformations, chiral transformations and local Lorentz transformations, all with field-dependent parameters as given.

The consequences of the <u>Bianchi-identities</u> are then

$$f_{\alpha\beta} = 0 \quad ; \quad f_{\alpha a} = \frac{1}{4}(\gamma_5\gamma_a\gamma_b\mathcal{R}^b)_\alpha \tag{30a,b}$$

$$R_{\alpha\beta}{}^{cd} = 2i\,\varepsilon^{abcd}(\gamma_a C)_{\alpha\beta} V_b \tag{30c}$$

$$R_{\alpha abc} = -\frac{1}{2}\varepsilon_{abcd}\,\gamma_5\mathcal{R}^d + 2i\,(\gamma_a C)_{\alpha\beta} T_{bc}{}^\beta \tag{30d}$$

$$\delta_S V_a = -\frac{1}{2}\bar{\zeta}\gamma_5\mathcal{R}_a \quad ; \quad \mathcal{D}_a V^a = 0 \tag{30e,f}$$

where $\mathcal{R}_\alpha{}^a$ is a vector-spinor, defined from $T_{ab}{}^\alpha$ by

$$\mathcal{R}_\alpha{}^a \equiv i\,\varepsilon^{abcd}(\gamma_b\gamma_5 C)_{\alpha\beta} T_{cd}{}^\beta \tag{31}$$

We can now collect the supersymmetry transformation laws for the entire multiplet from (20) and (30e). Together with chiral transformations and Lorentz transformations (parameters $\alpha$ and $\lambda^{ab}$), they are

$$\begin{aligned}
\delta e_\mu{}^a &= -2i\,\bar{\zeta}\,\gamma^a\,\psi_\mu - \lambda^{ab}\,e_{\mu b} \\
\delta\psi_\mu &= \mathcal{D}_\mu\zeta - \gamma_5\zeta\,e_\mu{}^a\,V_a - \tfrac{i}{2}\,\sigma_{\mu a}\zeta\,V^a + \alpha\,\gamma_5\psi_\mu + \tfrac{i}{4}\,\lambda^{ab}\,\sigma_{ab}\psi_\mu \\
\delta A_\mu &= \tfrac{1}{4}\,\bar{\zeta}\gamma_5\gamma_\mu\gamma\cdot\mathcal{R} + \partial_\mu\alpha \\
\delta V^a &= -\tfrac{1}{2}\,\bar{\zeta}\,\gamma_5\mathcal{R}^a - \lambda^{ab}\,V_b
\end{aligned} \tag{32}$$

Again from general considerations, we know that the algebra will be

$$[\,\delta(\zeta_1),\ \delta(\zeta_2)] = \delta\big(\zeta_1^B\ \zeta_2^A\ (T_{AB} + R_{AB})\ \big)$$

or here

$$[\,\delta_S(\zeta_1),\ \delta_S(\zeta_2)] = \delta_G(\xi^\mu) + \delta_S(\zeta) + \delta_C(\alpha) + \delta_L(\lambda^{ab}) \tag{33}$$

with the following field-dependent parameters for the general coordinate transformation $\delta_G$ and the gauge transformations:

$$\begin{aligned}
&\xi^\mu \equiv e_a{}^\mu\,\xi^a = 2i\,e_a{}^\mu\,\bar{\zeta}_1\,\gamma^a\,\zeta_2 \quad ; \qquad \zeta = -\,\xi^\mu\,\psi_\mu \\
&\alpha = -\,\xi^\mu\,A_\mu \quad ; \qquad \lambda^{ab} = -\,\xi^\mu\,\omega_\mu{}^{ab} + \varepsilon^{abcd}\,\xi_c\,V_d
\end{aligned} \tag{33'}$$

Further algebraic relations are ($\zeta$ not to be differentiated):

$$[\mathcal{D}_a,\ \delta_S(\zeta)] = -\,\delta_S(\bar{\zeta}^\beta\,T_{a\beta}{}^\alpha) + \delta_C(\bar{\zeta}^\alpha\,f_{\alpha a}) + \delta_L(\bar{\zeta}^\alpha\,R_{\alpha a}{}^{cd}) \tag{34a}$$

$$[\mathcal{D}_a\ ,\ \mathcal{D}_b] = -\,\delta_S(T_{ab}{}^\alpha) - \delta_C(f_{ab}) - \delta_L(R_{ab}{}^{cd}) \tag{34b}$$

$$[\,\delta_S(\zeta),\ \delta_C(\alpha)] = \delta_S(\alpha\,\gamma_5\zeta) \tag{35}$$

There remains the problem of the constraint (30f) on the axial-vector $V_a$. Fortunately, this constraint can - as in the rigid case - be solved in terms of an antisymmetric tensor $a_{\mu\nu}$:

$$V^a = e^{-1}\,e_\mu{}^a\,\varepsilon^{\mu\nu\kappa\lambda}\,\big(\,\tfrac{1}{4}\,\partial_\nu a_{\kappa\lambda} - \tfrac{i}{2}\,\bar{\psi}_\nu\gamma_\kappa\psi_\lambda\,\big) \tag{36}$$

where $e = \det e_\mu{}^a$. The transformation law for $a_{\mu\nu}$,

$$\delta_S a_{\mu\nu} = 2i\ \bar{\zeta}\ (\gamma_\mu\psi_\nu - \gamma_\nu\psi_\mu) \tag{37}$$

will represent the algebra (33) only modulo an additional "vector" gauge transformation

$$[\delta_1, \delta_2]\ a_{\mu\nu} = \ldots + \partial_\mu\ \beta_\nu - \partial_\nu\ \beta_\mu \tag{38}$$

with $\beta_\mu = \xi_\mu + a_{\mu\nu}\ \xi^\nu$.

Anticipating one of the results of the tensor calculus, we consider the supergravity Lagrangian

$$\begin{aligned}\mathcal{L}_{SG} = -\tfrac{1}{2}\ e\ e_a{}^\mu\ e_b{}^\nu\ R_{\mu\nu}{}^{ab} + 2i\ \varepsilon^{\mu\nu\kappa\lambda}\ \bar{\psi}_\mu\gamma_\nu\gamma_5\mathcal{D}_\kappa\psi_\lambda \\ - 3\ e\ V_a\ V^a - \varepsilon^{\mu\nu\kappa\lambda}\ A_\mu\ \partial_\nu\ a_{\kappa\lambda}\end{aligned} \tag{39}$$

The field equations for $a_{\mu\nu}$ and $A_\mu$ are

$$0 = \varepsilon^{\mu\nu\kappa\lambda}\left(f_{\kappa\lambda} - 3\ \partial_\kappa\ (e_\lambda{}^a\ V_a)\right) = -4\ e\ e_a{}^\mu\ V^a \tag{40}$$

and they set to zero the gauge-invariant part of $A_\mu$ and express the gauge-invariant part of $a_{\mu\nu}$ in terms of $\bar{\psi}_\mu\gamma_a\psi_\nu e_\kappa{}^a$, such that $V_a = 0$. The auxiliary fields then drop out from the Lagrangian and we are left with the $\mathcal{L}_{SG}$ of on-shell N=1 supergravity (Freedman, van Nieuwenhuizen & Ferrara 1976; Deser & Zumino 1976).

## QUESTIONS RAISED

In the previous section, the new minimal off-shell formulation of N=1 supergravity was presented, minimal in the sense that it contains the smallest possible number (six) of auxiliary fields. There are already known to exist two other auxiliary field formulations of N=1 supergravity, the "old minimal" set of 12 + 12 fields ($e_\mu{}^a$; $\psi_{\mu\alpha}$; $M$, $N$, $b_a$) of Stelle & West (1978a) and Ferrara & van Nieuwenhuizen (1978a) and the non-minimal set of Breitenlohner's (1977a) 20 +20 fields ($e_\mu{}^a$; $\psi_{\mu\alpha}$, $\chi$; $A_a$, $U_a$, $T_a$, $A$, $B$; $\lambda$). An intermediate set of 16 + 16 fields ($e_\mu{}^a$; $\psi_{\mu\alpha}$; $A_\mu$, $V_a$, $M$, $N$; $\lambda$; $S$) with no conservation law for $V_a$ but with a chiral gauge transformation $A_\mu \to A_\mu + \partial_\mu\alpha$ has been studied by de Wit & van Nieuwenhuizen (1978) in a different context and will be re-examined in Sohnius & West (1981c).

This set can be derived by our method, using a general multiplet as trace-submultiplet of the currents, and it can be constrained to either of the two minimal sets. Without coupling to matter, however, its simplest invariant is inconsistent as a Lagrangian (there is a field equation det $e = 0$). Self-coupled to its own axial gauge-submultiplet, the model broke supersymmetry spontaneously and led to a cosmological constant.

The existence of different auxiliary field formulations naturally raises the question whether they have the same properties, i.e.

(i) are they the same pure supergravity theories?

(ii) do they have the same properties when coupled to matter?

The remainder of this article is devoted to answering mainly question (ii), in particular to the more restricted question whether the "new minimal" formulation couples to the same type of matter and in the same way as the old minimal one. We will in fact derive all the important couplings of the new minimal set, those of the old minimal set being well known (Ferrara & van Nieuwenhuizen 1978b,c; Stelle & West 1978b-d). We will not discuss coupling to the Breitenlohner fields, which is known for the chiral and the gauge multiplets (Breitenlohner 1977b).

Concerning (i), it is clear that at the classical level the two minimal theories are identical, provided one neglects possible topological effects. The question of their quantum equivalence will be discussed later.

The presence of the new set of auxiliary fields raises two general points in connection with the physical consequences of auxiliary fields. Given a set of auxiliary fields, we can deduce a current multiplet. It is this current multiplet which is important in the coupling of the given off-shell formulation to matter. Also, it is generally believed that different types of current multiplets lead to different physics. In fact, the attempts so far to find "physics" in the N=8 supergravity theory rely on the observed particles being bound states which belong to the representations in the current multiplet of supergravity (Ellis, Gaillard & Zumino 1980). Given different current multiplets we would, according to this strategy, get different bound states and it would therefore be interesting to see if the new type of current multiplet which corresponds to the new auxiliary fields generalizes to N=8 and which particle spectrum it would predict.

A related point is the observation that auxiliary fields can actually increase the symmetry of a theory. The new auxiliary fields belong to a formulation of supergravity with a *local* chiral *U(1)* whereas the old minimal set and indeed the on-shell theory (where the gauge has been fixed to $A_\mu = 0$) have only a *global* chiral *U(1)*. A similar situation is known from the *SU(2)* of N=2 supergravity (Breitenlohner & Sohnius 1981). This contrasts sharply with those off-shell formulations found by the introduction of an off-shell central charge (Sohnius, Stelle & West 1980a,b) where the global symmetry is reduced from *U(N)* or *SU(N)* to *USp(N)* off-shell.

Should an analogue of the new auxiliary fields exist for N=8, one can expect it to have a *local U(8)* symmetry corresponding to that of the superconformal group. This contrasts with the rigid *USp(8)* arising from the off-shell version with a central charge (Cremmer, Ferrara, Stelle & West 1980).

## Practical advantages of auxiliary fields

We are now faced with the task of constructing the matter couplings of our formulation of supergravity. In fact, this construction will demonstrate one of the practical advantages of auxiliary fields, a topic on which we shall now make a few comments.

The practical advantages of auxiliary fields can be summarized in the quotation "the only good symmetry is a manifest symmetry". A symmetry leads to a proliferation of fields which can make calculations very difficult to do if the fields are dealt with on an individual footing rather than as a member of the symmetry multiplet to which they belong. A good example of this is the Lorentz group. We all remember the complexity of Maxwell's equations when the electromagnetic field is expressed in terms of $\vec{E}$ and $\vec{B}$, as compared to the elegance of the four-vector notation. We have also seen old textbooks where, written out in components, they fill half a page! It is this development which we wish to copy in the case of the more complicated space-time symmetry of supergravity.

A manifestly supersymmetric formalism from which we can construct actions, requires a representation on fields which are not subject to their equations of motion. For this it is necessary to know the auxiliary field structure. There are, in fact, two entirely equivalent manifestly supersymmetric formalisms, superspace and tensor calculus. Both are easy to construct if the auxiliary fields are known, and it is also easy to find one from the other.

The superspace formalism (Salam & Strathdee 1974) is especially suitable for calculating Feynman diagrams, using the technique of super-Feynman rules (Salam & Strathdee 1975). The tensor calculus, on the other hand, is especially useful for constructing actions, in particular those of the couplings of supergravity to matter. We will work out the tensor calculus for our auxiliary field formulation of supergravity and use it to calculate the couplings of our fields to matter. In the following section we will review the tensor calculus in the context of rigid supersymmetry before tackling the more complicated local case.

## RIGID TENSOR CALCULUS

The tensor calculus of rigid supersymmetry was worked out in the early days of supersymmetry by Wess & Zumino (1974a,c). The technique works with fields over $x$-space, grouped into multiplets (called supermultiplets) which have well-defined transformations under supersymmetry. As for any other symmetry, we will find rules for multiplying together these representations of supersymmetry (supermultiplets) and ways of constructing invariants from them.

### Supermultiplets

The basic such multiplet is the general supermultiplet (called "vector multiplet" in Wess & Zumino 1974a) which has component field content

$$\mathcal{C} = ( C ; \chi_\alpha ; M , N , v_\mu ; \lambda_\alpha ; D ) \quad . \tag{41}$$

This supermultiplet has supersymmetry transformations (parameter $\zeta$) and chiral transformations (parameter $\alpha$):

$$\delta C = \bar{\zeta} \gamma_5 \chi \tag{42a}$$

$$\delta \chi = ( M + \gamma_5 N ) \zeta - i \gamma^\mu ( v_\mu + \gamma_5 \partial_\mu C ) \zeta - \alpha \gamma_5 \chi \tag{42b}$$

$$\delta M = \bar{\zeta} ( \lambda - i \not{\partial} \chi ) + 2 \alpha N \tag{42c}$$

$$\delta N = \bar{\zeta}\gamma_5 (\lambda - i\not{\partial}\chi) - 2\alpha M \tag{42d}$$

$$\delta v_\mu = i\bar{\zeta}\gamma_\mu \lambda + \bar{\zeta}\partial_\mu \chi \tag{42e}$$

$$\delta \lambda = -i\,\sigma^{\mu\nu}\zeta\,\partial_\mu v_\nu - \gamma_5\zeta D + \alpha\gamma_5\lambda \tag{42f}$$

$$\delta D = i\bar{\zeta}\gamma_5\not{\partial}\,\lambda \tag{42g}$$

The above transformations can be derived in the following way: we start with the pseudoscalar $C$, invariant under chiral transformations, and write down the most general form of $\delta C$, which is (42a). We then write the most general form for $\delta\chi$ which preserves the Majorana condition, namely

$$\delta\chi = \left(M + \gamma_5 N - i\gamma^\mu v_\mu - i\gamma^\mu\gamma_5 u_\mu + i\sigma^{\mu\nu} t_{\mu\nu}\right)\zeta - \alpha\gamma_5\chi' ,$$

and now enforce the algebra (4) and (35) on $\delta_1\delta_2 C$. We find as a consequence that $u_\mu = \partial_\mu C$, $t_{\mu\nu} = 0$ and $\chi' = \chi$, so that we are left with (42b) as the most general $\delta\chi$ compatible with the algebra. Next we make an ansatz for $\delta M$, $\delta N$ and $\delta v_\mu$ and enforce the algebra on $\chi$, and so on until we have derived transformation laws for the entire multiplet (42).

We shall see later that the same technique can be used to enforce the more complicated algebra (33) of supergravity on the multiplet.

While the chiral transformations are essential if they actually appear in $[\delta_1, \delta_2]$ , as in (33), we could have left them out for the purposes of rigid supersymmetry.

In superspace, the general multiplet corresponds to a general pseudoscalar superfield $C(x^\mu, \theta_\alpha)$ and the chiral transformations are $\theta \rightarrow \theta + \alpha\gamma_5\theta$.

The general supermultiplet is actually a reducible representation of supersymmetry. It contains two invariant submultiplets. One of these is a curl multiplet ("restricted vector multiplet" in Wess & Zumino 1974a)

$$\boldsymbol{\lambda} = (\lambda_\alpha ; v_{\mu\nu}, D) , \tag{43}$$

in which the antisymmetric tensor $v_{\mu\nu}$ has to fulfill the constraint $\varepsilon^{\mu\nu\kappa\lambda}\partial_\nu v_{\kappa\lambda} = 0$, as is indeed the case for the curl-submultiplet of $\mathbb{C}$ where $v_{\mu\nu} = \partial_\mu v_\nu - \partial_\nu v_\mu$. If we take this solution for the constraint seriously without referring back to the entire $\mathbb{C}$ and simply replace (42e) by

$$\delta v_\mu = i\bar{\zeta}\gamma_\mu\lambda , \tag{44}$$

then the algebra will close on $v_\mu$ only modulo a field-dependent gauge transformation $v_\mu \rightarrow v_\mu + \partial_\mu v$. We call the multiplet

$$\mathbf{V} = ( v_\mu ; \lambda ; D ) \tag{45}$$

the gauge multiplet. $\mathbf{V}$ is really a $\mathbf{C}$ in the Wess & Zumino (1974c) gauge.

The other submultiplet of $\mathbf{C}$ is a chiral multiplet ("scalar multiplet" in Wess & Zumino 1974a) of weight 2. A chiral multiplet

$$\mathbf{A} = ( A , B ; \varphi_\alpha ; F , G ) \tag{46}$$

has supersymmetry transformations as in (6) and chiral transformations which depend on a number $n$, the chiral weight:

$$\delta_C A \equiv \alpha n B \quad ; \quad \delta_C B = - \alpha n A \tag{47a,b}$$

$$\delta_C \varphi = \alpha (n-1) \gamma_5 \varphi \tag{47c}$$

$$\delta_C F = \alpha (2-n) G \quad ; \quad \delta_C G = - \alpha (2-n) F \tag{47d,e}$$

The chiral submultiplet of $\mathbf{C}$ is formed by $( M , N ; \lambda - i \partial\!\!\!/ \chi ; \partial^\mu v_\mu , D + \Box C )$. Generally, a chiral multiplet is defined by the condition that $\delta B$ with parameter $\zeta$ is the same as $\delta A$ with parameter $\gamma_5 \zeta$. This, together with the definition (47a) of the chiral weight, determines the multiplet completely.

If we restrict $\mathbf{C}$ by setting $\boldsymbol{\lambda} = 0$, we get $v_\mu = \partial_\mu v$, and the left-over components $( v , C ; \chi ; -M , -N )$ form a chiral multiplet of weight 0.

If we restrict $\mathbf{C}$ by setting the chiral submultiplet to zero, we are left with a $\mathbf{C}$ of the form $( C ; \chi ; 0 , 0 , v_\mu ; i \partial\!\!\!/ \chi ; - \Box C )$ with

$$0 = \partial^\mu v_\mu \tag{48}$$

Such a $\mathbf{C}$ is called a linear multiplet and we can also write it as

$$\mathbf{L} = ( C ; \chi ; v_\mu ) \tag{49}$$

### Combination of supermultiplets

We now wish to combine two supermultiplets to obtain a third one. This operation is analogous to the combination of two vectors to form a second-rank tensor. The combinations are as follows:

**(a)** We can combine two general supermultiplets in a symmetric way to find a third one,

$$\mathbf{C}_1 \cdot \mathbf{C}_2 = \mathbf{C}_3 \tag{50}$$

with components

$$\begin{aligned}
C_3 &\equiv C_1 C_2 \\
\chi_3 &= \chi_1 C_2 + \chi_2 C_1 \\
M_3 &= M_1 C_2 + M_2 C_1 - \tfrac{1}{2} \bar{\chi}_1 \gamma_5 \chi_2 \\
N_3 &= N_1 C_2 + N_2 C_1 - \tfrac{1}{2} \bar{\chi}_1 \chi_2 \\
v_{\mu 3} &= v_{\mu 1} C_2 + v_{\mu 2} C_1 + \tfrac{i}{2} \bar{\chi}_1 \gamma_\mu \gamma_5 \chi_2 \\
\lambda_3 &= \lambda_1 C_2 + \tfrac{1}{2}\left( N_1 + \gamma_5 M_1 - i \not{v}_1 \gamma_5 + i \not{\partial} C_1 \right)\chi_2 + (1 \longleftrightarrow 2) \\
D_3 &= D_1 C_2 + D_2 C_1 - M_1 M_2 - N_1 N_2 - \partial^\mu C_1 \partial_\mu C_2 - v_1^\mu v_{\mu 2} \\
&\qquad + \bar{\lambda}_1 \chi_2 + \bar{\lambda}_2 \chi_1 - \tfrac{i}{2} \bar{\chi}_1 \overleftrightarrow{\not{\partial}} \chi_2
\end{aligned} \tag{50'}$$

This result is obtained by varying $C_1 C_2$, which is defined to be $C_3$, under supersymmetry to obtain the next component

$$\delta(C_1 C_2) = \bar{\zeta} \gamma_5 \, ( C_1 \chi_2 + C_2 \chi_1 )$$

The operation is repeated on $C_1\chi_2 + C_2\chi_1$ to find the next components, and so on.

(b) We can combine two chiral multiplets in a symmetric way to form a third one of chiral weight $n_3 = n_1 + n_2$:

$$\mathbf{A}_1 \cdot \mathbf{A}_2 = \mathbf{A}_3 \tag{51}$$

with components

$$\begin{aligned}
A_3 &\equiv A_1 A_2 - B_1 B_2 \\
B_3 &\equiv A_1 B_2 + B_1 A_2 \\
\varphi_3 &= ( A_1 - \gamma_5 B_1 ) \varphi_2 + ( A_2 - \gamma_5 B_2 ) \varphi_1
\end{aligned} \tag{51'}$$

$$F_3 = A_1 F_2 + B_1 G_2 + A_2 F_1 + B_2 G_1 + \bar{\varphi}_1 \varphi_2$$

$$G_3 = A_1 G_2 - B_1 F_2 + A_2 G_1 - B_2 F_1 - \bar{\varphi}_1 \gamma_5 \varphi_2$$

Multiplication of $\mathbb{A}$ with the two constant multiplets

$$\mathbb{1}_+ = (1, 0; 0; 0, 0) \text{ and } \mathbb{1}_- = (0, 1; 0; 0, 0) \tag{52}$$

gives $\mathbb{A} \cdot \mathbb{1}_+ = \mathbb{A}$ and $\mathbb{A} \cdot \mathbb{1}_- = (-B, A; -\gamma_5\varphi; G, -F)$, the "parity flipped" chiral multiplet.

(c) We can combine two chiral multiplets of the same chiral weight, $n_1 = n_2$ in a symmetric way to form a general multiplet

$$\mathbb{A}_1 \times \mathbb{A}_2 = \mathbb{C} \tag{53}$$

with components

$$\begin{aligned}
C &\equiv A_1 A_2 + B_1 B_2 \\
\chi &= (B_1 - \gamma_5 A_1)\varphi_2 + (B_2 - \gamma_5 A_2)\varphi_1 \\
M &= -F_1 B_2 - F_2 B_1 - A_1 G_2 - A_2 G_1 \\
N &= F_1 A_2 + F_2 A_1 - B_1 G_2 - B_2 G_1 \\
v_\mu &= B_1 \overleftrightarrow{\partial}_\mu A_2 + B_2 \overleftrightarrow{\partial}_\mu A_1 + i\bar{\varphi}_1 \gamma_\mu \gamma_5 \varphi_2 \\
\lambda &= -(G_1 + \gamma_5 F_1)\varphi_2 - (G_2 + \gamma_5 F_2)\varphi_1 \\
&\quad + i\not{\partial} B_2 \varphi_1 + i\not{\partial} B_1 \varphi_2 + i\gamma_5 \not{\partial} A_2 \varphi_1 + i\gamma_5 \not{\partial} A_1 \varphi_2 \\
D &= -2F_1 F_2 - 2G_1 G_2 - 2\partial_\mu A_1 \partial^\mu A_2 - 2\partial_\mu B_1 \partial^\mu B_2 - i\bar{\varphi}_1 \overleftrightarrow{\not{\partial}} \varphi_2
\end{aligned} \tag{53'}$$

(d) The particular combination

$$\mathbb{A}_1 \wedge \mathbb{A}_2 \equiv (\mathbb{1}_- \cdot \mathbb{A}_1) \times \mathbb{A}_2 \tag{54}$$

is antisymmetric in the fields of $\mathbb{A}_1$ and $\mathbb{A}_2$ and has as lowest component $C = A_1B_2 - A_2B_1$.

In superspace these rules correspond to multiplications of superfields: $\phi_1\phi_2 = \phi_3$, $i\phi = \phi_-$, $\mathrm{Re}\ \bar{\phi}_1\phi_2 = C$ and $\mathrm{Im}\ \bar{\phi}_1\phi_2 = \mathrm{Re}\ \overline{i\phi_1}\phi_2$ .

One further element of tensor calculus is based on the

observation that for any $\mathbb{A}$ the components $F$ and $G$ are the start for a new chiral multiplet, called $\mathrm{T}\mathbb{A}$, with weight $2-n$ and components

$$\mathrm{T}\mathbb{A} = (F, G; \, i\partial\!\!\!/\varphi; \, -\Box A, \, -\Box B) \tag{55}$$

Note that $\mathrm{TT} = -\Box$. $\mathrm{T}\mathbb{A}$ is sometimes called the kinetic multiplet.

### Action formulas

Having found multiplets of supersymmetry and rules for combining them, it remains to find how to construct invariant actions. We do this separately for the general and the chiral multiplets.

Given a general supermultiplet, the integral over its highest component, denoted by

$$\int d^4x \, |\mathbb{C}|_D \equiv \int d^4x \, D \tag{56}$$

is invariant under supersymmetry transformations, since $D$ varies into a divergence and therefore $\delta\int d^4x \, |\mathbb{C}|_D = 0$.

Given a chiral multiplet, we can construct an invariant by integrating over its $F$-component:

$$\int d^4x \, |\mathbb{A}|_F \equiv \int d^4x \, F \tag{57}$$

since $F$ varies into a divergence. Note, however, that $|\mathbb{A}|_F$ will only be *chirally* invariant if the chiral weight of $\mathbb{A}$ is 2.

In superspace, these operations correspond to integration over the whole superspace in the case of the $D$-action formula,

$$\int d^4x \, |\mathbb{C}|_D \sim \int d^4x \, d^4\theta \, C(x,\theta)$$

and integration over the chiral subspace in the case of the $F$-action formula,

$$\int d^4x \, |\mathbb{A}|_F \sim \int d^4x \, d^2\theta \, \phi + \text{h.c.}$$

### An example - the Wess-Zumino model

We now show how to construct the model of Wess & Zumino (1974b) using the rigid tensor calculus. Consider a chiral supermultiplet, eqs. (46), where $A$ and $B$ have dimension 1, i.e. the canonical dimension for bosonic fields. Then $\varphi$ has dimension $\frac{3}{2}$, as a spinor should. $F$ and $G$

have dimension 2 and we expect them to be auxiliary fields. We can use our rules to construct other supermultiplets out of $\mathbb{A}$, namely

$\mathbb{A} \times \mathbb{A}$ a general supermultiplet

$\mathbb{A} \cdot \mathbb{A}$ a chiral multiplet

$\mathbb{A} \cdot \mathbb{A} \cdot \mathbb{A}$ a chiral multiplet.

As an action we take the most general dimensionless invariant $\int d^4x\, \mathcal{L}_{WZ}$ which can be constructed without use of coupling constants with negative dimensions:

$$\mathcal{L}_{WZ} = -\tfrac{1}{4}\,|\mathbb{A}\times\mathbb{A}|_D - \tfrac{m}{2}\,|\mathbb{A}\cdot\mathbb{A}|_F - \tfrac{g}{3}\,|\mathbb{A}\cdot\mathbb{A}\cdot\mathbb{A}|_F\,. \tag{58}$$

A term $|\mathbb{A}\cdot T\mathbb{A}|_F$ gives nothing new since it differs only by a divergence from $-\tfrac{1}{2}\,|\mathbb{A}\times\mathbb{A}|_D$, and a term $|\mathbb{A}\,|_F$ can always be shifted away by combinations of shifts $\mathbb{A} \to \mathbb{A} + \alpha\,\mathbb{1}_+ + \beta\,\mathbb{1}_-$ and constant chiral transformations

Upon evaluating this expression according to the rules (51') and (53'), we find the Wess-Zumino Lagrangian, the first term in (58) being the kinetic Lagrangian,

$$-\tfrac{1}{4}\,|\mathbb{A}\times\mathbb{A}\,|_D = \tfrac{1}{2}\left(\partial_\mu A\,\partial^\mu A + \partial_\mu B\,\partial^\mu B + i\,\bar{\varphi}\,\partial\!\!\!/\,\varphi + F^2 + G^2\right),$$

the second and third being the mass and interaction terms, respectively. This kinetic Lagrangian confirms the suspicion, born of simple dimensional analysis, that $A$, $B$ and $\varphi$ propagate while $F$ and $G$ are auxiliary fields.

The requirement of *chiral invariance* would allow $F$-action formulas only for certain chiral weights. Then we could either have a free massive theory ($m \neq 0$, $g = 0$) if $\mathbb{A}$ has chiral weight 1, or an interacting massless one ($m = 0$, $g \neq 0$) if $\mathbb{A}$ has chiral weight $\tfrac{2}{3}$, but never $m \neq 0 \neq g$.

An alternative description (Siegel 1979) of a theory having the same states as the free massless Wess-Zumino model is in terms of a linear multiplet with Lagrangian

$$\mathcal{L} = \tfrac{1}{2}\,|\mathbb{L}\cdot\mathbb{L}|_D\quad. \tag{59}$$

LOCAL TENSOR CALCULUS

The local tensor calculus follows the same pattern as the rigid one, but now we must find representations and invariants of all the local symmetries. These are general coordinate transformations, local supersymmetry transformations, local chiral and Lorentz transformations, and the vector-gauge transformation on $a_{\mu\nu}$. Their actions on the supergravity fields are given in eqs. (32) and (37) and the corresponding algebra in eqs. (33), (35) and (38). Since the vector-gauge transformation has come into the scheme only as a consequence of the solution (36) for the constraint (30f) on $V_a$, it will not play a role in the tensor calculus for matter fields, since these only see the geometrical quantities $e_\mu{}^a$, $\psi_\mu$, $A_\mu$ and $V_a$.

However, the local tensor calculus is different from both the rigid one and the one for the old minimal fields, in that the chiral transformations

$$\delta_C A_\mu = \partial_\mu \alpha$$

$$\delta_C \psi_\mu = \alpha \gamma_5 \psi_\mu$$

must be taken into account since they appear in the commutator of two supersymmetry transformations. In the rigid algebra and the local algebra for the old minimal fields the (global) chiral transformations can be included or left out at will. In the new auxiliary field algebra, the presence of local chiral transformations in the commutator of two supersymmetries means we cannot leave them out without incurring a non-closing algebra.

The tensor calculus and the corresponding matter couplings have been worked out for the old auxiliary fields in Stelle & West (1978b,c,d) and in Ferrara & van Nieuwenhuizen (1978b,c). We proceed to work them out for the new minimal auxiliary fields.

Supermultiplets

We now wish to find the local analogues of the general and chiral supermultiplets. In doing so, we can follow exactly the procedure outlined above for the construction of the rigid multiplets, except that this time we enforce the more complicated algebra (33) instead of (4).

This will have the consequence that wherever $\partial^\mu$ appeared in the rigid transformation laws, we will now have $\mathcal{D}_a + \delta_L(\epsilon_{ab}{}^{cd} V^b)$. Further complications are introduced by having to vary expressions like $\mathcal{D}_a \varphi$, and by the fact that $\mathcal{D}_a$ and $\mathcal{D}_b$ don't commute anymore. Both of these problems can be dealt with with the help of eqs. (34).

An alternative procedure works order by order in the gravitational constant $\kappa$. We start with the rigid transformations and let the supersymmetry parameter $\zeta_\alpha$ become space-time dependent, $\zeta_\alpha(x)$; we then modify the transformations so as to give rise to the closing local algebra of equation (33) up to zeroeth order in $\kappa$. The process is repeated until one obtains a representation of the local algebra to all orders in $\kappa$. We now outline some of the major features of this approach.

Replacing $\zeta_\alpha$ by $\zeta_\alpha(x)$ in the transformations of the rigid general supermultiplet yields

$$\delta C = \bar{\zeta}(x)\, \gamma_5\, \chi$$

$$\delta\chi_\alpha = \big((M + \gamma_5 N - i \not{v} - i \not{\partial}\gamma_5 C)\zeta(x)\big)_\alpha$$

$$\delta M = \bar{\zeta}(x)\ (\lambda - i \not{\partial}\chi)$$

$$\vdots$$

The commutator of two supersymmetry transformations on $\chi_\alpha$ yields

$$[\delta_1, \delta_2]\, \chi = \ldots + i\gamma_5\not{\partial}(\bar{\zeta}_1\gamma_5\chi)\, \zeta_2 - (1 \longleftrightarrow 2) .$$

This commutator has a term of order $\kappa^0$ which contains a derivative of $\zeta_\alpha$. Such a term is not present in the algebra. This term is avoided if we replace $\partial_\mu C$ in $\delta\chi$ by

$$\partial_\mu C \longrightarrow \partial_\mu C - \bar{\psi}_\mu\gamma_5\chi$$

since then to order $\kappa^0$

$$\delta(\partial_\mu C - \bar{\psi}_\mu\gamma_5\chi) = \bar{\zeta}(x)\gamma_5\partial_\mu\chi$$

Clearly, this procedure is necessary wherever derivatives occur in the rigid transformations. The general replacement of a space-time derivative by a "supercovariantised" one is given by

$$\partial_\mu \longrightarrow \partial_\mu - \bar{\psi}^\alpha_\mu \frac{\delta_S}{\delta\bar{\zeta}^\alpha}$$

A similar problem occurs with the local chiral transformations. The behaviour of the general multiplet under local chiral transformations is given in eqs. (42), only that now $\alpha = \alpha(x)$. The commutator of a supersymmetry and a chiral transformation gives, e.g.

$$[\delta_C, \delta_S]\, M = \ldots\ldots + i\, \bar{\zeta}(x)\, \not\partial\, (\, \alpha(x)\, \gamma_5 \chi\,)$$

Again the appearance of a $\partial_\mu \alpha(x)$ term, which does not appear in the algebra, can be avoided by the replacement of the $\partial_\mu \chi$ term in $\delta M$ by

$$\partial_\mu \chi \longrightarrow \partial_\mu \chi + A_\mu\, \gamma_5 \chi$$

This procedure must be carried out for all space-time derivatives on fields which transform under chiral transformations, and a similar one for all those on fields which are not Lorentz scalars.

In this way, one achieves a modification to the rigid transformations such that they now form a representation of the local algebra (33) to order $\kappa^0$. Further modifications to the transformations allow one to obtain a general supermultiplet that forms a representation of the local algebra to order $\kappa^1$, and so on to all orders of $\kappa$.

Our result for the transformation laws of a general multiplet is

$$\begin{aligned}
\delta_S C &= \bar{\zeta}\gamma_5 \chi \\
\delta_S \chi &= (M + \gamma_5 N)\zeta - i\gamma^a (v_a + \gamma_5 \mathcal{D}_a C)\zeta \\
\delta_S M &= \bar{\zeta}\left(\lambda - i\not{\mathcal{D}}\chi - \tfrac{i}{2}\gamma^a \gamma_5 \chi V_a\right) \\
\delta_S N &= \bar{\zeta}\gamma_5\left(\lambda - i\not{\mathcal{D}}\chi - \tfrac{i}{2}\gamma^a \gamma_5 \chi V_a\right) \qquad\qquad (60)\\
\delta_S v_a &= i\bar{\zeta}\gamma_a \lambda + \bar{\zeta}\mathcal{D}_a \chi + \bar{\zeta}\left(\gamma_5 V_a - \tfrac{i}{2}\sigma_{ab}\gamma_5 V^b\right) \\
\delta_S \lambda &= -i\sigma^{ab}\zeta \mathcal{D}_a v_b - \gamma_5 \zeta D - \tfrac{i}{2}\sigma^{ab}\zeta\, T_{ab}{}^\alpha \chi_\alpha \\
\delta_S D &= i\bar{\zeta}\gamma_5 \not{\mathcal{D}}\lambda - \tfrac{3i}{2}\bar{\zeta}\gamma^a \lambda\, V_a \quad ;
\end{aligned}$$

the chiral transformations remain unchanged from eqs. (42), except that their parameter $\alpha$ is now $x$-dependent.

Similarly, a chiral multiplet of chiral weight $n$ has still the chiral transformation properties of eqs. (47), but modified supersymmetry transformations

$$\begin{aligned}
\delta_S A &= \bar\zeta\,\varphi \\
\delta_S B &= \bar\zeta\,\gamma_5\,\varphi \\
\delta_S \varphi &= -\,(F+\gamma_5 G)\,\zeta - i\,\not{\mathcal{D}}\,(A+\gamma_5 B)\,\zeta \qquad (61)\\
\delta_S F &= i\,\bar\zeta\not{\mathcal{D}}\,\varphi + \frac{i}{2}\bar\zeta\,\gamma^a\gamma_5\,\varphi\,V_a + \frac{in}{4}\bar\zeta\,(A+\gamma_5 B)\,\gamma\cdot\mathcal{R} \\
\delta_S G &= i\,\bar\zeta\,\gamma_5\not{\mathcal{D}}\,\varphi + \frac{i}{2}\bar\zeta\,\gamma^a\varphi\,V_a + \frac{in}{4}\bar\zeta\,\gamma_5\,(A+\gamma_5 B)\,\gamma\cdot\mathcal{R}
\end{aligned}$$

The curl submultiplet of $\mathbf{C}$ is

$$\boldsymbol{\lambda} = (\,\lambda\,;\ v_{ab}\,,\ D\,) \qquad (62)$$

with

$$\begin{aligned}
v_{ab} &\equiv \mathcal{D}_a v_b - \mathcal{D}_b v_a + T_{ab}{}^{\alpha}\chi_\alpha \\
\delta_S v_{ab} &= i\,\bar\zeta\,(\mathcal{D}_a\gamma_b - \mathcal{D}_b\gamma_a)\,\lambda - i\,\varepsilon_{abcd}\,\bar\zeta\,\gamma^c\lambda\,V^d \qquad (62 \\
&\quad + \frac{i}{2}\bar\zeta\,(\gamma_a V_b - \gamma_b V_a)\,\gamma_5\lambda
\end{aligned}$$

The constraint $\boldsymbol{\lambda} = 0$ has exactly the same solution as in the rigid case, namely $\mathbf{C} = (\,C\,;\ \chi\,;\ M\,,\ N\,,\ \mathcal{D}_a v\,;\ 0\,;\ 0\,)$ where the left-over fields form a chiral multiplet $(\,v\,,\ C\,;\ \chi\,;\ -M\,,\ -N\,)$ of weight 0.

It proves useful to define a world vector $v_\mu$,

$$\begin{aligned}
v_\mu &\equiv e_\mu{}^a\,v_a + \bar\psi_\mu\,\chi \\
\delta_S v_\mu &= i\,\bar\zeta\,\gamma_\mu\,\lambda + \partial_\mu\,(\,\bar\zeta\,\chi\,) \qquad (63)\\
v_{\mu\nu} &\equiv \partial_\mu v_\nu - \partial_\nu v_\mu = e_\mu{}^a\,e_\nu{}^b\,v_{ab} + i\,(\,\bar\psi_\mu\gamma_\nu - \bar\psi_\nu\gamma_\mu\,)\,\lambda\ ,
\end{aligned}$$

such that the constraint $\boldsymbol{\lambda} = 0$ implies $v_\mu = \partial_\mu v$. The gauge multiplet is then

$$\mathbf{V} = (\,v_\mu\,;\ \lambda\,;\ D\,) \qquad (64)$$

The chiral submultiplet of $\mathbb{C}$ has weight 2 and components

$$A = M \quad ; \quad B = N \quad ; \quad \varphi = \lambda - i \not{\mathcal{D}} \chi - \frac{i}{2}\gamma^a \gamma_5 \chi V_a$$

$$F = \mathcal{D}^a v_a - 2V^a \mathcal{D}_a C + \frac{i}{4} \bar{\chi} \gamma \cdot \mathcal{R} \tag{65}$$

$$G = D + \mathcal{D}^a \mathcal{D}_a C + 2V^a v_a + \frac{i}{2} \bar{\chi} \gamma_5 \gamma \cdot \mathcal{R}$$

Constraining this submultiplet to zero leaves us with the linear multiplet with a modified transversality condition on the vector:

$$0 = \mathcal{D}^a v_a - 2 V^a \mathcal{D}_a C + \frac{i}{4} \bar{\chi} \gamma \cdot \mathcal{R} \tag{66}$$

It proves useful to define a contravariant vector density

$$v'^\mu \equiv e\, e_a{}^\mu (v^a - 2V^a C) - i e \bar{\psi}_\nu \sigma^{\nu\mu} \chi - i \varepsilon^{\mu\nu\kappa\lambda} \bar{\psi}_\nu \gamma_\kappa \psi_\lambda C$$

$$\delta_S v'^\mu = ie \bar{\zeta} \gamma^\mu \varphi - 2i e \bar{\zeta} \sigma^{\mu\nu} (A + \gamma_5 B) \psi_\nu - 2i \varepsilon^{\mu\nu\kappa\lambda} \partial_\nu (\bar{\zeta} \gamma_\kappa \psi_\lambda C)$$
$$+ i \partial_\nu (e \bar{\zeta} \sigma^{\mu\nu} \chi) \tag{67}$$

$$\partial_\mu v'^\mu = e \left(\mathcal{D}^a v_a - V^a \mathcal{D}_a C + \frac{i}{4} \bar{\chi} \gamma \cdot \mathcal{R}\right) ,$$

such that the linear multiplet constraint (66) is just $\partial_\mu v'^\mu = 0$ , solved by

$$v'^\mu = \varepsilon^{\mu\nu\kappa\lambda} \partial_\nu b_{\kappa\lambda} \tag{68}$$

The corresponding transformation law for $b_{\mu\nu}$,

$$\delta_S b_{\mu\nu} = - \frac{i}{2} \bar{\zeta} \sigma_{\mu\nu} \gamma_5 \chi - i \bar{\zeta} (\gamma_\mu \psi_\nu - \gamma_\nu \psi_\mu) C \tag{69}$$

represents the algebra (33) only modulo yet another vector-gauge transformation

$$[\delta_1, \delta_2]\, b_{\mu\nu} = \ldots + \partial_\mu \beta_\nu - \partial_\nu \beta_\mu \quad \text{with} \quad \beta_\mu = - \tfrac{1}{2} \xi_\mu C + b_{\mu\nu} \xi^\nu . \tag{70}$$

The <u>linear multiplet</u> is then

$$\mathbb{L} = (C ; \chi ; v'^\mu) . \tag{71}$$

Combinations of local supermultiplets

Given the local supermultiplets of the preceding subsection, we can, as in the rigid case, find rules to multiply them to form new supermultiplets. The calculation proceeds in exactly the same way as in the rigid case: after having decided which combination of component fields is to be the first component of the composite supermultiplet, we apply successive supervariation and identify the higher components. The type of multiplet which results will become apparent in the calculation. The results are:

**(a)** Two general multiplets can be multiplied symmetrically to form a third one,

$$\mathbf{C}_1 \cdot \mathbf{C}_2 = \mathbf{C}_3 \tag{72}$$

The components are as in the rigid case, eq.(50), except that every derivative $\partial_\mu$ has to be replaced by $\mathcal{D}_a$ and that there is an additional term in $D_3$,

$$D_3 = \ldots - \frac{3i}{2}\bar{\chi}_1 \gamma^a \gamma_5 \chi_2 V_a \tag{72'}$$

(b) Two chiral multiplets can be combined in three distinct ways. First, the symmetric product

$$\mathbf{A}_1 \cdot \mathbf{A}_2 = \mathbf{A}_3 \tag{73}$$

gives a new chiral multiplet with weight $n_3 = n_1 + n_2$ and components exactly as in the rigid case, eq.(51). Second, the symmetric product of two chiral multiplets with the same weight $n$,

$$\mathbf{A}_1 \times \mathbf{A}_2 = \mathbf{C} \tag{74}$$

gives a general multiplet with components as in the rigid case, eq.(53), except that every derivative $\partial_\mu$ is to be replaced by $\mathcal{D}_a$ and that there are additional terms in $\lambda$ and $D$,

$$\begin{aligned} \lambda &= \ldots - \frac{in}{2}\gamma_5 \gamma\cdot\mathcal{R}\, C \\ D &= \ldots - 3i\,\bar{\varphi}_1 \gamma^a \gamma_5 \varphi_2 V_a - \frac{n}{2} C\,(R + 6V^2) - \frac{in}{2}\bar{\chi}\gamma_5\gamma\cdot\mathcal{R}\,. \end{aligned} \tag{74'}$$

Third, the antisymmetric product of two chiral multiplets with the same weight $n$, defined as in the rigid case by

$$\mathbb{A}_1 \wedge \mathbb{A}_2 \equiv ( \mathbb{1}_- \cdot \mathbb{A}_1 ) \times \mathbb{A}_2 \tag{75}$$

We notice that the multiplication rules for $\mathbb{A}_1 \times \mathbb{A}_2$ and hence for $\mathbb{A}_1 \wedge \mathbb{A}_2$ depend on the chiral weight $n$.

The quantity $R$ in (74') is the fully covariantized Ricci scalar, defined from $R_{ab}{}^{cd}$ via

$$R \equiv R_a{}^a \quad ; \quad R_{ab} \equiv R_{acb}{}^c \tag{76}$$

The kinetic multiplet $\mathrm{T}\mathbb{A}$ for a chiral multiplet $\mathbb{A}$ of weight $n$ has weight $2-n$ and components

$$\begin{aligned}
\mathrm{T}A &= F \quad ; \quad \mathrm{T}B = G \\
\mathrm{T}\varphi &= \mathrm{i}\,\mathcal{D\!\!\!/}\varphi + \tfrac{\mathrm{i}}{2}\gamma^a\gamma_5\varphi\, V_a + \tfrac{\mathrm{i}n}{4}\,(A+\gamma_5 B)\,\gamma\cdot\mathcal{R}, \\
\mathrm{T}F &= -\mathcal{D}^2 A + 2V^a\mathcal{D}_a B + \tfrac{n}{4} A\,(R+6V^2) + \tfrac{\mathrm{i}}{4}(n-1)\,\bar{\chi}\gamma\cdot\mathcal{R} \\
\mathrm{T}G &= -\mathcal{D}^2 B - 2V^a\mathcal{D}_a A + \tfrac{n}{4} B\,(R+6V^2) + \tfrac{\mathrm{i}}{4}(n-1)\,\bar{\chi}\gamma_5\gamma\cdot\mathcal{R} \;.
\end{aligned} \tag{77}$$

## Local action formulas

The action formulas of rigid supersymmetry require considerable modification to become invariant under local supersymmetry. The local action formulas can be found order by order in the gravitational coupling constant $\kappa$. So far we had set $\kappa = 1$, but it is easy to include $\kappa$ through dimensional analysis.

We now outline the general features of the calculation of action formulas on the example of the $D$-action formula for the general multiplet. The rigid $D$-formula is no longer invariant once the constant parameter $\zeta$ has been made $x$-dependent, namely

$$\int \mathrm{d}^4x\ \delta D = \int \mathrm{d}^4x\ ( -\mathrm{i}\,\partial_\mu\bar{\zeta}\,\gamma_5\gamma^\mu\lambda )$$

We obtain an action formula invariant to order $\kappa^0$ by replacing $\int \mathrm{d}^4x\ D$ by

$$\int \mathrm{d}^4x\ e\ ( D - \mathrm{i}\,\bar{\psi}_\mu\gamma^\mu\gamma_5\lambda )$$

Upon supersymmetry variations, this will give terms of order $\kappa$ from the variation of $\psi_\mu$. These terms are partly cancelled by non-linear terms in $\delta D$ and $\delta\lambda$, and by the variations of det $e_\mu{}^a$ and of the vierbein $e_a{}^\mu$ implicit in $\gamma^\mu$. However, it is necessary to add new terms to the action formula in order to cancel all terms of order $\kappa$. The calculation proceeds in this manner until an action formula has been found which is invariant to all orders of $\kappa$.

The action for the general multiplet is $\int d^4x \, |\mathbf{C}|_D$ where

$$|\mathbf{C}|_D = e\,D - i\,e\,\bar{\psi}\cdot\gamma\,\gamma_5\,\lambda + \tfrac{1}{2}\,\varepsilon^{\mu\nu\kappa\lambda}\,v_\mu\,\partial_\nu\,a_{\kappa\lambda} \quad . \tag{78}$$

The vector $v_\mu$ is defined in (63). In superspace this would be $\int d^4x\, d^4\theta \,\mathrm{Ber}\, E_M{}^A \; C(x,\theta)$ .

This simple formula is to be compared with the equivalent for the old minimal tensor calculus (Stelle & West 1978c,d). The obvious difference is that here the fields $C$, $\chi$, $M$, $N$ and the longitudinal part of $v_\mu$ do not appear. The action formula (78) is therefore gauge invariant under the transformations $\mathbf{C} \to \mathbf{C} + \mathbf{C}'$ where $\boldsymbol{\lambda}(\mathbf{C}') = 0$.

In particular, we find for the unit multiplet

$$\mathbb{1} = \mathbf{1}_+ \times \mathbf{1}_+ = (\;1\,;\;0\,;\;0\,,\;0\,,\;0\,;\;0\,;\;0\;) \tag{79}$$

that

$$|\mathbf{1}|_D = 0 \;, \tag{80}$$

a result which in superspace would read $\int d^4x\; d^4\theta \;\mathrm{Ber}\; E_M{}^A = 0$ .

In the context of an abelian gauge theory coupled to supergravity, the action formula (78) allows a gauge invariant generalization of the Fayet-Iliopoulos term of rigid supersymmetry (Fayet & Iliopoulos 1974). For on-shell supergravity such a term was given by Freedman (1977).

The action formula for a chiral multiplet of weight 2 is

$$|\mathbf{A}|_F = e\;F + i\,e\,\bar{\psi}\cdot\gamma\,\varphi - i\,e\,\bar{\psi}_\mu\,\sigma^{\mu\nu}\,(\,A + \gamma_5\,B\,)\,\psi_\nu \tag{81}$$

The restriction to weight 2 is necessary even to ensure chiral invariance of $|\mathbf{A}|_F$, since $F$ rotates into $G$ with weight $2-n$. This condition, not necessary for the old $F$-action formula (Ferrara & van Nieuwenhuizen 1978b), will place restrictions on the type of Lagrangians that are invariant within our framework.

It would also be possible to deduce the tensor calculus from the superconformal tensor calculus by making the appropriate superconformal transformation. This technique was used to find some of the old minimal tensor calculus (Ferrara, Grisaru & van Nieuwenhuizen 1978) and has been used here to check some of the results given.

## THE SUPERGRAVITY ACTION

As a first example of the use of tensor calculus, we construct the supergravity action itself. We observe that the supermultiplet

$$\mathbf{E} = ( A_\mu ; \tfrac{i}{4}\gamma_5 \gamma\cdot\mathcal{R} ; \tfrac{1}{4}(R + 6V^2) ) \tag{82}$$

transforms like a gauge multiplet (with the chiral transformation as gauge transformation). We call this multiplet the Einstein multiplet, in analogy with Ferrara & van Nieuwenhuizen (1978c).

Because of the gauge invariance of our $D$-action formula, we can use it on $\mathbf{E}$ and get the supergravity Lagrangian

$$\mathcal{L}_{SG} = -2\,|\mathbf{E}|_D = -\frac{e}{2}(R + 6V^2) + \frac{e}{2}\,\bar{\psi}\cdot\gamma\,\gamma\cdot\mathcal{R} - \varepsilon^{\mu\nu\kappa\lambda} A_\mu \partial_\nu a_{\kappa\lambda} \tag{83}$$

If we replace the fully covariantized $R_{ab}{}^{cd}$ and $T_{ab}{}^{\alpha}$ which appear in the definitions of $R$ and $\mathcal{R}$, eqs.(76) and (31), by the curls $R_{\mu\nu}{}^{cd}$ and $T_{\mu\nu}{}^{\alpha}$, eqs.(28), then this "decovariantization" leads to the form of the Lagrangian given earlier in eq.(39).

In superspace, a gauge multiplet, and thus $\mathbf{E}$, corresponds to a general gauge superfield (the prepotential) in the Wess-Zumino gauge. In order to write our supergravity action in superspace, we would therefore first have to complete the chiral gauge freedom to a full supermultiplet worth of gauge freedoms, and then introduce a prepotential $V(x,\theta)$ for this "superchiral" invariance. The supergravity Lagrangian (83) would then be proportional to $d^4\theta$ Ber $E_M{}^A\, V(x,\theta)$, and gauge invariant as discussed after eq. (78).

Unlike in Einstein's Lagrangian ("N=0 supergravity") and the old minimal formulation, we cannot add an explicit cosmological term to the action. A cosmological term would have to include the term $\bar{\psi}_\mu \sigma^{\mu\nu} \psi_\nu$ which is not chirally invariant. Expressed in the language of our tensor calculus, the chiral unit multiplet has weight 0 and no invariant action can be constructed from $|\mathbf{1}_+|_F$.

A cosmological term could arise in a spontaneous breaking of supersymmetry where either $\langle D \rangle \neq 0$ or $\langle F \rangle \neq 0$ for some multiplet.

This is the first example of essential differences between the two minimal formulations of N=1 supergravity.

THE WESS-ZUMINO MODEL

Given a chiral supermultiplet $\mathbb{A} = (A, B; \varphi; F, G)$ of chiral weight $n$, we can form a general supermultiplet $\mathbb{A}\times\mathbb{A}$, a chiral multiplet $\mathbb{A}\cdot\mathbb{A}$ of weight $2n$ and a chiral multiplet $\mathbb{A}\cdot\mathbb{A}\cdot\mathbb{A}$ of weight $3n$. We can obtain the coupling of the Wess-Zumino model to supergravity by inserting these multiplets into the appropriate action formulas.

The kinetic Lagrangian is

$$\begin{aligned}\mathcal{L}_0 &= -\tfrac{1}{4}\,|\mathbb{A}\times\mathbb{A}|_D \\ &= \frac{e}{2}\Big(\mathcal{D}_a A\mathcal{D}^a A + \mathcal{D}_a B\mathcal{D}^a B + i\,\bar{\varphi}\gamma^a\mathcal{D}_a\varphi + F^2 + G^2 \\ &\qquad + \frac{3i}{2}\,\bar{\varphi}\gamma^a\gamma_5\varphi\,V_a - i\,\bar{\varphi}\gamma^\mu(F+\gamma_5 G)\psi_\mu - \bar{\varphi}\,\mathcal{D}\!\!\!/\,(A+\gamma_5 B)\gamma\cdot\psi\Big) \\ &\quad + \tfrac{1}{4}\,\varepsilon^{\mu\nu\kappa\lambda}\,(A\overleftrightarrow{\partial}_\mu B - \tfrac{i}{2}\,\bar{\varphi}\gamma_\mu\gamma_5\varphi)\,\partial_\nu a_{\kappa\lambda} \\ &\quad + \frac{in}{4}\,e\,\bar{\varphi}(A+\gamma_5 B)\gamma\cdot\mathcal{R} - \frac{n}{4}\,(A^2+B^2)\,\mathcal{L}_{SG}\;.\end{aligned} \tag{84}$$

This is an "invariant" Lagrangian for all values of the chiral weight $n$. The value $n = 0$ corresponds to the original coupling of the Wess-Zumino model to supergravity, found without auxiliary fields (Ferrara et.al.1977). The value $n = \frac{2}{3}$ corresponds to the superconformal coupling (Kaku & Townsend 1978).

The mass term is given by

$$\mathcal{L}_m = -\frac{m}{2}\,|\mathbb{A}\cdot\mathbb{A}|_F \tag{85}$$

but is invariant only if $\underline{n = 1}$, because only then $\mathbb{A}\cdot\mathbb{A}$ has weight 2 as required for an $F$-action formula. Similarly, the interaction term

$$\mathcal{L}_g = -\frac{g}{3}\,|\mathbb{A}\cdot\mathbb{A}\cdot\mathbb{A}|_F \tag{86}$$

is invariant only for $\underline{n = \frac{2}{3}}$. A term $\mathcal{L}_{FI} = -\lambda\,|\mathbb{A}|_F$ would be invariant only for $n = 1$ and cannot be eliminated by a shift as in the rigid case.

We note that, unlike in the old minimal formulation, we cannot couple the Wess-Zumino model in its most general form of arbitrary parameters $m$, $g$ and $\lambda$. This difference is again caused by the chiral transformations under which our actions must be invariant.

As has been pointed out to us by D.Z. Freedman (private communication), the use of several chiral multiplets with different weights allows an interaction term and mass terms for some of the multiplets.

The field equations of the matter fields are non-covariant linear combinations of the covariant equations

$$\mathbf{T}\mathbf{A} = m\mathbf{A} + g\,\mathbf{A}\cdot\mathbf{A} + \lambda\,\mathbf{1}_+ \tag{87}$$

where, as pointed out, at most one of the parameters on the right-hand side can be different from zero, depending on the chiral weight. The presence of $\mathcal{L}_{FI}$ means $\langle F \rangle = \lambda$ and thus spontaneous breaking of supersymmetry, resulting in a negative cosmological term $-\frac{e}{2}\kappa^2\lambda^2$.

The relation between currents and auxiliary fields discussed earlier allows the above couplings to yield explicit examples of the new type of trace multiplet $(v_\mu\,;\ -i\gamma\cdot J\,;\ 2\Theta_\mu{}^\mu)$. These are given by the matter field combinations which appear on the right-hand sides of the equations of motion for the gravitational fields in the presence of the matter and its coupling.

The spectrum of the massless Wess-Zumino model with $g = 0$, but coupled to supergravity, is the same as that obtained from the Lagrangian

$$\mathcal{L} = \tfrac{1}{2}\,|\mathbf{L}\cdot\mathbf{L}|_D \tag{88}$$

for the linear multiplet.

MAXWELL EINSTEIN THEORY

From the gauge multiplet $\mathbf{V} = (v_\mu\,;\ \lambda\,;\ D)$ we can obtain a chiral multiplet $\mathbf{M}$ of weight 2 and with components

$$\begin{aligned}\mathbf{M} = \Big(\tfrac{1}{4}\bar\lambda\lambda,\ \tfrac{1}{4}\bar\lambda\gamma_5\lambda\,;\ -\tfrac{1}{2}\gamma_5\lambda D + \tfrac{i}{4}\sigma^{ab}\lambda v_{ab}\,;\\ -\tfrac{1}{4}v^{ab}v_{ab} + \tfrac{1}{2}D^2 + \tfrac{i}{2}\bar\lambda\mathcal{D}\lambda,\\ \tfrac{1}{8}\varepsilon^{abcd}v_{ab}v_{cd} + \tfrac{i}{4}\mathcal{D}^a(\bar\lambda\gamma_a\gamma_5\lambda)\Big)\ .\end{aligned} \tag{89}$$

The Lagrangian

$$\mathcal{L} = |\mathbf{M}|_F \tag{90}$$

is the kinetic Lagrangian for the photon $v_\mu$ and the photino $\lambda$, coupled to supergravity, and agrees after elimination of auxiliary fields with the previously known results derived either without auxiliary fields (Ferrara, Scherk & van Nieuwenhuizen 1976) or with the old minimal set (Stelle & West 1978b, Ferrara & van Nieuwenhuizen 1978b).

The coupling of super-QED (two chiral multiplets rotating into each other) to supergravity is straightforward (compare Stelle & West 1978d). The Lagrangian is, as in the rigid case (Wess & Zumino 1974c),

$$\mathcal{L} = |\mathbb{M}|_F - \tfrac{1}{8}\left|(\mathbb{A}_1 \times \mathbb{A}_1 + \mathbb{A}_2 \times \mathbb{A}_2)\cdot(e^{2g\mathbb{C}} + e^{-2g\mathbb{C}})\right|_D - \tfrac{1}{4}\left|(\mathbb{A}_1 \wedge \mathbb{A}_2)\cdot(e^{2g\mathbb{C}} - e^{-2g\mathbb{C}})\right|_D \tag{91}$$

where $\mathbb{C}$ is the general gauge multiplet before a Wess-Zumino gauge has been chosen. The chiral weights of $\mathbb{A}_1$ and $\mathbb{A}_2$ must be equal but are otherwise arbitrary.

## COUNTERTERMS

Given the additional local chiral invariance of the new formulation, one may wonder if this new symmetry will rule out all higher invariants present in the old formulation. This is in fact not the case, as the following dicussion demonstrates.

The field $W_{ab}$, defined from the supercovariant curl $T_{ab\alpha}$ of $\psi_{\mu\alpha}$ by

$$T_{ab} \equiv W_{ab} - \tfrac{1}{8}\gamma_c\sigma_{ab}\mathcal{R}^c - \tfrac{1}{24}\sigma_{ab}\gamma\cdot\mathcal{R} \tag{92}$$

transforms under supersymmetry like

$$\delta_S W_{ab} = -\tfrac{1}{8}(\sigma^{cd}\sigma_{ab} + \tfrac{1}{3}\sigma_{ab}\sigma^{cd})\gamma_5\zeta\,(3\mathcal{D}_c V_d + f_{cd}) - \tfrac{i}{4}\sigma^{cd}\zeta\, C_{abcd}\,. \tag{93}$$

Here $C_{abcd}$ is the Weyl tensor

$$C_{abcd} \equiv \tfrac{1}{2}R_{abcd} - \tfrac{1}{8}\varepsilon_{aba'b'}\varepsilon_{cdc'd'}R^{a'b'c'd'} - \tfrac{1}{12}(\eta_{ac}\eta_{bd} - \eta_{bc}\eta_{ad})R\,. \tag{94}$$

The combinations $\bar{W}_{ab}W^{ab}$ and $\tfrac{1}{2}\varepsilon^{abcd}\bar{W}_{ab}W_{cd}$ are the lowest components of a chiral multiplet $\mathbb{W}^2$ of weight 2, from which we can therefore construct a Weyl action by an $F$-action formula:

$$\mathcal{L}_{\text{Weyl}} \sim |\mathbf{W}^2|_F \quad . \tag{95}$$

The $F$-component of $\mathbf{W}^2$ will contain the square of the Weyl tensor. Just as $\frac{1}{2} F^2$ appeared in the kinetic action for the chiral mutlplet, we will get a fourth power of the Weyl tensor in

$$\mathcal{L}_{3\text{-loop}} \sim |\mathbf{W}^2 \times \mathbf{W}^2|_D \tag{96}$$

and generally a $4k$-th power in

$$\mathcal{L}_{(4k-1)\text{-loop}} \sim |(\mathbf{W}^2 \times \mathbf{W}^2)^k|_D \quad . \tag{96'}$$

Since $C_{abcd}$ does not vanish on-shell (nor does $W_{ab}$), these are candidates for counterterms. Note, however, that actions like

$$|\mathbf{W}^2 \cdot \mathbf{W}^2 \cdot \ldots |_F$$

are ruled out.

This discussion implicitly assumes that the local chiral symmetry is preserved in the quantum theory. As was pointed out earlier (Sohnius & West 1981b), a breaking of this symmetry by a chiral anomaly destroys the Fermi-Bose balance and, as a consequence, breaks off-shell supersymmetry. It is certainly true that if one ignores the auxiliary fields and in a straightforward way takes into account only the spin-$\frac{3}{2}$ field, then, according to Duff & Christensen (1978) there will be a chiral anomaly. There is, however, considerable uncertainty as to the status of anomalies in supersymmetry, and it is not clear if the results remain the same if auxiliary fields are taken into account.

### FAYET-ILIOPOULOS MECHANISM

As a final example of coupling to supergravity, we construct the local analogue of the rigid mechanism that breaks simultaneously gauge symmetry and supersymmetry. In the rigid case, this is achieved by adding to the coupling of super-QED a mass term for the matter and a term $\xi D$ (Fayet & Iliopoulos 1974):

$$\mathcal{L} = \mathcal{L}_{\text{SQED}} + \frac{m}{2} |\mathbb{A}_1 \cdot \mathbb{A}_1 + \mathbb{A}_2 \cdot \mathbb{A}_2|_F + \xi \, |\mathbf{C}|_D \tag{97}$$

This leads to a massive spin-1 and a massless Goldstone spinor corresponding to the spontaneous breaking of gauge-invariance and rigid supersymmetry. Our tensor calculus allows us to interpret the above Lagrangian in curved space if the chiral weights of $\mathbf{A}_1$ and $\mathbf{A}_2$ are both $n = 1$. Furthermore, the "Maxwell" gauge-invariance is not broken by the term $\xi \; |\mathbf{C}|_D$, since our $D$-action formula does not involve gauge-dependent components of $\mathbf{C}$.

This model will lead to a massive spin-1 and a massive spin-$\frac{3}{2}$ corresponding to the spontaneous breaking of "Maxwell" gauge invariance and supersymmetry. The exact spectrum is under study and will be presented elsewhere.

This construction is in stark contrast to the corresponding situation with the old auxiliary fields. In the old minimal tensor calculus, the $D$-action formula involves all fields of $\mathbf{C}$ and thus has no gauge-invariance. Despite this, it was found possible to write a Fayet-Iliopoulos term in the form $|e^{\mathbf{C}}|_D$ (Stelle & West 1978d), and the Maxwell gauge-invariance was brought back by using the local conformal invariance of the action. This, however, ruled out non-superconformally invariant terms in the action like the mass term in (97).

Here we have encountered another example of the differences in coupling between the two minimal auxiliary field formulations.

## CONCLUSION

We have demonstrated how the "new minimal" formulation of N=1 supergravity can sustain a full tensor calculus for all types of matter multiplets. We have pointed out where essential differences to the old minimal formulation arise, and that they are always related to the requirement of chiral gauge-invariance.

In particular the existence - and gauge-invariance - of a Fayet-Iliopoulos term and the non-existence of an (explicit) cosmological constant (the one with the positive sign!) make it desirable to re-examine super-Higgs and other models in this new light.

It also remains to be seen whether the new formulation can survive quantization, i.e. whether the chiral anomaly remains absent for supergravity. In the matter sector, this would require continued absence of a chiral anomaly in the presence of conformal anomalies and thus invalidate one of the assumptions which make the finiteness of rigid

N=4 Yang-Mills theory plausible (Sohnius & West 1981a).

## REFERENCES

Akulov, V.P., Volkov, D.V. & Soroka, V.A. (1977). Generally covariant theories of gauge fields on superspace. Theor. Math. Phys. 31, p.12/285 (Russian/English).
Breitenlohner, P. (1977a). A geometric interpretation of local supersymmetry. Phys. Lett. 67B, p.49.
Breitenlohner, P. (1977b). Some invariant Lagrangians for local supersymmetry. Nucl. Phys. B124, p.500.
Breitenlohner, P. & Sohnius, M.F. (1980). Superfields, auxiliary fields and tensor calculus for N=2 extended supergravity. Nucl. Phys. B165, p.483.
Breitenlohner, P. & Sohnius, M.F. (1981). An almost simple off-shell version of SU(2) Poincaré supergravity. Nucl. Phys. B178,p.151.
Cremmer, E,, Ferrara, S., Stelle, K.S. & West, P.C. (1980). Off-shell N=8 supersymmetry with central charges. Phys. Lett. 94B, p.349.
Deser, S. & Zumino, B. (1976). Consistent supergravity. Phys. Lett. 62B, p. 335.
de Wit, B. & van Nieuwenhuizen, P. (1978). The auxiliary field structure in chirally extended supergravity. Nucl. Phys. B139, p. 216.
Duff, M. & Christensen (1978). Axial and conformal anomalies for arbitrary spin in gravity and supergravity. Phys. Lett. 76B, p. 571.
Ellis, J., Gaillard, M.K. & Zumino, B. (1980). A grand unified theory obtained from broken supergravity. Phys. Lett. 94B, p. 343.
Fayet, P. & Iliopoulos, J. (1974). Spontaneously broken supergauge symmetries and Goldstone spinors. Phys. Lett. 51B, p. 461.
Ferrara, S., Freedman, D.Z., van Nieuwenhuizen, P., Breitenlohner, P., Gliozzi, F. & Scherk, J. (1977). Scalar multiplet coupled to supergravity. Phys. Rev. D15, p. 1013.
Ferrara, S., Grisaru, M. & van Nieuwenhuizen, P. (1978). Poincaré and conformal supergravity models with closed algebras. Nucl. Phys. B 138, p. 430.
Ferrara, S., Scherk, J. & van Nieuwenhuizen, P. (1976). Locally supersymmetric Maxwell-Einstein theory. Phys. Rev. Lett. 37, p. 1035.
Ferrara, S. & van Nieuwenhuizen, P. (1978a). The auxiliary fields of supergravity. Phys. Lett. 74B, p. 333.
Ferrara, S. & van Nieuwenhuizen, P. (1978b). Tensor calculus for supergravity. Phys. Lett. 76B, p. 404.
Ferrara, S. & van Nieuwenhuizen, P. (1978c). Structure of supergravity. Phys. Lett. 78b, p. 573.
Ferrara, S. & Zumino, B. (1975). Transformation properties of the supercurrent. Nucl. Phys. B87, p. 207.
Freedman, D.Z. (1977). Supergravity with axial gauge invariance. Phys. Rev. D15, p. 1013.
Freedman, D.Z., van Nieuwenhuizen, P. & Ferrara, S. (1976). Progress towards a theory of supergravity. Phys.Rev. D13, p. 3214.
Kaku, M. & Townsend, P. (1978). Poincaré supergravity as broken superconformal gravity. Phys. Lett 76B, p. 54.
Ogievetsky, V. & Sokatchev, E, (1977). On a vector superfield generated by the supercurrent. Nucl. Phys. B124, p. 309.

Salam, A. & Strathdee, J. (1974). Supergauge transformations. **Nucl. Phys.** B 76, p.477.

Salam, A. & Strathdee, J. (1975). On superfields and Fermi-Bose symmetry. Phys. Rev. D11, p. 521.

Siegel, W. (1979). Gauge spinor superfield as scalar multiplet. Phys. Lett. 85B, p. 333.

Sohnius, M. (1981). Gauge algebras with field-dependent structure. In preparation.

Sohnius, M.F., Stelle, K.S. & West, P.C. (1980a). Off-mass-shell formulation of extended supersymmetric gauge theories. Phys. Lett. 92B, p 123.

Sohnius, M.F., Stelle, K.S. & West, P.C. (1980b). Dimensional reduction by Legendre transformation generates off-shell supersymmetric Yang-Mills theories. Nucl. Phys. B173, p.127.

Sohnius, M.F. & West, P.C. (1981a). Conformal invariance in N=4 supersymmetric Yang-Mills theory. Phys. Lett. 100b, p. 245.

Sohnius, M.F. & West, P.C. (1981b). An alternative minimal off-shell version of N=1 supergravity. Phys. Lett. 105B, p. 353.

Sohnius, M.F. & West, P.C. (1981c). In preparation.

Stelle, K.S. & West. P.C. (1978a). Minimal auxiliary fields for supergravity. Phys. Lett. 74B, p. 330.

Stelle, K.S. & West, P.C. (1978b). Matter coupling and BRS transformations with auxiliary fields in supergravity. Nucl. Phys. B 140, p.285.

Stelle, K.S. & West, P.C. (1978c). Tensor calculus for the vector multiplet coupled to supergravity. Phys. Lett. 77B, p.376.

Stelle, K.S. & West, P.C. (1978d). Relation between vector and scalar multiplets and gauge invariance in supergravity. Nucl. Phys. B 145, p. 175.

Wess, J. (1978). Supersymmetry-Supergravity. In Lecture Notes in Physics, vol. 77, p. 81.

Wess, J. & Zumino, B. (1974a). Supergauge transformations in four dimensions. Nucl. Phys. B70, p.39.

Wess, J. & Zumino, B. (1974b). A Lagrangian model invariant under supergauge transformations. Phys. Lett. 49B, p. 52.

Wess, J. & Zumino, B. (1974c). Supergauge invariant extension of Quantum Electrodynamics. Nucl. Phys. B 78, p. 1.

Wess, J. & Zumino, B. (1977). Superspace formulation of supergravity. Phys. Lett. 66B, p. 361.

# A NEW DETERIORATED ENERGY-MOMENTUM TENSOR†)

M.J. Duff
Physics Dept., Imperial College, London
P.K. Townsend, CERN, Geneva

## 1 INTRODUCTION

It is well known that the stress-tensor $T_{\mu\nu}$ of a scalar field theory is not unique because of the possibility of adding an "improvement" term, e.g. for the model

$$\mathcal{L} = -\frac{1}{2}(\partial_\mu\phi)^2 \tag{1.1}$$

we have the family of stress-tensors

$$T_{\mu\nu}(\xi) = -\left[\partial_\mu\phi\partial_\nu\phi - \frac{1}{2}\eta_{\mu\nu}(\partial\phi)^2\right] - \xi(\Box\eta_{\mu\nu} - \partial_\mu\partial_\nu)\phi^2 \tag{1.2}$$

parametrized by the constant $\xi$, which is the coefficient of an identically conserved addition to $T_{\mu\nu}$. The $\xi$-term does affect the trace $T^\mu_\mu$, however, and for $\xi = 1/6$ we have $T^\mu_\mu = 0$. The $T_{\mu\nu}$ for this value of $\xi$ is called the "new-improved" stress tensor (Callan et al. 1970). We will refer to it as the conformally improved stress-tensor and the parameter $\xi$ as the improvement coefficient. If $\xi \neq 1/6$, so that $T^\mu_\mu \neq 0$, we refer to the corresponding $T_{\mu\nu}$ as a "deteriorated" stress-tensor, hence the title of this talk. The gravitational coupling of the scalar field $\phi$ with stress-tensor $T_{\mu\nu}(\xi)$ corresponds to an extra term $\xi R\phi^2/2$ in the curved space Lagrangian. The action is then Weyl (local scale) invariant for the special value $\xi = 1/6$ and it was in this context that the $\xi$ parameter was first introduced (Penrose 1965).

In supersymmetric field theories $T_{\mu\nu}$ will appear in a supercurrent multiplet along with the supersymmetry current $S_\mu$. In particular, for the Wess-Zumino (W-Z) model for a chiral field $\phi$ (Wess & Zumino 1974)

---

†) Delivered by P.K. Townsend.

with field equations $D^2\phi = 4\mu\bar{\phi}$, the supercurrent superfield of Ferrara & Zumino (1975) is

$$V_{\alpha\dot{\alpha}} = D_\alpha\phi\bar{D}_{\dot{\alpha}}\bar{\phi} + 2i\bar{\phi}\overleftrightarrow{\partial}_{\alpha\dot{\alpha}}\phi \quad . \tag{1.3}$$

(The conventions are such that $\{D_\alpha,\bar{D}_{\dot{\alpha}}\} = 2i\partial_{\alpha\dot{\alpha}}$.) An expansion in components yields a conserved tensor $T_{\mu\nu}$ whose trace vanishes with the mass $\mu$. Hence in the $\mu \to 0$ limit it is the conformally improved $T_{\mu\nu}$ that appears in $V_{\alpha\dot{\alpha}}$. When $\mu \neq 0$ the trace $T^\mu_\mu$ appears in a sub-multiplet whose superfield is

$$\bar{D}^{\dot{\alpha}}V_{\alpha\dot{\alpha}} = \mu D_\alpha(\phi^2) \ , \tag{1.4}$$

i.e. the basic "trace-multiplet" is determined by the chiral superfield $\phi^2$. The complete supercurrent multiplet, for $\mu \neq 0$, couples straightforwardly to the usual minimal supergravity fields (Ferrara & van Nieuwenhuizen 1978; Stelle & West 1978).

However there are other possibilities for the trace-multiplet, as was first pointed out in the context of anomalies in $T^\mu_\mu$, and their corresponding multiplets, by Clark et al. (1978). Recently this other possibility has also been discussed in the context of supergravity auxiliary fields by Sohnius & West (1981). Their work implies the existence of a "new-minimal" supercurrent $V_{\alpha\dot{\alpha}}$ differing from that of Eq. (1.3). Which supercurrent is the relevant one may be a matter of choice or it may depend on the model under consideration. For example, consider the dual form of the massless W-Z model,

$$\mathcal{L} = \int d^4\theta G^2 \ , \quad G \equiv D^\alpha\psi_\alpha + \bar{D}_{\dot{\alpha}}\bar{\psi}^{\dot{\alpha}} \ , \tag{1.5}$$

for the gauge chiral spinor $\psi_\alpha$ (Siegel 1979). This has the same spectrum as the massless W-Z model but with a gauge antisymmetric tensor replacing the usual pseudoscalar. This model, when coupled to supergravity, cannot couple through a conformally improved stress-tensor because there is no analogue of the improvement term for an antisymmetric tensor. (This does not mean that no interacting conformal invariant model for the linear superfield G cannot be found. In fact there is such a model for which the scalar field $G|_{\theta=0}$ acts as a conformal "compensating" field for the antisymmetric tensor kinetic term (de Wit & Roček 1981). Coupling this model to the conformal supergravity fields leads directly (de Wit & Roček 1981; Howe, Stelle & Townsend 1981; Gates, Roček & Siegel 1981) to the new-minimal version of Poincaré supergravity. See also Sohnius & West, this volume.) There is still one improvement coefficient for the scalar, and one expects that the supercurrent multiplet will simplify for special choices. But what are these choices, and what are the corresponding multiplets? This point is relevant to extended supergravity theories with a cosmological constant because the radiative corrections to the cosmological constant, which vanish for $N \geq 5$, depend directly on the improvement coefficients for the scalars (Christensen & Duff 1980; Christensen et al. 1980), and we would like to know how the presence of gauge antisymmetric tensors (if possible) affects this result.

This brings us to the general question of the supercurrent multiplet for arbitrary deteriorated stress tensors, and how these are related to supercurrent multiplets for models with gauge antisymmetric tensors. We will answer this question in the following for various models of N = 1, 2, and 4 supersymmetry. For N = 1 the work presented here was a precursor to a more general approach to matter-coupling in supergravity theories developed in Howe et al. (1981b). The work on N = 2 and N = 4 was done in collaboration with P. Howe and K. Stelle.

## 2 N = 1

Consider the Wess-Zumino model,

$$\mathcal{L} = -\frac{1}{2}(\partial_\mu A)^2 - \frac{1}{2}(\partial_\mu B)^2 - \frac{1}{2}\bar{\lambda}\not\partial\lambda - \frac{1}{2}\mu^2(A^2+B^2) - \frac{1}{2}\mu\bar{\lambda}\lambda , \quad (2.1)$$

which has a two-parameter family of stress tensors:

$$T_{\mu\nu} = -\left[\partial_\mu A\partial_\nu A - \frac{1}{2}\eta_{\mu\nu}(\partial A)^2\right] - \left[\partial_\mu B\partial_\nu B - \frac{1}{2}\eta_{\mu\nu}(\partial B)^2\right] - \bar{\lambda}\gamma_{(\mu}\partial_{\nu)}\lambda$$
$$+ \frac{1}{2}\eta_{\mu\nu}\left[\mu^2(A^2+B^2) + \mu\bar{\lambda}\lambda\right] - \xi(\eta_{\mu\nu}\Box - \partial_\mu\partial_\nu)A^2$$
$$- \xi'(\eta_{\mu\nu}\Box - \partial_\mu\partial_\nu)B^2 . \quad (2.2)$$

The trace is

$$T^{\mu}_{\mu} = \mu^2(A^2+B^2) + \mu\bar{\lambda}\lambda + \frac{1}{2}\left[1 - 3(\xi+\xi')\right] \Box(A^2+B^2)$$
$$+ \frac{3}{2}(\xi-\xi')\Box(A^2-B^2) \qquad (2.3)$$

This is the $\theta = 0$ component of the following superfield:

$$T = \frac{\mu}{16}\left[D^2(\phi^2) + \bar{D}^2(\bar{\phi}^2)\right] + \Box\left\{\frac{[1 - 3(\xi+\xi')]}{2}\bar{\phi}\phi + \frac{3}{4}(\xi-\xi')(\phi^2+\bar{\phi}^2)\right\}, \qquad (2.4)$$

with $\phi|_{\theta=0} = A + iB$. Since the trace multiplet is what distinguishes the various possibilities for the supercurrent, we may concentrate our attention on it. Equation (2.4) establishes the existence of such a multiplet for all values of $\xi$ and $\xi'$, but we generally expect that $T^{\mu}_{\mu}$ will appear as a higher component of a lower dimension superfield, e.g. $V_{\alpha\dot{\alpha}}$. We will therefore impose, from now on, a "minimality" requirement, namely that all components of the supercurrent multiplet, and its trace, be found within the single superfield $V_{\alpha\dot{\alpha}}$. A candidate for such a supercurrent superfield is

$$V_{\alpha\dot{\alpha}} = D_{\alpha}\phi\bar{D}_{\dot{\alpha}}\bar{\phi} + ia\bar{\phi}\overleftrightarrow{\partial}_{\alpha\dot{\alpha}}\phi + ib\partial_{\alpha\dot{\alpha}}(\phi^2-\bar{\phi}^2), \qquad (2.5)$$

where the coefficient of the first term has been scaled to unity. This is possible except when it vanishes, i.e. when the spinor has zero chiral charge. In that case we must consider separately

$$V_{\alpha\dot{\alpha}} = i\bar{\phi}\overleftrightarrow{\partial}_{\alpha\dot{\alpha}}\phi + ib'\partial_{\alpha\dot{\alpha}}(\phi^2-\bar{\phi}^2), \qquad (2.6)$$

where the chiral charge of the scalars has now been scaled to unity. [We could add $\partial_{\alpha\dot{\alpha}}(\phi^2+\bar{\phi}^2)$ or $\partial_{\alpha\dot{\alpha}}(\bar{\phi}\phi)$ to Eqs. (2.5) and (2.6), but the former violates parity while the latter is easily ruled out by the considerations to follow.]

In order to qualify as genuine supercurrents, these candidates must contain a conserved vector spinor and a conserved symmetric tensor that can be identified with the supersymmetry current and the stress-tensor, respectively. The latter follows from the former, which is easier to check because the spinor current has lower dimension. The most general spinor current that can be constructed locally of the right dimension is

$$\psi_{\dot\beta\alpha\dot\alpha} = \bar D_{\dot\beta} V_{\alpha\dot\alpha} + C\,\varepsilon_{\dot\alpha\dot\beta}\bar D^{\dot\gamma} V_{\alpha\dot\gamma}\ , \qquad (2.7)$$

with C an arbitrary constant. The divergence of this quantity is

$$\partial^{\alpha\dot\alpha}\psi_{\dot\beta\alpha\dot\alpha} = -2\mu\partial^{\alpha}_{\dot\beta}D_{\alpha}(\phi^2)\left[1 - \frac{(3-a)}{2}C\right] + 2i\Box\bar D_{\dot\beta}(\bar\phi\phi)\,C\left(1-\frac{a}{2}\right)$$
$$+ 2ib\Box\bar D_{\dot\beta}(\bar\phi^2)(1+C)\ . \qquad (2.8)$$

Requiring that it vanish leads to the following possibilities:

i) $\mu \neq 0$ , $a = 2$ , $b = 0$ , $C = 2$

ii) $\mu = 0$ , $b \neq 0$ , $a = 2$ , $C = -1$

iii) $\mu = 0$ , $b = 0$ , and $a = 2$ , C arbitrary ,
or $a \neq 2$ , $C = 0$ .

In the case that the supercurrent is not of the form (2.5) but rather of the form (2.6) we find the additional possibility:

iv) $b' = 0$ , $C = 0$ and $\mu = 0$ or $\mu \neq 0$ .

It now remains to translate these restrictions on a, b, C, and $\mu$ to restrictions on the coefficients $\xi$ and $\xi'$. In the massive case, one can show that

$$\frac{1}{16(3-a)}\left[D^{\alpha},\bar D^{\dot\alpha}\right]V_{\alpha\dot\alpha} = \frac{\mu}{16}\left[D^2(\phi^2) + \bar D^2(\bar\phi^2)\right]$$
$$+\Box\left\{\frac{(a-2)}{4(3-a)}\bar\phi\phi - \frac{b}{4(3-a)}(\phi^2+\bar\phi^2)\right\} \qquad (2.9)$$

which agrees with T of Eq. (2.4) provided

$$\left(\frac{a-2}{3-a}\right) = 2\left[1 - 3(\xi+\xi')\right]\ , \qquad \frac{b}{3-a} = -3(\xi-\xi') \qquad (2.10)$$

which fixes a and b in terms of $\xi$ and $\xi'$ such that for the allowed values of the former $\xi = \xi' = 1/6$; i.e. case (i) is the "conformal point" corresponding to the usual minimal stress tensor. In the case that the chiral charge of the spinor vanishes, a similar computation yields (since $b' = 0$),

$$-\frac{1}{16}\left[D^{\alpha},\bar{D}^{\dot{\alpha}}\right]V_{\alpha\dot{\alpha}} = \frac{\mu}{16}\left[D^2(\phi^2) + \bar{D}^2(\bar{\phi}^2)\right] + \Box\left\{-\frac{1}{4}\bar{\phi}\phi\right\} . \tag{2.11}$$

This agrees with T of Eq. (2.4) provided

$$\xi = \xi' = \frac{1}{4} , \tag{2.12}$$

i.e. the improvement coefficients are $3/2$ times their values at the conformal point. The structure of the supercurrent multiplet in this case turns out to be of "new-minimal" form; i.e. $T^{\mu}_{\mu}$ appears in a multiplet with $\gamma\cdot S$ and a conserved antisymmetric tensor $t_{\mu\nu}$, but the axial current $j^5_{\mu}$ remains conserved. This is only possible for $\mu \neq 0$ if the chiral charge of the spinor vanishes. These results can be represented by a graph of the allowed values for $\xi$ and $\xi'$ in the $\xi,\xi'$ plane, as shown in Fig. 1 When $\mu = 0$, we have

$$-\frac{1}{16}\left[D^{\alpha},\bar{D}^{\dot{\alpha}}\right]V_{\alpha\dot{\alpha}} = \Box\left\{\frac{(2-a)}{4}\bar{\phi}\phi + \frac{b}{4}(\phi^2+\bar{\phi}^2)\right\} \tag{2.13}$$

for $V_{\alpha\dot{\alpha}}$ of the form (2.5). [The form (2.6) yields nothing new for $\mu = 0$, so we need not consider it.] This agrees with T of Eq. (2.4) provided

$$\frac{(2-a)}{4} = \frac{1-3(\xi+\xi')}{2} , \quad \frac{b}{4} = \frac{3}{4}(\xi-\xi') . \tag{2.14}$$

Of course we would still have agreement for some choice of a and b if Eq. (2.13) were scaled by some constant. What fixes the scale is the requirement that this quantity be the trace of the full $T_{\mu\nu}$ whose scale is separately fixed. Cases (ii) and (iii) above can now be translated to conditions on $\xi$ and $\xi'$. These are given by two allowed lines in the $\xi,\xi'$ plane as shown in Fig. 2. The two lines cross at the conformal point. On the line $\xi+\xi' = 1/3$ the supercurrent multiplet is of the conventional minimal form. It is not quite the same as the usual one however, because the way in which $T_{\mu\nu}$ is extracted from $V_{\alpha\dot{\alpha}}$, represented by the constant C(= -1), is different from the usual case (where C = 2). Thus, although the components are the same the transformation rules differ. On the line $\xi = \xi'$ the supercurrent is of new-minimal form, which is not surprising since the chiral symmetry A ↔ B is preserved if $\xi = \xi'$ and so the axial current will be conserved.

Whenever one of the lines in Fig. 2 crosses an axis, we expect that a duality transformation could be performed to convert a (pseudo)scalar to an antisymmetric tensor. Thus, for the model of Eq. (1.5) we would anticipate two distinct (and discrete) solutions for $V_{\alpha\dot{\alpha}}$. Indeed this is the case. If we start from the candidate supercurrent

$$V_{\alpha\dot{\alpha}} = D_\alpha G \bar{D}_{\dot{\alpha}} G - \frac{a'}{2} G \left[ D_\alpha, \bar{D}_{\dot{\alpha}} \right] G \, , \tag{2.15}$$

then repeating the previous analysis leads to the two possibilities $a' = 2$ or $a = 0$, which imply $\xi = 1/3$ or $\xi = 0$, respectively. We can plot these possibilities on the $\xi$-line as in Fig. 3.

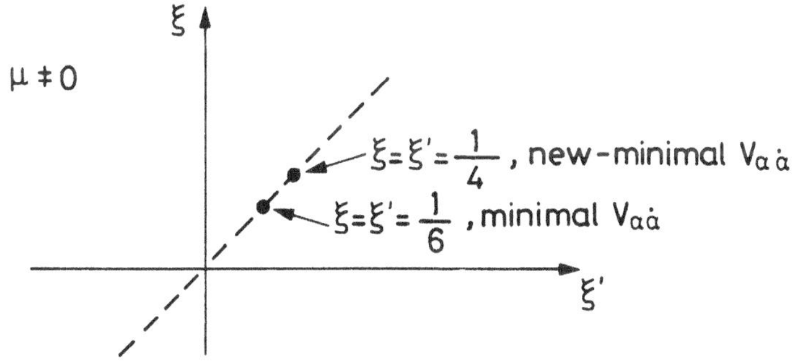


FIG .1

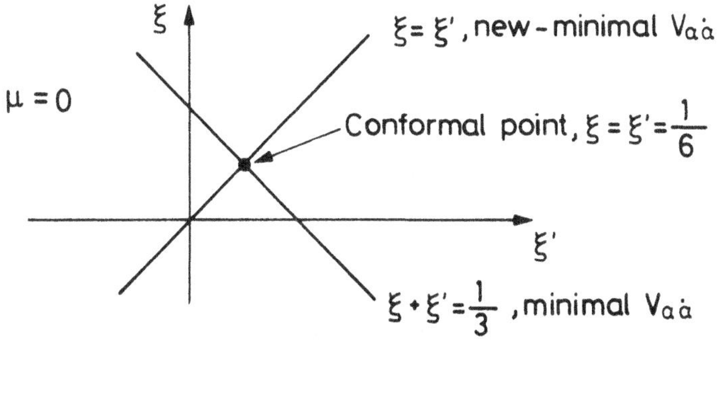


FIG. 2

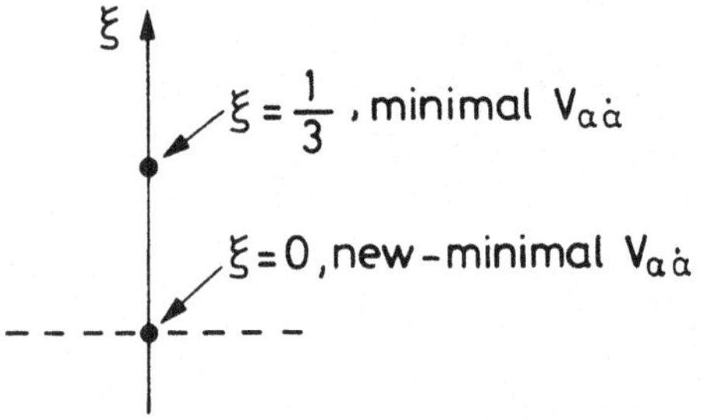


FIG. 3

This is obviously the $\xi' = 0$ slice of Fig. 2. For $\xi = 1/3$ we have the minimal stress-tensor multiplet, while for $\xi = 0$ we have the new-minimal one. One point that is worth mentioning is that since the renormalization of the cosmological constant in scalar theories depends on $\xi$ via $\sum_i (\xi_i - 1/6)^2$, where i runs over all the scalars, the two values $\xi = 1/3$ and $\xi = 0$ yield the same result for this renormalization, i.e. the renormalization is not affected by the choice of auxiliary fields (minimal or new-minimal). This is a gratifying result; however, it is only for $\xi_i = 1/6$ for all i that the renormalization of the cosmological constant vanishes for $N \geq 5$ gauged supergravity (Christensen et al. 1980) so that the occurrence of any antisymmetric tensors in these models (if possible) would spoil this cancellation. This is curious because antisymmetric tensors are needed for a cancellation of the trace anomaly in these theories (Duff & van Nieuwenhuizen 1980; Siegel 1980; Duff 1980; Nicolai & Townsend 1981).

Before leaving N = 1 we want to mention that there is a version of the W-Z model with two antisymmetric tensors (Freedman 1977). This would be expected to have a unique supercurrent multiplet of new-minimal form because no improvement of $T_{\mu\nu}$ is possible. This model has some rather pathological features, however, arising from the fact that the supersymmetry algebra fails to close off-shell on the bosons as well as on the fermions (Townsend & van Nieuwenhuizen 1981). This model is therefore a counter-example to the often-made claim that this cannot happen.

## 3 N = 2

The analogue of $V_{\alpha\dot{\alpha}}$ for N = 2 is V, a real scalar superfield, which if $\mu = 0$ and $T^{\mu}_{\mu} = 0$, satisfies the conservation condition (Howe et al. 1981; Gates & Siegel 1981):

$$D^{ij}V = 0\ , \quad D^{ij} \equiv D^{\alpha i}D_{\alpha}{}^{j} \equiv D^{ji}\ . \tag{3.1}$$

A superfield V satisfying Eq. (3.1) is an irreducible conformal supercurrent. A non-conformal V can be obtained either by adding a mass to the underlying model which produces a 8+8 component trace multiplet (Sohnius 1979) or by deteriorating the stress-tensor. To investigate the latter possibility we choose the simplest N = 2 model with four scalars and two Majorana spinors. We will assign the scalars to an SU(2) triplet $A_{ij}$ and an SU(2) singlet B, and the spinors to an SU(2) doublet $\lambda^i$. The fields are related by

$$D^i_\alpha A_{jk} = \delta^i_{(j}\lambda_{\alpha k)} , \quad D^i_\alpha B = \lambda^i_\alpha , \tag{3.2}$$

and since we put the model on-shell we have also

$$D^i_\alpha \lambda_{\alpha j} = 0 , \quad \bar{D}_{\dot\beta i}\lambda_{\alpha j} = i\partial_{\dot\beta\alpha}B\varepsilon_{ij} + 2i\partial_{\dot\beta\alpha}A_{ij} . \tag{3.3}$$

Now, since V must have dimension 2 and since we preserve SU(2), there is only one candidate for V, namely

$$V = aA_{ij}A^{ij} + bB^2 , \tag{3.4}$$

which can be written as

$$V = \frac{(3a+2b)}{8} V^{conf} + \frac{3}{8}(2a-b)V^{tr} , \tag{3.5}$$

where $V^{conf}$ and $V^{tr}$ are two irreducible multiplets

$$V^{conf} = 2A_{ij}A^{ij} + B^2 , \quad V^{tr} = B^2 - \frac{2}{3}A_{ij}A^{ij} , \tag{3.6}$$

satisfying

$$\begin{aligned} &D^{ij}V^{conf} = 0 \\ &D_{\alpha\beta}V^{tr} = 0 ; \quad D_{\alpha\beta} \equiv D^i_\alpha D_{\beta i} \\ &\bar{D}_{ij}D^{ij}V^{tr} = -16\,\Box V^{tr} . \end{aligned} \tag{3.7}$$

A component expansion of $V^{tr}$ shows that it has 24+24 components and maximum spin 1. The spin 1 current, however, is not conserved. Table 1 gives the spin, SL(2,C), and SU(2) content and dimensions of the components of $V^{tr}$.

However, before we can say that $V^{tr}$ is a genuine trace multiplet for N = 2, we must check that V still contains a conserved vector spinor. As before, we construct from V a vector spinor of the right dimension in the most general way,

$$\psi^{k}_{\dot{\beta}\alpha\beta} = \bar{D}^{k}_{\dot{\beta}} D_{\alpha\beta} V + \frac{2i}{3} C' \partial_{\dot{\beta}(\alpha} D^{k}_{\beta)} V + C\varepsilon_{\beta\alpha} \bar{D}_{\dot{\beta}\ell} D^{\ell k} V \; . \tag{3.8}$$

This is conserved provided the following matrix equation is satisfied

$$\begin{bmatrix} 2(a-2b) & 2b \\ -(a-2b) & a \end{bmatrix} \begin{bmatrix} C \\ C' \end{bmatrix} = -(3a+2b) \begin{bmatrix} 1 \\ 1 \end{bmatrix} \tag{3.9}$$

This has a solution, so V of Eq. (3.4) is an acceptable supercurrent. (If $3a+2b = 0$ the solution is $C = C' = 0$, and since $D_{\alpha\beta}V$ vanishes when $3a+2b = 0$ the entire spinor current vanishes, so this special case is obviously excepted; it corresponds to the choice $V = V^{tr}$.)

We now want to relate these results to the improvement coefficients for $A_{ij}$ and B. With the normalization fixed by Eq. (3.3) the Lagrangian is

$$\mathcal{L} = \partial_{\alpha\dot{\alpha}} A_{ij} \partial^{\alpha\dot{\alpha}} A^{ij} + \frac{1}{2} \partial_{\alpha\dot{\alpha}} B \partial^{\alpha\dot{\alpha}} B + \ldots \; . \tag{3.10}$$

Defining the improvement coefficients as for N = 1, with $\xi$ being the coefficient for $A_{ij}$ and $\xi'$ that for B, we find for the trace of the stress-tensor

$$T^{\mu}_{\mu} = \Box(1-6\xi) A_{ij} A^{ij} + \Box \frac{(1-6\xi')}{2} B^2 \; , \tag{3.11}$$

with the relative factor of ½ explained by Eq. (3.10). But from the form of $V^{tr}$ in Eq. (3.6) we can read off the trace contained in V:

$$T \sim \Box(B^2 - \frac{2}{3} A_{ij} A^{ij}) \; . \tag{3.12}$$

Table 1

| Spin | SU(2) | SL(2,C) | Dimension |
|---|---|---|---|
| 1 | $\underset{\sim}{3}$ | $(\frac{1}{2},\frac{1}{2}) + (0,0)$ | 1 |
| $\frac{1}{2}$ | $\underset{\sim}{4}$ | $(\frac{1}{2},0) + (0,\frac{1}{2})$ | $\frac{3}{2}$ |
| $\frac{1}{2}$ | $\underset{\sim}{2}$ | $(\frac{1}{2},0) + (0,\frac{1}{2})$ | $\frac{1}{2}$ |
| 0 | $\underset{\sim}{5}$ | $(0,0)$ | 2 |
| 0 | $\underset{\sim}{3} + \underset{\sim}{3}$ | $(0,0)$ | 1 |
| 0 | $\underset{\sim}{1}$ | $(0,0)$ | 0 |

Equations (3.11) and (3.12) agree if

$$\xi' + 3\xi = \frac{2}{3}\,, \tag{3.13}$$

which we can plot in the $\xi,\xi'$ plane as shown in Fig. 4. This is just the N = 2 version of the line $\xi + \xi' = \frac{1}{3}$ of Fig. 1. The slope of the line in Fig. 2 is $-\frac{1}{3}$ instead of $-1$ simply because there are three fields in $A_{ij}$. Both lines are characterized by the fact that the sum of the improvement coefficients over all scalars is constant and equal to its value at the conformal point. It seems that N = 2 is more restrictive than N = 1, but of course we have chosen a rather restrictive "minimality" condition, namely that all components of the supercurrent be found in a single scalar superfield. Since the N = 2 Poincaré supergravity prepotential is a spinor superfield (Sokatchev 1980; Gates & Siegel 1981) this may be too stringent a requirement.

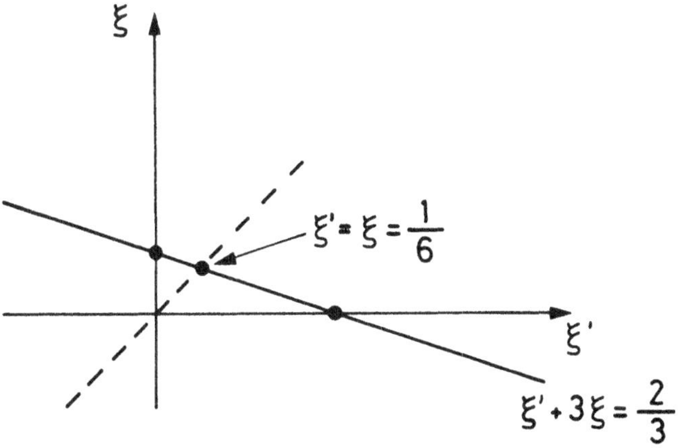


FIG. 4

There is a dual version of the above model based on the N = 2 linear superfield $L_{ij}$ satisfying $D^i_\alpha L_{jk} = \delta^i_{(j}\lambda_{\alpha k)}$. The unique candidate scalar supercurrent $V = L_{ij}L^{ij}$ satisfies (Stelle 1981)

$$D^{ij}V = D_{k\ell}\{L^{(ij}L^{k\ell)}\} , \tag{3.14}$$

where the trace multiplet $L^{(ij}L^{k\ell)}$ is again a 24+24 component multiplet, differing from the one given in Table 1 by the assignment of dimensions of the scalars. It is clear from this result that in order to couple supergravity to the kinetic action for the N = 2 linear multiplet, a new set of N = 2 auxiliary fields is needed that includes a 24+24 trace submultiplet, i.e. that differs from the usual set of Fradkin & Vasiliev (1979) and de Wit & van Holten (1979).

## 4 N = 4

The N = 4 super-Maxwell theory is described by a scalar superfield $\phi_{ij}$ in the real $\underset{\sim}{6}$ representation of SU(4) and satisfying (Sohnius 1978)

$$D^k_\alpha \phi_{ij} = \delta^k_{(i}\lambda_{\alpha j)} , \tag{4.1}$$

which also implies the field equations. If we insist on SU(4) invariance then there is only one improvement coefficient $\xi$, and the trace of the SU(4) invariant deteriorated stress-tensor is therefore

$$T \sim (1-6\xi)\,\Box\,\phi_{ij}\phi^{ij} . \tag{4.2}$$

Hence the basic trace-multiplet is $\phi_{ij}\phi^{ij}$. In the non-interacting theory this is an irreducible multiplet with maximum conserved spin 4 (Howe et al. 1981a). In the presence of interactions, i.e. for N = 4 super-Yang-Mills theory, the high-spin current cannot be conserved and so the trace-multiplet will itself be partially locally reducible. In the N = 4 case it is not clear which superfield will contain both the trace multiplet and the spin-2 conformal supercurrent multiplet (Bergshoeff et al. 1981) whose superfield is the Lorentz scalar $V^{ijk\ell}$ in the $\underset{\sim}{20}'$ representation of SU(4). It is possible that the full supercurrent superfield could be a Lorentz and SU(4) scalar V with some weak conservation condition on it, because such a superfield does contain both a superspin 0 (i.e. maximum spin 2) in a $\underset{\sim}{20}'$ of SU(4) and a singlet superspin 2 (i.e. maximum spin 4).

Of course, if the assumption of SU(4) invariance is relaxed to USp(4) or O(4) then there is a more general deteriorated stress-tensor. In fact, as for N = 2, there are then two possible terms in the trace $T^{\mu}_{\mu}$ corresponding to a split of the six scalars into a $\underset{\sim}{5} \oplus \underset{\sim}{1}$ of USp(4) or a $\underset{\sim}{3} \oplus \underset{\sim}{3}$ of O(4). In this case the superfield T can again be separated into a linear combination of two irreducible multiplets one of which has maximum spin 2 and the other maximum spin 4. For some linear combination, only the spin-2 multiplet remains. In the case of USp(4) symmetry this is just the 128+128 non-conformal supercurrent multiplet of Howe & Lindström (1981).

While deterioration of the stress-tensor allows us to rederive multiplets for USp(4) symmetry that can be obtained by other methods, it also allows us to discuss the trace multiplet for SU(4) symmetry. In this respect the method of deterioration gives us information that cannot be obtained in other ways. These other ways of obtaining a non-conformal supercurrent multiplet are i) adding mass terms, ii) going to dimensions other than four, iii) shifting scalar fields when the potential has a valley; but all of them imply a breakdown of SU(4) to USp(4) in the N = 4 case.

Conversely, by choosing to break SU(4) by the improvement coefficients, one could arrange to have a supercurrent multiplet that includes the divergences of any number of the SU(4) vector currents according to which group SU(4) is broken down. In the extreme case SU(4) can be completely broken, thus generating supercurrents that may be applicable for coupling to N = 4 Poincaré supergravity. This remains a task for the future.

## REFERENCES


Bergshoeff, E., de Roo, M. & de Wit, B. (1981). Extended conformal supergravity. Nucl. Phys. B 182, 173.

Callan, C.G., Coleman, S. & Jackiw, R. (1970). A new improved energy momentum tensor. Ann. Phys. (NY) 59, 42.

Christensen, S.M. & Duff, M.J. (1980). Quantizing gravity with a cosmological constant. Nucl. Phys. B 170, 480.

Christensen, S.M., Duff, M.J., Gibbons, G.W. & Roček, M. (1980). Vanishing one-loop $\beta$-function in gauged $N > 4$ supergravity. Phys. Rev. Lett. 45, 161.

Clark, T.E., Piguet, O. & Sibold, K. (1978). Supercurrents, renormalization and anomalies. Nucl. Phys. B 143, 445.

de Wit, B. & Roček, M. (1981). Improved tensor multiplets. NIKHEF preprint H/81-28.

de Wit, B. & van Holten, J.W. (1979). Multiplets of linearized SO(2) supergravity. Nucl. Phys. B 155, 530.

Duff, M.J. (1980). Antisymmetric tensors and supergravity. In Superspace and Supergravity, eds. S.W. Hawking & M. Roček (Cambridge University Press).

Duff, M.J. & van Nieuwenhuizen, P. (1980). Quantum inequivalence of different field representations. Phys. Lett. B 94, 179.

Ferrara, S. & van Nieuwenhuizen, P. (1978). The auxiliary fields of supergravity. Phys. Lett. B 74, 333.

Ferrara, S. & Zumino, B. (1975). Transformation properties of the supercurrent. Nucl. Phys. B 87, 207.

Fradkin, E.S. & Vasiliev, M. (1979). Minimal set of auxiliary fields and S matrix for extended supergravity. Lett. Nuovo Cimento 25, 79; Minimal set of auxiliary fields in SO(2) extended supergravity. Phys. Lett. B 85, 47.

Freedman, D.Z. (1977). Gauge theories of antisymmetric tensor fields, unpublished.

Gates, S.J. & Siegel, W. (1981). Superprojectors. CalTech. preprint; Linearized $N = 2$ superfield supergravity; CalTech preprint.

Gates, S.J. Jr., Roček, M. & Siegel, W. (1981). Solution to constraints for $n = 0$ supergravity. Cal. Tech. preprint CALT-68-868.

Howe, P. & Lindström, U. (1981). The supercurrent in five dimensions. Phys. Lett. B 103, 422.

Howe, P., Stelle, K.S. & Townsend, P.K. (1981a). Supercurrents. To appear in Nucl. Phys. B.

Howe, P., Stelle, K.S. & Townsend, P.K. (1981b). Matter coupling and auxiliary field choices in superspace supergravity, in preparation.

Howe, P., Stelle, K. & Townsend, P. (1981c). The vanishing volume of $N = 1$ superspace. Phys. Lett. B107, 420.

Nicolai H. & Townsend, P.K. (1981). $N = 3$ supersymmetry multiplets with vanishing trace anomaly: building blocks of the $N > 3$ supergravities. Phys. Lett. B 98, 257.

Penrose, R. (1965). Zero rest-mass fields including gravitation:asymptotic behaviour. Proc. Roy. Soc. 284A, 159.

Siegel, W. (1979a). Gauge spinor superfield as scalar multiplet. Phys. Lett. B 85, 333.

Siegel, W. (1980). Quantum equivalence of different field representations. Phys. Lett. B 138, 107.

Sohnius, M. (1979). The multiplet of currents for $N = 2$ extended supergravity. Phys. Lett. B 81, 8.

Sohnius, M. (1978). Supersymmetry and central charges. Nucl. Phys. B 138, 109.
Sohnius, M. & West, P.C. (1981). An alternative minimal off-shell version of N = 1 supergravity. Imperial College preprint ICTP 80-81/37.
Sokatchev, E. (1980). A superspace action for N = 2 supergravity. Phys. Lett. B 100, 107.
Stelle, K.S. (1981). See talk in this volume.
Stelle, K.S. & West, P.C. (1978). Auxiliary fields for supergravity. Phys. Lett. B 74, 330.
Townsend, P.K. & van Nieuwenhuizen, P. (1981). Which field representations are allowed in interacting locally supersymmetric field theories? To be published in Physics Letters B.
Wess, J. & Zumino, B. (1974). Supergauge transformations in four dimensions. Nucl. Phys. B 70, 39.

# OFF-SHELL N = 2 AND 4 SUPERGRAVITY IN FIVE DIMENSIONS

P. Howe
CERN, 1211 Geneva 23, Switzerland

## 1 INTRODUCTION

Currently available x-space Lagrangians for supergravity theories with N ≥ 3 (Ferrara et al. 1977; Freedman & Schwarz 1978; Cremmer et al. 1978; Cremmer & Julia 1979) do not contain auxiliary fields so that the supersymmetry algebra only closes if the field equations are satisfied. In order to quantize these theories it is therefore desirable to know the auxiliary fields and, if possible, to have a superspace formulation, since in the latter case one might hope to be able to use supergraph techniques. In this talk I shall discuss the auxiliary field problem for N = 2 and 4 supergravities in five dimensions from a superspace point of view.

Five dimensions is of interest for several reasons. First, five-dimensional supergravity theories have some elegant formal features (Cremmer 1980), in particular, the x-space Lagrangians for N ≥ 4 have local Sp(N) and global non-compact invariances (e.g., in N = 8 the latter is $E_6$) as opposed to the four-dimensional case where the global symmetry group is an invariance only of the equations of motion. Secondly, five-dimensional theories may be dimensionally reduced in a non-standard way to yield four-dimensional theories with broken supersymmetry (Scherk & Schwarz 1979; Cremmer et al. 1979). Thirdly, matter multiplets in five dimensions are not conformally invariant even when they are massless. It is this last feature we shall exploit here; it implies that the supercurrents constructed as bilinears from underlying on-shell matter multiplets contain stress-tensors which are not traceless, so that the corresponding multiplets of supergravity fields are Poincaré multiplets.

The five-dimensional N = 2 and 4 supercurrents have recently been constructed (Howe & Lindström 1981) and the N = 4 supercurrent multiplet turns out to have the same number of components (128 + + 128) as the four-dimensional N = 4 conformal supercurrent (Bergshoeff et al. 1981). Hence for N = 4 in five-dimensions, the supercurrent is still of minimal size but no conformal compensating multiplet is needed. Furthermore, although the N = 4 conformal supergravity multiplet has a complex dimension-zero spin-zero field (Bergshoeff et al. 1981), this is not the case for the N = 4 five-dimensional multiplet which couples to the supercurrent (Howe & Lindström 1981). Indeed, there are no dimension one-half spinors either and this leads to considerable simplifications in the superspace theory compared with four-dimensional conformal supergravity in superspace (Howe 1981 a,b).

After briefly reviewing superspace and the five-dimensional supercurrents, I shall use information furnished by the latter to derive the superspace torsion constraints in a simple way. The supergravity multiplets corresponding to these constraints are not adequate for constructing Lagrangians - they are the analogues of the four-dimensional N = 2 minimal Poincaré multiplet (Breitenlohner & Sohnius 1980) - and in order to remedy this situation further multiplets are needed. In four-dimensional N = 2 there are two known possibilities; one may either use a linear multiplet (Breitenlohner & Sohnius 1980; Bergshoeff et al. 1980) or a scalar multiplet (de Wit et al. 1980), the latter having an off-shell central charge. N = 2 five-dimensional supergravity follows the same pattern, not surprisingly, but for N = 4 it seems that a central charge is obligatory. A candidate multiplet for the N = 4 case is exhibited in the final section.

## 2 SUPERSPACE

The superspaces we shall use have co-ordinates $z^M = (x^m, \theta^{\underline{\mu}})$ where the five $x^m$ are commuting co-ordinates corresponding to ordinary space-time and the $\theta^{\underline{\mu}}$ are 4N anticommuting co-ordinates which are pseudo-real. The tangent space group is SO(1,4) × Sp(N) under which a tangent-space co-vector $V_A = (V_a, V_{\underline{\alpha}}) = (V_a, V_{\alpha i})$ transforms as follows:

$$\delta V_A = -L_A{}^B V_B \tag{2.1}$$

where

$$L_A{}^B = \text{diag.}(L_a{}^b, L_{\underline{\alpha}}{}^{\underline{\beta}}) \tag{2.2}$$

$$L_{\underline{\alpha}}{}^{\underline{\beta}} = L_{\alpha i}{}^{\beta i} = \delta_\alpha{}^\beta L_i{}^j + \delta_i{}^j L_\alpha{}^\beta$$

$$L_{\alpha\beta} = -\frac{i}{4}(\gamma^{ab})_{\alpha\beta} L_{ab}$$

$$L_{ij} \equiv L_i{}^k \eta_{kj} = L_{ji} = -\eta_{ik}\eta_{jl}\bar{L}^{kl}$$

Here, a,b... are vector indices, $\alpha,\beta$... four-component spinor indices and i,j... (= 1..N) internal indices. The $\gamma$ matrices are

$$\eta_{\alpha\beta},\ (\gamma^a)_{\alpha\beta},\ (\gamma^{ab})_{\alpha\beta} = (\gamma^{[a}\gamma^{b]})_{\alpha\beta} \tag{2.3}$$

where $\eta_{\alpha\beta} = \eta^{\alpha\beta}$ is the charge conjugation matrix. $\eta_{\alpha\beta}$ and $(\gamma^a)_{\alpha\beta}$ are antisymmetric while $(\gamma^{ab})_{\alpha\beta}$ is symmetric (the normal $\gamma$'s are those with the second index raised by $\eta^{\alpha\beta}$). $\eta_{ij} = \eta^{ij} = -\eta_{ji}$ is the Sp(N) metric and all spinors satisfy the Sp(N) Majorana condition

$$\bar{\psi}^{\alpha i} = \eta^{\alpha\beta}\eta^{ij}\psi_{\beta j} \tag{2.4}$$

The basic geometrical variables are the vielbein $E_M{}^A$ and the connection $\Omega_{MA}{}^B$, the latter transforming inhomogeneously under (2.2):

$$\delta\Omega_{M,A}{}^B = -\partial_M L_A{}^B - L_A{}^C\Omega_{MC}{}^B + \Omega_{MA}{}^C L_C{}^B \tag{2.5}$$

so that it has the same symmetry properties as $L_A{}^B$ on the indices A and B. Using $\Omega$ we can construct a covariant derivative $D_A = E_A{}^M D_M$ which acts on fields transforming tensorially under (2.2), e.g.,

$$D_A V_B = E_A{}^M\{\partial_M V^B - \Omega_{MB}{}^C V_C\} \tag{2.6}$$

Taking the graded commutator of two covariant derivatives one finds

$$[D_A, D_B\} V_C = -T_{AB}{}^D D_D V_C - R_{AB,C}{}^D V_D \tag{2.7}$$

where $T_{AB}{}^C$ is the torsion and $R_{ABC}{}^D$ the curvature which is SO(1,4) $\times$ Sp(N) Lie algebra valued on its last two indices. The Bianchi identity is

$$\sum_{(ABC)} [[D_A,\ D_B, D_C\}\} = 0 \tag{2.8}$$

where the summation is graded cyclic. It is known that the Bianchi identities may be used to solve for $R_{ABC}{}^D$ algebraically in terms of $T_{AB}{}^C$ and its derivatives (Dragon 1979) so that the torsion is the fundamental tensor in superspace geometry.

The vielbein and the connection define a highly reducible representation of supersymmetry and it is therefore necessary to impose constraints. This may be done covariantly by constraining $T_{AB}{}^{C}$ thereby avoiding the complication of Wess-Zumino gauges associated with superspace general co-ordinates and tangent space transformations. The non-zero torsion components may be associated with component fields by evaluating them at $\theta = 0$; there is therefore no need to perform any $\theta$ expansions and this way of identifying component fields also allows one to read off their x-space supersymmetry transformation properties in a simple way (Wess & Zumino 1978). So for any tangent space tensor V, one defines

$$\delta V = \xi^{A} D_{A} V \tag{2.9}$$

where $\xi^{A} = \xi^{M} E_{M}{}^{A}$ is taken to be field independent and $\xi^{M}$ is the superspace general co-ordinate parameter. Then for supersymmetry transformations of components one has

$$\delta V|_{\theta=0} = i\, \xi^{\alpha i}|_{\theta=0} (D_{\alpha i} V)_{\theta=0} \tag{2.10}$$

Gauge fields themselves will not be located in the torsions but rather in the vielbein and the connection. We make the following identifications:

$$\begin{aligned} E_{m}{}^{a}|_{\theta=0} &= e_{m}{}^{a}(x) \\ E_{m}{}^{\underline{\alpha}}|_{\theta=0} &= \psi_{m}{}^{\underline{\alpha}}(x) \\ \Omega_{m,\underline{\alpha}}{}^{\underline{\beta}}|_{\theta=0} &= -\frac{i}{4}(\gamma^{ab})_{\alpha}{}^{\beta}\omega_{m,ab}(x)\delta_{i}{}^{j} + \delta_{\alpha}{}^{\beta} V_{m,i}{}^{j}(x) \end{aligned} \tag{2.11}$$

where $e_{m}{}^{a}$ is the fünfbein, $\psi_{m}{}^{\underline{\alpha}}$ the N gravitino fields, $\omega_{m,ab}(x)$ the x-space Lorentz connection and $V_{mi}{}^{j}$ the Sp(N) gauge fields. In both N = 2 and N = 4 supergravity there is also a central charge gauge field which is not contained in $E_{M}{}^{A}$ or $\Omega_{M,A}{}^{B}$ and it is necessary to introduce a superspace Abelian gauge field $A_{M}$ with its corresponding field strength $F_{AB}$

$$F_{AB} = D_{A}A_{B} - (-)^{AB} D_{B}A_{A} + T_{AB}{}^{C}A_{C};\; A_{A} = E_{A}{}^{M}A_{M} \tag{2.12}$$

and Bianchi identity

$$\sum_{(ABC)} (D_A F_{BC} + T_{AB}{}^D F_{DC}) = 0 \tag{2.13}$$

The x-space central charge gauge field may then be identified as $A_m|_{\theta=0}$. The supersymmetry transformation rules for the gauge fields may be computed with the aid of the following formulae:

$$\begin{aligned} \delta E_M{}^A &= D_M \xi^A + \xi^B T_{BM}{}^A \\ \delta \Omega_{MA}{}^B &= \xi^C R_{CM,A}{}^B \\ \delta A_M &= \xi^A F_{AM} \end{aligned} \tag{2.14}$$

In (2.9) and (2.14) we have ignored field dependent tangent space and central charge gauge transformations which arise when one converts a $\xi^M$ transformation into a $\xi^A$ transformation. The N = 4 theory also has an additional five Abelian vector fields but they do not occur in the minimal representation and we shall defer discussion of them until the final section.

## 3 SUPERCURRENTS

The supercurrents may be constructed using on-shell Abelian supersymmetric Yang-Mills for both N = 2 and N = 4; we shall give the details for the N = 4 case, the N = 2 results being derivable in exactly the same fashion. N = 4 Yang-Mills in five dimensions is described by a scalar superfield $W_{ij}$ which is in the $\underset{\sim}{5}$ representation of Sp(4) ($W_{ij} = -W_{ij}$; $\eta^{ij}W_{ij} = 0$; $\bar{W}^{ij} = \eta^{ik}\eta^{jl}W_{kl}$). It satisfies the following equation:

$$D_{\alpha i} W_{jk} = \eta_{ij}\lambda_{\alpha k} - \eta_{ik}\lambda_{\alpha j} + \frac{1}{2}\eta_{jk}\lambda_{\alpha i} \tag{3.1}$$

where the $\lambda_{\alpha i}$ are the $\underset{\sim}{4}$ spin-1/2 fields. They vary into the field strength of the vector field:

$$D_{\alpha i}\lambda_{\beta j} = \frac{i}{2}\eta_{ij}(\gamma^{ab})_{\alpha\beta}F_{ab} - i(\gamma^a)_{\alpha\beta}\partial_a W_{ij} \tag{3.2}$$

All fields are on-shell (from (3.1)):

$$\Box W_{ij} = \partial\!\!\!/ \lambda_i = \partial^a F_{ab} = \partial_{[a}F_{bc]} = 0 \tag{3.3}$$

The supercurrent is (Howe & Lindström 1981)

$$J_{ij,kl} = W_{ij}W_{kl} - \frac{1}{20}(\eta_{ij}\eta_{kl} - 2\epsilon_{ijkl})W^{mn}W_{mn} \tag{3.4}$$

and transforms as a $\underset{\sim}{14}$ under Sp(4). The supercurrent has 128 + 128 components which is the same number as for the four-dimensional N = 4 conformal supercurrent (Bergshoeff et al. 1981) but the two are not identical. Indeed, if five-dimensional super Yang-Mills is reduced à la Scherk & Schwarz (1979) so that it becomes massive spin-1 with a central charge in four dimensions, then it is not possible to re-arrange the supercurrent into its conformal form. (This reduction requires that the original five-dimensional multiplet be doubled, but this does not alter the supercurrent.)

The supercurrent satisfies the following constraint:

$$(D_{\alpha i}J_{jk,lm})\ \underset{\sim}{40} = 0 \tag{3.5}$$

i.e., the $\underset{\sim}{14}$ can vary into a $\underset{\sim}{40}$ and a $\underset{\sim}{16}$, but only the latter is present. It couples to the linearized supergravity prepotential, V, via a Noether coupling:

$$I_{Noether} = \int d^5x\ d^{16}\theta\ J_{ijkl}V^{ij,kl} \tag{3.6}$$

so that V is also a $\underset{\sim}{14}$. From (3.5) the following gauge transformation leaves (3.6) invariant,

$$V_{ijkl} = D_\alpha{}^m\Lambda^\alpha{}_{ij,kl,m} \tag{3.7}$$

where $\Lambda$ is in the $\underset{\sim}{40}$ dimensional representation of Sp(4). Unlike the four-dimensional conformal case there is no simple field strength superfield for the supergravity fields in V, but the latter may easily be read off from the explicit component form of the supercurrent. Similar arguments apply in the N = 2 case and we find the following gravitational multiplets:

$$\begin{array}{ll} \text{dim. } 0 & h_{ab},\ A_a \\ \text{dim. } 1/2 & \psi_{a\alpha i}\ (\underset{\sim}{2},\underset{\sim}{4}) \\ \text{dim. } 1 & V_a{}^{ij}\ (\underset{\sim}{3},\underset{\sim}{10});\ S_{ij}\ (\underset{\sim}{3},\underset{\sim}{10});\ N_{ij}{}^{ab}\ (\underset{\sim}{1},\underset{\sim}{5}) \end{array} \tag{3.8}$$

$$\text{dim. } 3/2 \quad N = 2,\ \chi_{\alpha i}\ (\underset{\sim}{2});\ N = 4\ \chi_{\alpha i,jk}\ (\underset{\sim}{16})$$

$$\text{dim. } 2 \quad N = 2,\ C;\ N = 4,\ C_{ij,kl}\ (\underset{\sim}{14}) \qquad (3.8) \text{ cont.}$$

where the numbers in brackets indicate the Sp(N) representations of the fields for N = 2 and 4 respectively. $h_{ab}$ is the graviton field, $A_a$ the central charge gauge field, $\psi_{a\alpha i}$ the gravitino fields and $V_a{}^{ij}$ the Sp(N) gauge fields. The $N^{ab}$'s are non-gauge antisymmetric tensors. We have used geometrical dimensions so, for example, the real linearized graviton field is $1/\kappa\ h_{ab}$, but since $\kappa$ appears in the expansion of the vielbein, these are the appropriate dimensions to be used in the superspace torsions. We note that for N = 4 there is a dimension zero scalar, five-dimension 0 vectors and four-dimension 1/2 spinors absent which are present in the physical spectrum, while for both N = 2 and 4 the fields V, S, N, $\chi$ and C are auxiliary. Hence it is clear that these "minimal" multiplets cannot be used to write a Lagrangian; even in N = 2 where the physical fields are already present there is no possibility of writing terms for the auxiliary fields $\chi$ and C.

## 4 SUPERSPACE CONSTRAINTS

We shall now use the knowledge of the linearized supergravity multiplets (3.8) to derive the superspace constraints. The philosophy is simple: we place the covariant component fields in the torsions wherever they will fit on dimensional and symmetry grounds and declare all other components of the torsions to be zero, any arbitrary coefficients being adjusted by use of the Bianchi identities (2.8) and (2.13). We may also use conventional constraints to choose many torsions to vanish (we recall that conventional constraints are those that correspond to the freedom to redefine the vielbein and the connection). One advantage of five dimensions over four-dimensional conformal supergravity is now apparent, namely the fact that we do not need to worry about conformal invariance which has the consequence that no fields corresponding to Wess-Zumino gauges for super-Weyl transformations will appear in the torsions as they do in four-dimensional conformal supergeometry (Howe 1981 b). Furthermore, it turns out that the dimension zero and one-half constraints (together with the conventional constraints at dimension one) specify the geometry completely. Since there are no dimension zero covariant fields in (3.8) there can be no bilinears at dimension one-half in the torsions, so that the linearized constraints will be identical to the full non-linear ones. Hence the full non-linear multiplet corresponding to (3.8) can be straightforwardly computed using the Bianchi identities (although to work out the supersymmetry transformations of the high dimension fields $\chi$ and C the Bianchi's do not suffice; one must also use the Ricci identity (2.7) - indeed, C does not appear in the torsion or the curvature at $\theta = 0$).

We are now in a position to state the torsion constraints. At dimension zero there are no covariant supergravity fields in (3.8) but we can have constants:

$$T_{\alpha i\beta j}{}^{c} = -i\eta_{ij}(\gamma^{c})_{\alpha\beta};\ F_{\alpha i\beta j} = -i\eta_{ij}\eta_{\alpha\beta} \tag{4.1}$$

At dimension one-half there are again no covariant fields, hence

$$T_{\alpha i\beta j}{}^{\gamma k} = T_{\alpha ib}{}^{c} = F_{\alpha ib} = 0 \tag{4.2}$$

At dimension one we may choose as conventional constraints

$$T_{ab}{}^{c} = T_{a\beta(j}{}^{\beta}{}_{k)} = 0 \tag{4.3}$$

We then find

$$T_{a\beta j\gamma k} = -\frac{i}{2}\eta_{ij}(\gamma^{b})_{\alpha\beta}F_{ab} + i(\gamma_{a})_{\alpha\beta}S_{ij} + i(\gamma^{bc})_{\alpha\beta}M_{abcij}$$

$$M^{ab}_{ij} \equiv \frac{1}{2}\epsilon^{abcde}M_{cdeij} = \begin{cases} N^{ab}_{ij} + \frac{3}{8}\eta_{ij}F^{ab}, & N = 4 \\ \eta_{ij}N^{ab}, & N = 2 \end{cases} \tag{4.4}$$

where $F_{ab}$ is the central charge field strength, $S_{ij}$ symmetric, and $N^{ab}_{ij}$ is skew on a,b and i,j and Sp(4) traceless for N = 4. The remaining (dimension 3/2) torsion is

$$T_{ab\alpha i} = \Psi_{ab\alpha i} + 2(\gamma_{[a}\Psi_{b]})_{\alpha i} + (\gamma_{ab}\Psi)_{\alpha i} \tag{4.5}$$

where the $\Psi$'s are $\gamma$-traceless and are the irreducible parts of the covariantized gravitino field strength. The variations of the dimension one fields are given by

$$D_{\alpha i}F_{ab} = -T_{ab\alpha i}$$

$$D_{\alpha i}N_{abjk} = -\frac{3}{2}(\eta_{ij}\Psi_{ab\alpha k} - \eta_{ik}\Psi_{ab\alpha j} + \frac{1}{2}\eta_{jk}\Psi_{ab\alpha i})$$
$$+\frac{3}{4}\{(\gamma_a)_\alpha{}^\beta(\eta_{ij}\Psi_{b\beta k} - \eta_{ik}\Psi_{b\beta j} + \frac{1}{2}\eta_{jk}\Psi_{b\alpha i}) - a \leftrightarrow b\}$$
$$+\frac{1}{2}(\gamma_{ab})_\alpha{}^\beta(\chi_{\beta j,ki} - \chi_{\beta k,ji}),\ N = 4$$
$$D_{\alpha i}N_{ab} = \frac{1}{8}\Psi_{ab\alpha i} - \left(\frac{3}{4}(\gamma_a)_\alpha{}^\beta\Psi_{b\beta i} - a \leftrightarrow b\right) \tag{4.6}$$
$$+ (\gamma_{ab})_\alpha{}^\beta\chi_{\beta i},\ N = 2$$

$$D_{\alpha i}S_{jk} = -(\eta_{ij}\Psi_{\alpha k} + \eta_{ik}\Psi_{\alpha j}) + \chi_{\alpha i,jk},\ N = 4$$

$$D_{\alpha i}S_{jk} = \eta_{ij}(\frac{2}{3}\chi_{\alpha k} - \frac{3}{4}\Psi_{\alpha k}) + (j, \leftrightarrow k),\ N = 2$$

where $\chi_{i,jk}$ is in a $\underset{\sim}{16}$ of Sp(4) for N = 4. Thus the only new field appearing at dimension 3/2 is the $\chi$ field, as required. This completes the discussion of the superspace geometry since the curvatures are determined as functions of the torsion as we remarked in Sect. 2. We have now located all the fields of the multiplets (3.8) except for the C field, which has dimension two and arises in the variation of $\chi$. The gauge fields (fünfbein, gravitini, Sp(N) gauge fields and the central charge vector) are to be found in the superspace potentials, as we discussed in Sect. 2.

## 5 LAGRANGIAN REPRESENTATIONS

The supergravity multiplets of (3.8), which correspond to the superspace constraints of the last section, are, as we have already remarked, not suitable for use in actions. In particular, these multiplets contain gauge fields corresponding to local Sp(N) which are of auxiliary dimension, and it is necessary to introduce compensating multiplets for these invariances. In superspace we can do this straightforwardly by splitting the connection into its SO(1,4) and Sp(N) parts. We denote the SO(1,4) × × Sp(N) connection by $\hat{\Omega}$, the SO(1,4) connection by $\Omega$ and the Sp(N) connection by $\Sigma$, so

$$\hat{\Omega}_{AB}{}^C = E_A{}^M\hat{\Omega}_{MB}{}^C = \Omega_{AB}{}^C + \Sigma_{AB}{}^C \tag{5.1}$$

where

$$\Omega_{A\underline{\beta}}{}^{\underline{\gamma}} = \delta_j{}^k \Omega_{A\beta}{}^\gamma$$
$$\Sigma_{A\underline{\beta}}{}^{\underline{\gamma}} = \delta_\beta{}^\gamma \Sigma_{Aj}{}^k \quad (\Sigma_{Ab}{}^c = 0) \tag{5.2}$$

The torsions and curvatures computed with respect to $\hat{\Omega}$ and $\Omega$ are related by

$$\hat{T}_{AB}{}^c = T_{AB}{}^c + \Sigma_{AB}{}^c - (-1)^{AB}\Sigma_{BA}{}^c$$
$$\hat{R}_{ABc}{}^d = R_{ABc}{}^d$$
$$\hat{R}_{ABk}{}^\ell = D_A\Sigma_{Bk}{}^\ell - (-1)^{AB}D_B\Sigma_{Ak}{}^\ell + T_{AB}{}^c\Sigma_{ck}{}^\ell \tag{5.3}$$
$$- i\Sigma_{Ak}{}^m\Sigma_{Bm}{}^\ell + (-1)^{AB} i\Sigma_{Bk}{}^m\Sigma_{Am}{}^\ell$$

the $\hat{T}$'s being the torsions of the last section. This is, of course, a trivial re-writing, but now fields previously contained in the Sp(N) connection will appear in the torsions of the Lorentz superspace, the local Sp(N) invariance appearing as an additional invariance of the constraints on $T_{AB}{}^C$. If we now impose constraints on $\Sigma_{AB}{}^C$ (or equivalently $T_{AB}{}^C$), this invariance will be reduced from 1/2N(N + 1) scalar superfields to a smaller multiplet, which is essentially the compensating multiplet. This procedure is similar to choosing "type (iii)" constraints (Gates et.al. 1980) and is also related to the pre-curvature method (Breitenlohner & Sohnius 1981).

To see how it works let us consider N = 2 and impose

$$\Sigma_{\alpha(ijk)} = 0 \rightarrow \Sigma_{\alpha ijk} = \eta_{ij}\lambda_{\alpha k} + \eta_{ik}\lambda_{\alpha j} \tag{5.4}$$

But under Sp(2) = SU(2), we have

$$\delta\Sigma_{\alpha i,kl} = -D_{\alpha i}L_{jk} - iL_i{}^m\Sigma_{\alpha m,jk} + i\Sigma_{\alpha ij}{}^m L_{mk} + i\Sigma_{\alpha ik}{}^m L_{mj} \tag{5.5}$$

Hence, (5.4) will be invariant under a restricted SU(2) transformation where the parameter satisfies

$$D_{\alpha(i}L_{jk)} = 0 \tag{5.6}$$

But (5.6) defines an irreducible multiplet - the linear multiplet. In flat space its components are a triplet of scalars, a doublet of spinors, a conserved vector and another scalar. In the present context this vector is no longer conserved but its divergence is related to the C field as in the four-dimensional N = 2 case (Breitenlohner & Sohnius 1981). Hence the Lorentz superspace will now contain a multiplet which is the sum of the minimal multiplet and the linear multiplet. Its components are: $(e_m{}^a, A_m; \psi_{m\alpha i}, \lambda_{\alpha i}; V_m^{ij}, S_{ij}, P, V'_a, N_{ab}; \chi_{\alpha i})$, where $\lambda_{\alpha i}$ is the doublet from the linear multiplet, P the scalar, $V'_a$ the vector whose divergence is C and the $V_m^{ij}$ are no longer gauge fields as they absorb the dimension zero triplet of the linear multiplet. This is a (48 +48) component multiplet which is the five-dimensional version of the four-dimensional N = 240 + 40 of de Wit & van Holten (1979) and Fradkin & Vasiliev (1979). The extra 8 + 8 components arise because five-dimensional N = 2 supergravity becomes four-dimensional N = 2 supergravity together with an N = 2 Yang-Mills multiplet upon dimensional reduction.

If we try to repeat the above procedure for N = 4 we encounter problems. The analogue of (5.4) would be

$$\Sigma_{\alpha(ijk)} = 0 \rightarrow \Sigma_{\alpha i,jk} = \lambda_{\alpha i,jk} + \eta_{ij}\zeta_{\alpha k} + \eta_{ik}\zeta_{\alpha j} \tag{5.7}$$

This looks encouraging because λ is a $\underset{\sim}{16}$-plet and hence could pair with χ, and $\zeta_{\alpha i}$ could be the missing physical fermion field. The residual Sp(4) invariance is again given by (5.6), but in N = 4 the multiplet defined by this constraint contains a spin 3 field of dimension three (Howe et al. 1981) so the resulting combination of supergravity and linear multiplets would not be satisfactory from the point of view of constructing Lagrangians. It is difficult to see any modification of the constraint (5.6) that would work; for example, any N = 4 multiplet has maximum spin ≥ 2, but there are no spin 2 fields of dimension one in the Lorentz superspace torsions. Indeed, it has been argued (Rivelles & Taylor 1981; see also the contribution by J.G. Taylor in this book) that there are no auxiliary fields for supergravity theories with N ≥ 3 in the absence of central charges, so we now turn to the possibility of including them.

To do this superspace needs to be enlarged by the introduction of an additional commuting co-ordinate y. Correspondingly, there will be an extra direction in the tangent space, so that a tangent space vector will now have components $(V_a, V_{\alpha i}, V_5)$, where $V_5$ is inert under the tangent space group which is still taken to be SO(1,4) × Sp(N). The minimal supergravity multiplet may still be described in this superspace if we identify

$$T_{AB}{}^{5} = F_{AB} \tag{5.8}$$

and impose

$$T_{5A}{}^{B} = T_{5A}{}^{5} = R_{5BC}{}^{D} = 0 \tag{5.9}$$

so that

$$D_5 T_{AB}{}^{C} = 0 \tag{5.10}$$

We split up the SO(1,4) × Sp(N) connection as before, but there will now be more flexibility in choosing constraints due to the presence of the central charge. Consider first N = 2. We again impose (5.4) so that the reduced SU(2) parameter satisfies (5.6). However, this does not as yet define an irreducible multiplet and a further constraint must be imposed. If we choose $D_5 L_{ij} = 0$ we will be back with the 48 + 48 already discussed, but we may instead impose the higher order constraint

$$D_5\lambda_{\alpha i} = -\tfrac{i}{4}D_{\alpha i}(D^{\beta j}\lambda_{\beta j}) - \tfrac{1}{2}\,(D^{\beta}_{i}\lambda_{\beta j} + i\lambda^{\beta}{}_{i}\lambda_{\beta j} - 6S_{ij})\lambda^{j}_{\alpha} \tag{5.11}$$

which reduces $L_{ij}$ to a scalar multiplet ($\underset{\sim}{1} + \underset{\sim}{3}$ dimension zero and dimension one scalars, a doublet of dimension one-half spinors). To see this, consider the one-form

$$P = E^a P_a + E^5 P_5 + iE^{\alpha i}\lambda_{\alpha i} \tag{5.12}$$

where

$$\begin{aligned} P_5 &= -\tfrac{i}{4}D^{\alpha i}\lambda_{\alpha i} \\ P_a &= -\tfrac{i}{4}D^{\alpha i}(\gamma_a)_{\alpha}{}^{\beta}\lambda_{\beta i} \end{aligned} \tag{5.13}$$

Then (5.11) ensures that

$$dP = 0 \rightarrow P = dA \tag{5.14}$$

Hence the resultant combined multiplet has components ($e_m{}^a$, $A_m$ A; $\lambda_{\alpha i}$, $\psi_{m\alpha i}$; $P_5$, $S_{ij}$, $B_{ij}$, $N_{ab}$, $V_m^{ij}$; $\chi_{\alpha i}$; C), where $B_{ij}$ is the triplet of scalars arising from $\Sigma_{5ij}$ and $V_m^{ij}$ are again non-gauge. This multiplet has 48 + 48 components and is the five-dimensional version of the alternative four-dimensional N = 2 40 + 40 of de Wit et al. (1980).

For N = 4 we can now impose the constraint (5.7) without having to go all the way up to spin 3. We can, in fact, choose additional constraints such that $L_{ij}$ will describe five off-shell Yang-Mills multiplets with a central charge (Sohnius et al. 1980). The (linearized) components of this multiplet are

$$\begin{array}{ll} \text{dim. } 0 & A\ (\underset{\sim}{1}),\ A_{ij}\ (\underset{\sim}{10}),\ A_{ij,kl}\ (\underset{\sim}{14}),\ A_{ab}^{ij}\ (\underset{\sim}{5}) \\ \text{dim. } 1/2 & \lambda_{\alpha i,jh}\ (\underset{\sim}{16}),\ \xi_{\alpha i}\ (\underset{\sim}{4}) \\ \text{dim. } 1 & B\ (\underset{\sim}{1}),\ B_{ij}\ (\underset{\sim}{10}),\ B_{ij,kl}\ (\underset{\sim}{14}) \end{array} \tag{5.15}$$

where the $A_{ab}$'s are gauge antisymmetric tensors (describing spin 1 in five dimensions) and A and $\zeta_{\alpha i}$ the other missing physical fields. $\lambda_{\alpha i,jk}$ and $A_{ij,kl}$ are of the right dimension to pair with $\chi_{\alpha i,jk}$ and $C_{ij,kl}$ of (3.8) whilst $A_{ij}$ gets absorbed into the Sp(4) gauge fields. Hence the multiplet formed by combining (3.8) with (5.15) is a 208 + 208 component multiplet which is a candidate for a Lagrangian multiplet.

## 6 CONCLUDING REMARKS

We have seen that for both N = 2 and N = 4 supergravities in five dimensions there are minimal Poincaré multiplets which have 40 + 40 and 128 + 128 components, respectively. Lagrangian multiplets for N = 2 may be constructed by combining the minimal multiplet with either a linear multiplet or a centrally charged scalar multiplet, but for N = 4 there seems to be no simple way of finding a Lagrangian multiplet without a central charge. There is a candidate compensating multiplet in the centrally charged case for N = 4, however, and it remains to be seen whether or not one can construct a Lagrangian (Howe 1981 c). There is also the question of consistency in the presence of a central charge at the non-linear level.

If this set of auxiliary fields does work, it is interesting to note that it would not yield the auxiliary fields for four-dimensional N = 4 supergravity as one would be unable to separate off the additional four-dimensional Yang-Mills multiplet because of the presence of the $\underset{\sim}{5}$

antisymmetric tensors. Therefore we unfortunately do not gain any insight into the four-dimensional N = 4 Yang-Mills problem, although the fact that one might only be able to go off-shell with more than one physical multiplet is perhaps worthy of note.

## REFERENCES

Bergshoeff, E., de Roo, M., van Holten, J.W., de Wit, B. & van Proeyen, A. (1980). In Supergravity and Superspace, eds S.W. Hawking & M. Roček. Cambridge: Cambridge University Press.
Bergshoeff, E., de Roo, M., & de Wit, B. (1981). Nucl. Phys., B182, 173.
Breitenlohner, P. & Sohnius, M. (1980). Nucl. Phys., B165, 483.
Breitenlohner, P. & Sohnius, M. (1981). Nucl. Phys., B178, 151.
Cremmer, E. (1980). In Supergravity and Superspace, eds S.W. Hawking & M. Roček. Cambridge: Cambridge University Press.
Cremmer, E., Ferrara, S. & Scherk, J. (1978). Phys. Lett., 74B, 61.
Cremmer, E. & Julia, B. (1979). Nucl. Phys., B159, 141.
Cremmer, E., Scherk, J. & Schwarz, J. (1979). Phys. Lett., 84B, 83.
de Wit, B. & van Holten, J.W. (1979). Nucl. Phys., B158, 130.
de Wit, B., van Holten, J.W. & van Proeyen, A. (1980). Nucl. Phys., B167 186.
Dragon, N. (1979). Z. Phys., C2, 29.
Ferrara, S., Scherk, J. & Zumino, B. (1977) Phys. Lett., 66B, 35.
Fradkin, E. & Vasiliev, M.A. (1979). Phys. Lett., 85B. 47
Freedman, D.Z. & Schwarz, J. (1978). Nucl. Phys., B137, 333.
Gates, S.J., Stelle, K.S. & West, P.C. (1980). Nucl. Phys., B169, 347.
Howe, P. (1981a). Phys. Lett., 100B, 389.
Howe, P. (1981b). CERN preprint TH. 3117.
Howe, P. (1981c). Lagrangian for N = 4 supergravity in five dimensions, in preparation.
Howe, P. & Lindström, U. (1981). Phys. Lett., 103B, 422.
Howe, P., Stelle, K.S. & Townsend, P.K. (1981). CERN preprint TH. 3098 to be published in Nucl. Phys. B.
Rivelles, V.O. & Taylor, J.G. (1981). Phys. Lett., 104B, 131.
Scherk, J. & Schwarz, J.H. (1979). Nucl. Phys., B153, 61.
Sohnius, M., Stelle, K.S. & West, P.C. (1980). In Supergravity and Superspace. eds S.W. Hawking & M. Roček. Cambridge: Cambridge University Press.
Wess, J. & Zumino, B. (1978). Phys. Lett., 79B, 394.

P. van Nieuwenhuizen
Institute for Theoretical Physics
State University of New York at Stony Brook
Stony Brook, Long Island, New York 11794

At the conference work was reported done in collaboration with E. Bergshoeff, M. de Roo and B. de Wit (sections 3,4) and B. de Wit and A. van Proeyen (section 2). To avoid duplication of articles which are (to be) published in the literature, we will concentrate below on paedagogical reviews and introductions into the subjects and refer to the articles for the working out.

In section 2, we quantize supergravity in d=11 dimensions. For some time there have been suggestions that without auxiliary fields (these are still unknown in d=11) one would need not only the familiar four-ghost couplings of d=4, but also six-ghost couplings and higher. Our results put to rest these nightmares: the special properties of d=11 supergravity terminate the ghost couplings at the four-ghost level, just as in d=4. From this point of view one has an analogy between (the auxiliary fields of ?) d=11 and d=4. If one would know the auxiliary fields of d=11, one could obtain those of the N=8 model in d=4 by what is called dimensional reduction. Let us recall why auxiliary fields are important. Only with them one can find representations in terms of fields of the local gauge algebra; without them this is only true on-shell.

After constructing in section 3 (by dimensional reduction and the Noether method) the N=1 gauge action and the supersymmetric Yang-Mills action in d=10, we use in section 4 the method of currents to find information about the d=10 auxiliary fields. The usual approach is to consider a matter system which is: (i) massless to avoid central charges acting on the matter fields and hence on the matter currents and (ii) on-shell in order to have a closed algebra. Its currents (the energy-momentum tensor

$T_{\mu\nu}$ and others which follow by successively varying) are coupled to fields, and by requiring this linearized action to be invariant, one finds the global linearized supersymmetry transformation rules of supergravity (the nonlinear extensions then follow from the Noether method[1]). If the currents are gauge-invariant, one will not find central charges acting on the gauge fields off-shell. Thus one obtains an ordinary Clifford algebra and hence a finite representation.

Let us explain this procedure by assuming that we know already the complete action of a matter system coupled to supergravity. This action reads

$$\mathcal{L} = \mathcal{L}_G(g) + \mathcal{L}_M(m,g) \;, \quad \delta g = \delta g(g) \;, \quad \delta m = \delta m(m,g)$$

Expanding the pure gauge action $\mathcal{L}_G$ and the coupling terms $\mathcal{L}_M$ into terms with 2, 3, ... fields one has

$$\mathcal{L}_G(g) = \mathcal{L}^2_{gg} + \mathcal{L}^3_{ggg} + \dots$$

$$\mathcal{L}_M(m,g) = \mathcal{L}^2_{mm} + \mathcal{L}^3_{mmg} + \mathcal{L}^3_{mgg} + \dots$$

The transformation rules can be expanded likewise

$$\delta g = \delta^0 + \delta^1(g) + \delta^2(g) + \dots \;, \quad \delta m = \delta^1(m) + \delta^2(m,g) + \delta^2(m,m) + \dots$$

Because we have (assumed that we know) the auxiliary fields, $\delta_g$ is independent of m. If we consider global parameters the $\delta^0$ terms (such as $\delta\psi_\mu = \partial_\mu \varepsilon$) vanish. If the free matter system is on-shell, i.e., the free field equation for the fields m due to $\mathcal{L}^2_{mm}$ are satisfied, there are no terms linear in m in $\mathcal{L}_M$ (m, g=0). For example, in the globally supersymmetric matter action in flat space, there are no terms linear in, for example, the auxiliary field F. Thus, in this case, all matter couplings involve at least two m-fields and $\mathcal{L}^3_{mgg}=0$. The invariance of the whole action under the complete transformation rules means in particular that in the varied action the terms with two m fields and

and one g field cancel. Where do these variations come from? They can only come from $\mathcal{L}^3_{mmg}$ if one uses $\delta^1(m)$ and $\delta^1(g)$ because the other candidate, $\mathcal{L}^2_{mm}$ yields a variation proportional to the free m-field equations which are zero. Thus $\mathcal{L}^3_{mmg}$ is invariant under global linearized transformations. The coefficients of the fields g are the (Noether) currents mentioned before, and we now see how the method of currents works: varying the known currents under $\delta^1(m)$, we can deduce what $\delta^1(g)$ must be in order that $\int \mathcal{L}^3_{mmg}$ be invariant.

For didactical reasons we first work out the current analysis for the massless Wess-Zumino system, both with and without improved currents. The massive case (as well as the massless spin (½ , 1) case) was considered long ago by Ferrara and Zumino, who could have found in this way the auxiliary fields (S, P, $A_m$). After this didactical introduction (the results of which seem strangely enough not available in the literature but were also obtained at this conference by (at least) Townsend, Stelle, West, Sohnius) we will compare with the results of a similar analysis in d=10.

## 1. SUPERGRAVITY IN THE HIGHEST POSSIBLE DIMENSION: d=11

Beyond d=11 dimensions lies a world without supergravity(1) and beyond d=4 a world without conformal supergravity. In d=11 dimensions only N=1 pure ordinary supergravity exists(2). It contains a graviton, a gravitino and an antisymmetric 3-index photon $A_{\mu\nu\rho}$. The need for $A_{\mu\nu\rho}$ easily follows if one counts states

graviton (transverse and traceless) : $\frac{1}{2}\, 9 \times 10 - 1 = 44$

gravitino (transverse and $\gamma\cdot\psi = 0$ ) : $(9-1) \times 32 \times \frac{1}{2} = 128$

photon (tranverse) : $\binom{9}{3} = 84$

One could also try $A_{\mu 1 \cdots \mu 6}$ because $\binom{9}{6}=\binom{9}{3}$, but for not understood reasons there is no corresponding Lagrange field theory. (Global supersymmetry transformations of $e_\mu^{\ m}$, $\psi_\mu$, $A_{\mu 1 \cdots \mu 6}$ with the usual on-shell algebra exist, but at the point where one needs the coupling FFA(3) the 3-index photon works while the 6-index photon does not.)

Since in general for supergravity theories one finds that

high N low d = low N high d

it comes as no surprise that N=8 d=4 theory is equivalent to N=1 d=11 theory. Indeed, in d=11 the Dirac matrices have 32 components, and by piling the eight gravitini on top of each other, one can make a d=11 spinor: eight 4-component spinors = one 32-component spinor.

It can be shown that in d=11

(i) There are two inequivalent irreducible higher-than-one dimensional representations of (the finite group generated by) the Dirac matrices, which are in fact faithful. Our conventions are

$$\{\Gamma_\mu, \Gamma_\nu\} = 2\delta_{\mu\nu} \left(\mu,\nu = 1,11 \text{ and } \Gamma_\mu \text{ all hermitian}\right) \quad (1)$$

They differ by $\Gamma_{2n+1} \to -\Gamma_{2n+1}$, hence from a physicist point of view the Dirac matrices are unique (in fact, in any dimension).

(ii) The charge conjugation matrix $C\Gamma_\mu C^{-1} = \pm\, \Gamma_\mu^{\ T}$ satisfies in d=11

$$C\Gamma_\mu C^{-1} = -\Gamma_\mu^{\ T} \qquad (d=11) \qquad (2)$$

Clearly it is unique up to equivalence transformations. (These read $\Gamma_\mu \to U^{-1}\Gamma_\mu U$ and $C \to U^T C U$ with U unitary to keep $\Gamma_\mu$ hermitian. Clearly C is antisymmetric in all representations since it is so in a Majorana representation),

(iii) One can define Majorana spinors

$$\bar{\Psi}_D \equiv \psi^\dagger \Gamma(\text{time}) = \bar{\Psi}_M \equiv \psi^T C \qquad (d=11) \tag{3}$$

Similarly we will use later that in d=10

(i) There is only one irreps of the Dirac matrices.

(ii) There are two charge conjugation matrices, since in $C\Gamma_\mu C^{-1} = \pm \Gamma_\mu{}^T$ both signs are possible. This is possible, since we consider massless spinors[1]

(iii) both C's allow Majorana representations.[4]

In an explicit representation the meaning of this is clear. One can choose in d=11 a Majorana representation with $\Gamma(\text{time}) = I(16) \otimes \sigma_2$. Let "$\Gamma_5$" (or, better, $\Gamma_{11}$) be given by $I(16) \otimes \sigma_3$, then

$$C_-(d=10) = I(16) \otimes \sigma_2 \text{ and } C_+(d=10) = I(16) \otimes \sigma_1. \tag{4}$$

In the first case $\psi$ is real, in the second case the first 16 components have opposite reality properties from the last 16 components (an anti-Majorana spinor)

(iv) The Weyl condition

$$\tfrac{1}{2}(1+\Gamma_{11})\psi = \psi \qquad (\text{hence } \Gamma_{11}\psi = \psi) \qquad (d=10) \tag{5}$$

or anti-Weyl condition $\Gamma_{11}\psi = -\psi$ puts in the special Majorana representation the first (last) 16 components equal to zero, and relates 16 components to the 16 remaining components in a general representation. Since $\Gamma_{11} = i\Gamma_1 \ldots \Gamma_{10}$ is real in the Majorana representation, spinors in d=10 can be both Majorana and Weyl.[5]

The action of d=11 supergravity reads

$$\mathcal{L} = -\frac{e}{2}R(e,\omega) - \frac{e}{2}\bar{\Psi}_\mu \Gamma^{\mu\rho\sigma} D_\rho\left(\frac{\omega+\hat{\omega}}{2}\right)\psi_\sigma - \frac{e}{48}F_{\mu\nu\rho\sigma}^2$$
$$-\frac{3}{4}De\left[\bar{\Psi}_\mu \Gamma^{\mu\alpha\beta\gamma\delta\nu}\psi_\nu + 12\bar{\psi}^\alpha\Gamma^\beta\psi^\gamma\right]\left(F_{\alpha\beta\gamma\delta} + \hat{F}_{\alpha\beta\gamma\delta}\right)$$
$$+ C\epsilon^{\mu_1\cdots\mu_{11}} F_{\mu_1\cdots\mu_4} F_{\mu_5\cdots\mu_8} A_{\mu_9\cdots\mu_{11}} \tag{6}$$

$$C = -i\sqrt{2}/(72\times 48) \quad \text{and} \quad D = \sqrt{2}/(8\times 36)$$

The photon curl $F_{\mu\nu\rho\sigma}$ equals $\partial_\mu A_{\nu\rho\sigma}$ + 23 terms, and $\Gamma^{\alpha\beta\gamma}$ has strength one. The action is invariant under

$$\delta e_\mu{}^m = \frac{1}{2}\bar{\epsilon}\Gamma^m\psi_\mu \;, \; \delta A_{\mu\nu\rho} = E\bar{\epsilon}\Gamma_{[\mu\nu}\psi_{\rho]} \text{ with } E = -\frac{\sqrt{2}}{8}$$
$$\delta\psi_\mu = D_\mu(\hat{\omega})\epsilon + D\left(\Gamma^{\alpha\beta\gamma\delta}{}_\mu - 8\delta^\alpha_\mu\Gamma^{\beta\gamma\delta}\right)\epsilon\,\hat{F}_{\alpha\beta\gamma\delta} \qquad (7)$$

The symbol [ ] means antisymmetrization with strength one, and $\hat{\omega}$ is the supercovariant spin connection (in form equal to its d=4 counterpart) but $\omega$ is the solution of its own field equation when one replaces in the action all $\hat{\omega}$ by $\omega$ and solves according to Palatini. Clearly

$$\omega_{\rho mn} = \hat{\omega}_{\rho mn} + \alpha\,\bar{\psi}^\alpha\,\Gamma_{\alpha\rho mn\beta}\psi^\beta \qquad (8)$$

The constant $\alpha$ follows without much work by taking the completely antisymmetric part of

$$\bar{\psi}_\mu\Gamma^{\mu\rho\sigma}\Gamma^{mn}\psi_\sigma = \bar{\psi}_\mu\Gamma^{\mu\rho\sigma mn}\psi_\sigma +$$
$$+ 2\left(\bar{\psi}\cdot\Gamma\psi^m e^{n\rho} - \bar{\psi}\cdot\Gamma\psi^n e^{m\rho} - \bar{\psi}^m\Gamma^\rho\psi^n\right) \qquad (9)$$

and comparing with the completely antisymmetric part of $\omega_{\rho mn}$, using

$$\hat{\omega}_{\rho mn} = \omega_{\rho mn}(e) + \frac{1}{4}\left(\bar{\psi}_\rho\gamma_m\psi_n - \bar{\psi}_\rho\gamma_n\psi_m + \bar{\psi}_m\gamma_\rho\psi_n\right) \qquad (10)$$

Clearly $\alpha = -\frac{1}{8}$

In d=4 N=1 or N=2, one does not find $\frac{1}{2}(\omega+\hat{\omega})$ in the gravitino action, while here one finds that this choice is needed to avoid explicit four-fermion terms. Why? To answer this, let us consider the gravitino field equation. According to 1.5 order formalism we replace $\frac{1}{2}(\omega+\hat{\omega})$ by $\omega - \frac{1}{2}(\omega-\hat{\omega})$ in the gravitino action and need only consider the $\psi$-dependence of $-\frac{1}{2}(\omega-\hat{\omega})$. Because the four-fermion terms in $\bar{\psi}\psi\hat{F}$ do not appear symmetrically, one needs compensating four-fermion terms, and these are given by $-\frac{1}{2}(\omega-\hat{\omega})$. In the d=4 N=2 model, the four-fermion terms in $\bar{\psi}\psi\hat{F}$ appear symmetrically and there one finds $\omega$ in the spin connection. (In d=4, $\omega = \hat{\omega}$ by coincidence).

The gauge algebra closes only on the bosons, in which case it reads

$$[\delta_Q(\epsilon), \delta_G(\eta^\mu)] = \delta_Q(\eta^\lambda \partial_\lambda \epsilon)$$

$$[\delta_Q(\epsilon), \delta_L(\lambda^{mn})] = \delta_Q(\tfrac{1}{4}\lambda^{mn}\Gamma_{mn}\epsilon)$$

$$[\delta_M(\Lambda_{\mu\nu}), \delta_G(\eta^\mu)] = \delta_M(\eta^\lambda \partial_\lambda \Lambda_{\mu\nu} + \eta^\lambda{}_{,\mu}\Lambda_{\lambda\nu} + \eta^\lambda{}_{,\nu}\Lambda_{\mu\lambda}) \tag{11}$$

In addition one has the usual spacetime algebra for the G(eneral coordinate) and L(orentz) transformations. As always, the Q-Q commutator is the only one with field-dependent structure functions. One finds

$$[\delta_Q(\epsilon_1), \delta_Q(\epsilon_2)] = \delta_G(\tfrac{1}{2}\bar\epsilon_2\Gamma^\mu\epsilon_1) + \delta_Q(-\tfrac{1}{2}\bar\epsilon_2\Gamma^\mu\epsilon_1\psi_\mu)$$
$$+\delta_L\left(\tfrac{1}{2}\bar\epsilon_2\Gamma^\mu\epsilon_1\hat\omega_{\mu mn} + D\bar\epsilon_2\{\Gamma^{mn\alpha\beta\gamma\delta} + 24 e^{m\alpha}e^{n\beta}\Gamma^{\gamma\delta}\}\epsilon_1\hat F_{\alpha\beta\gamma\delta}\right)$$
$$+\delta_M\left(-\tfrac{1}{2}\bar\epsilon_2\Gamma^\sigma\epsilon_1 A_{\sigma\mu\nu} + \tfrac{1}{3}E\,\bar\epsilon_2\Gamma_{\mu\nu}\epsilon_1\right) \tag{12}$$

The terms with $\xi^\mu = \frac{1}{2}\bar\varepsilon_2\Gamma^\Gamma\varepsilon_1$ are expected, they always appear and are due to the fact that P-gauge transformations and general coordinate transformations differ.[6]

Rather interesting are the last terms in $\delta_L$ and $\delta_M$. Already from $\delta e_\mu{}^m$ and from $\delta A_{\mu\nu\rho}$ one finds signals that there is a kind of algebra $\{Q, Q\} \sim \Gamma_m P^m + \Gamma_{mn}A^{mn}$, and this agrees with the above local (Q,Q) commutator. No $\Gamma_{mn}M^{mn}$ term can appear in d=11 since in d=11 no de-Sitter supergravity seems possible so that $A^{mn}$ must be interpreted as a photonic (central) rather than a Lorentz charge. However, if one starts from a Poincaré algebra with $[P_m, P_n] = 0$ (and hence $[Q,P] = 0$ according to Jacobi identities) the Haag-Lopuszanski-Sohnius theorem (of d=4! ) says that one cannot have $A^{mn}$ in the {Q,Q} bracket (in de-Sitter spaces one can find a $\{Q,Q\} \sim \Gamma_{mn}M^{mn}$ term, but this would lead to a $\bar\varepsilon_2\Gamma_{mn}\varepsilon_1$ term in $\lambda^{mn}$, not in $\Lambda_{\mu\nu}$). Perhaps one should redo the HLS theorem in d=11.

The $\hat F$ term in $\delta\psi$ is the d=11 echo from a similar "non-geometrical" term in the d=4 N=2 model.[7] In fact, as shown by the group-manifold people[8], this term is very geometrical: it comes from the relation between general (group-manifold) coordinate transformations and pure gauge transformations

$$\delta_G h^A = (D\xi)^A + \xi^B R_B{}^A . \tag{13}$$

Since as a result of solving the field equations in the group-manifold the Q-curvature $R(Q) = R(Q)_{mn} V^m \wedge V^n + \gamma^n (\Gamma^{kl} F_{kl})\psi \wedge V_n + \ldots$ ,projection onto x-space reproduces the $\delta\psi \sim F\varepsilon$ term. It is a challenge to the group-manifold approach to explain the geometry of d=11.

There exist speculations that the underlying super algebra in d=11 is Osp(1/32). The reason is that Sp(32) has as generators $\Gamma_m$, $\Gamma_{mn}$, $\Gamma_{mnrst}$ which should then correspond to vielbein, spin connection and ... six-index photon. However, as already stressed, there does not exist a non-contracted (de-Sitter) theory, so that the true group might be a non-semisimple superalgebra[4].

## 2. QUANTIZATION OF 11-DIMENSIONAL SUPERGRAVITY

In d=4 dimensions, the auxiliary fields of the N=1 model are $S, P, A_m$. Quantizing this model with a closed gauge algebra, i.e., in the presence of $S, P, A_m$, and eliminating $S, P, A_m$ from the quantum action, three new features arise

(i) a 4-ghost coupling is produced in the action

(ii) a 3-ghost term is found in the BRS gravitino law

(iii) a 4-ghost term is found in the BRS law for the Lorentz ghost field.

Historically, the correct BRS laws for theories with open gauge algebra were found first, and at least in the example of the N=2 model this led to a set of auxiliary fields. Although there are various sets of auxiliary fields for a given supergravity model, eliminating them should always lead to the same result, namely the result as obtained from requiring BRS invariance of the quantum action.

For d=11 supergravity the auxiliary fields are not known, and the question arises whether the correct BRS laws for this theory with open algebra are as in d=4, or whether for example six - and higher ghost couplings are needed in the action. In principle, such higher ghost couplings are needed, and it were only the particular properties of the d=4 supergravity transformation laws which prevented them . So let us first see how things work out, and then interpret the results.

If a classical action $I(c\ell)$ is invariant under classical gauge transformations $\delta\phi^j = R^j{}_\alpha \xi^\alpha$ (where $\xi^\alpha$ are the local parameters) with a closed gauge algebra (see (3)), the correct quantum action reads

$$I(qu) = I(cl) + \tfrac{1}{2} F_\alpha \gamma^{\alpha\beta} F_\beta + C^{*\alpha} F_{\alpha,j} R^j{}_\beta C^\beta \qquad (1)$$

where $F_\alpha$ are the gauge choices and $F_{\alpha,j}$ is the right derivative with respect to $\phi^j$. This action is invariant under the following BRS rules

$$\delta\phi^i = R^i{}_\alpha C^\alpha \Lambda \,, \quad \delta C^{*\alpha} = \Lambda F_\beta \gamma^{\beta\alpha} \,, \quad \delta C^\alpha = -\tfrac{1}{2} f^\alpha{}_{\beta\gamma} C^\gamma \Lambda C^\beta \qquad (2)$$

The $f^\alpha{}_{\beta\gamma}$ are the field-dependent structure functions of the classical gauge algebra and $\gamma^{\beta\alpha}$ is independent of $\phi^j$ (see below, however). For example, $\tfrac{1}{2} F_\alpha \gamma^{\alpha\beta} F_\beta = \tfrac{1}{4} \bar\psi \cdot \gamma \not\partial \gamma \cdot \psi$ with $F_\alpha = \gamma\cdot\psi$ . The BRS rules are nilpotent on $\phi^j$ (see (3)) and $C^\alpha$ (see (6) and (7)) when the gauge algebra closes (but on

$C^{*\alpha}$ only if one introduces a new ghost $d^{\alpha}$ which is BRS inert[9].

If the classical gauge algebra only closes on-shell, one has

$$[\delta(\zeta^{\beta}),\delta(\xi^{\alpha})]\phi^{i} = \delta(f^{\gamma}{}_{\alpha\beta}\zeta^{\beta}\xi^{\alpha})\phi^{i} + I(cl)_{,j}\,\eta^{ij}{}_{\alpha\beta}\zeta^{\beta}\xi^{\alpha} \qquad (3)$$

where $\eta$ is called the nonclosure function. Varying I(qu) in (1), the only variations which no longer cancel are $\delta(R^{j}{}_{\beta}C^{\beta})$, thus the double BRS variation of $\phi^{j}$. One is left according to (3) with a term proportional to $I(c\ell)$ times $\eta$, and hence one cancels this variation by adding a term to $\delta\phi^{j}$ proportional to $C^{*}\eta^{kj}C\Lambda C$. Let us call this term $\delta\phi^{j}$(extra). Since for $\eta=0$ according to (1) one obtains the correct ghost action by varying the gauge fixing term and sandwiching the result between an antighost and a ghost, it seems a good starting point to do the same in the case of open algebras. In this way one finds in the case of an open gauge algebra an extra four-ghost coupling[10]

$$I(4\text{-ghost}) = -\tfrac{1}{4}C^{*\alpha}F_{\alpha,i}C^{*\beta}F_{\beta,j}\eta^{ij}{}_{\gamma\delta}C^{\delta}C^{\gamma}(-)^{i+\gamma} \qquad (4)$$

Varying $I(c\ell)$ + I(4-ghost) under (2) plus $\delta\phi^{j}$ (extra), one finds four new variations which do not a priori cancel

(i) $\delta\phi^{j}$(extra) in I(2-ghost), (ii) $\delta\phi^{j}$ in I(4-ghost)

(iii) $\delta\phi^{j}$(extra) in I(4-ghost), (iv) $\delta C$ in I(4-ghost)

As in d=4, also in d=11, the nonclosure function $\eta^{ij}_{\alpha\beta}$ depends only on vielbeins, and again the indices i,j refer to gravitini, while $\alpha,\beta$ refer to local supersymmetry. Hence (4) is independent of gravitini, while $\delta\phi^{j}$ (extra) is nonvanishing only when j refers to the gravitino. Consequently, the contribution from (iii) vanishes. The contributions from (i) and (iv) are proportional to $\eta$, while those from (ii) contain terms proportional to $\delta\eta$. Since according to (3) $\eta$ has something to do with the gauge commutator of $\phi^{j}$, $\delta\eta$ is expected to appear in triple variations of $\phi^{j}$. Hence one is led to consider triple Jacobi identities[10]

$$\delta(\zeta)\,[\delta(\eta),\delta(\xi)] - [\delta(\eta),\delta(\xi)]\,\delta(\zeta) + \text{cyclic} = 0 \tag{5}$$

Evaluating these triple Jacobi identities and using $\delta(I,_k R^k{}_\alpha) \equiv 0$ to express $I,_{k\ell}$ into $I,_k$ , one finds the following identity

$$R^j{}_\lambda A^\lambda + I(c\ell)_{,i}\, B^{ij} = 0 \tag{6}$$

The function $A^\lambda$ is already found in the case of closed gauge algebras where it appears as the double BRS variation of the ghost $C^\alpha$. It is given by

$$A^\lambda = \left(-f^\lambda{}_{\gamma\delta} f^\delta{}_{\beta\alpha} + f^\lambda{}_{\gamma\beta,k} R^k{}_\alpha\right)\left(C^\alpha \Lambda C^\beta C^\gamma{}_{(-)}{}^\gamma\right) \tag{7}$$

The function $B^{ij}$, on the other hand, is precisely what is left of $\delta I(qu)$ (if one replaces the parameters $\xi, \eta, \zeta$ by three ghosts times $\Lambda$).

$$\delta I_{\text{total}} = -\frac{1}{4} C^{*\alpha} F_{\alpha,i}\, C^{*\beta} F_{\beta,j}\, B^{ji}{}_{(-)}{}^i \tag{8}$$

It would be gratifying if B would vanish, but things are more subtle.(11) Let us first restrict $\lambda$ by requiring that <u>on-shell</u> the gauge invariances are linearly independent: $R^i{}_\lambda \xi^\lambda = 0 \leftrightarrow \xi^\lambda = 0$. Any other gauge invariances (such as $\delta S = P$, $\delta P = -S$ (or $\delta S = \frac{1}{4}\bar{\epsilon}\gamma\cdot R$, $\delta\psi_\mu = \frac{1}{6} S\gamma_\mu\epsilon$) in the N=1 d=4 model) need not be fixed in the quantization process, since they are not needed to eliminate ghosts from the classical theory. (In the presence of such local gauge invariances, the kinetic matrix is still non-singular and one can still find the propagators). Under this restricted set, (3) still holds since at most one has to transfer some of these trivial gauge invariances from the first to the second term on the right hand side of (3).

Hence, on-shell $A^\lambda = 0$, so that $A^\lambda$ is proportional to $I(c\ell)_{,j}$. Extracting a factor $I(c\ell)_{,\ell}$ from (6), the remainder is either a gauge transformation of $\phi^j$ (because $I(c\ell)_{,\ell} R^\ell{}_\alpha \equiv 0$ vanishes identically) or it is proportional to $I(c\ell)_{,k}$ in such a way that upon multiplication by $I(c\ell)$, everything vanishes. Thus

$$(-)^{i\ell} R^{i}{}_{\lambda} A^{\lambda\ell} + B^{\ell i} = R^{\ell}{}_{\lambda} X^{\lambda i} + I(c\ell)_{,k} M^{k\ell i} \tag{9}$$

where

$$A^{\lambda} = I(c\ell)_{,\ell} A^{\lambda\ell} (-)^{\lambda\ell} \tag{10}$$

The function $\eta^{ij}_{\alpha\beta}$ has the symmetry $(-)^{ij+1}$ if one interchanges the indices ij. This follows from (3) if one multiplies on the left by $I(c\ell)_{,i}$. (The last term in (3) cannot be proportional to $R^{i}{}_{\alpha}$ since it is nonzero only if i refers to the gravitino). Inspection of the general form of $B^{ij}$ shows that it, too, has this symmetry.

Due to special properties of the d=11 model, one can show that

(i) $A^{\lambda\ell}$ does not contain field equations because

(ii) $A^{\lambda}$ can at most produce one field equation.

Let us explain this . Consider the second term in (7) first. It is the variation of a structure function. Since the only field-dependent structure functions are found in the commutator of two supersymmetry transformations, $\gamma,\beta$ are supersymmetric. Now the only way in which the variation of the structure functions can produce a field equation at all, is when one varies $\hat{\omega}, \psi$ or $\hat{F}$, but not for example $A_{\mu\nu\rho}$. These $\tilde{\omega}$ and $\hat{F}$ are only found in $f^{\lambda}{}_{\alpha\beta}$ when $\lambda$ is local Lorentz, while $\psi$ appears when $\lambda$ is supersymmetric (if k = gen.coord.one produces $\partial_{\mu}\psi_{\nu}$ terms).

(iii) Consider now the first term in (7). The $f^{\lambda}{}_{\gamma\delta}$ may contain derivatives which act on $f^{\delta}{}_{\beta\alpha}$ (or vice-versa) namely if $\gamma$ or $\delta$ refers to general coordinate invariance and $\lambda$(and hence $\delta$ ) to supersymmetry, while $f^{\delta}{}_{\beta\alpha}$ is field dependent if $\alpha,\beta$ are supersymmetric. If $\delta$ is supersymmetric one can find $\partial_{\mu}\psi_{\nu}$ terms and if $\delta$= Lorentz, one can find $\partial\hat{\omega}$ or $\partial\hat{F}$ terms. Finally if $\delta$= gen. coord. or $\delta$=Lorentz, one finds only $\partial e$ or $\partial A$ terms and hence one cannot produce field equations in these cases. However, the $\partial\psi$ terms cancel against the $\partial\psi$ terms found under ii.[12] Thus the only possibility for $A^{\lambda}$ to contain a field equation is when $\lambda$= Lorentz,

(iv) The function $B^{ij}$ does not contain field equations.

Thus, using the symmetry of $B^{\ell j}$ and considering (9) first on-shell one finds $R^{\ell}{}_{\lambda}\,(X-A)^{\lambda j} + (-)^{\ell j}\, R^{j}{}_{\lambda}(X-A)^{\lambda\ell} = 0$. Then one considers (9) off-shell and one finds

$$M^{k\ell i} = 0 \tag{11}$$

The first relation yields upon multiplication by a field equation $I(\ell)_{,j}$

$$(X-A)^{\lambda j} = 0 \tag{12}$$

so that $\chi^{\lambda\ell}$ is nonzero only if $\lambda$ refers to local Lorentz invariance (see ii).

Finally substituting $B^{\ell j}$ from (9) back into (8), one can cancel (8) by adding to $\delta C^{\lambda}$($\lambda$ = Lorentz) an extra contribution which yields the opposite from (8) when substituted in I(2-ghost). It does not contribute to I(4-ghost) because that only depends on supersymmetry ghosts.

Thus, as in d=4, there are in d=11 only four-ghosts but no higher ghost couplings. This is consistent with (but no proof of the existence of) auxiliary fields of the kind enumerated in (i), (ii), (iii), in the beginning of this section.

One remark. Up to now, we had always to take $\gamma^{\alpha\beta}$ independent of $\phi^{i}$, but recently Nielsen put forward BRS rules for $\gamma^{\alpha\beta}(\phi)$.[13] One could take, for example in Yang-Mills theory, as a check $\gamma \sim [1+(A^{\alpha}_{\mu})^{2}]^{2}$ so that one should recover the case $\gamma \neq \gamma(\phi)$ by redefining $\tilde{F}_{\alpha} = F_{\alpha}[\,1+(A_{\mu}{}^{\alpha})^{2}]$. This is under study.[14]

## 3. THE MAXWELL-EINSTEIN SUPERGRAVITY IN d=10.[15]

By dimensional reduction of N=1 pure supergravity in d=11, one finds N=2 d=10 pure supergravity. The d=11 Majorana gravitino $\psi_\mu^a$(a = 1,32)splits up into two Majorana-Weyl gravitini, one of which is $\frac{1}{2}(1+\Gamma_{11})\,\psi_\mu = \psi^L{}_\mu$ while the other is $\frac{1}{2}(1-\Gamma_{11})\psi_\mu = \psi_\mu{}^R$. Moreover, two spinors are spun off as well: $\psi^L{}_{\dot{1}\dot{1}} = \lambda^L$ and $\psi^R_{\dot{1}\dot{1}} = \lambda^R$. The elfbein splits into a tenbein $e_\mu{}^m$, one vector $e_\mu{}^{11}$. The Lorentz group O(10,1) goes into O(9,1), its off-diagonal part being used up to gauge $e_{\dot{1}\dot{1}}{}^m = 0$, and a scalar $e_{\dot{1}\dot{1}}{}^{11} \equiv \phi$(a dot indicating where necessary a curved index). The 3-index photon $A_{\mu\nu\rho}$ splits into $A_{\mu\nu\rho}$ and $A_{\mu\nu\dot{1}\dot{1}} \equiv A_{\mu\nu}$.

In d=10 only N=1 matter exists, namely the supersymmetric Yang-Mills $(A_\mu\ ,\ \chi)$. To truncate the N=2 d=10 gauge theory to the N=1 d=10 case, one puts $\psi_\mu^R = A_{\mu\nu\rho} = e_\mu^{\ 11} = \psi_\mu^R = 0$. From $\delta e_\mu{}^m$ one then finds that $\varepsilon = \varepsilon_L$, and from $\delta A_\mu$ that $\chi = \chi_L$ while one also finds in similar way that $\lambda^L = 0$. All these truncations are consistent, i.e., the variation of vanishing fields again vanish. After Weyl rescaling of e and $\psi$ (followed by $\psi \to k_2\psi + k_3\lambda$, $\lambda \to k_4\lambda + k_5\psi$. One puts $k_5 = 0$ to avoid $\delta\lambda \sim \not{\partial}\varepsilon$ ) one finds in d=10

(i) all kinetic actions in canonical form,

(ii) no kinetic cross term between gravitino and $\lambda$.

Having cast the action in canonical form, one proceeds to cast the transformation rules in standard form. By adding to the supersymmetry transformation (as obtained from d=11) a local Lorentz rotation, and by redefining the supersymmetry parameter, one obtains (i) $\delta e_\mu{}^m = \frac{1}{2}\,\bar{\eta}\,\Gamma^m\psi_\mu$ and (ii) $\delta\psi_\mu = D_\mu\eta$ + more.

As a check on the results one may verify that the $\psi$- and $\lambda$-field equations are supercovariant. This could have been used to determine the four-fermion couplings. As in d=5 N=8, but unlike in N=1 d=11 or N=1 d= 4, the four-fermion terms in the action cannot be removed by using $\frac{1}{2}(\omega+\hat{\omega})$ and $\frac{1}{2}(F+\hat{F})$.

The final action is a sum of Einstein, Rarita-Schwinger, Dirac ($\lambda$), Klein-Gordon ($\phi$) and Maxwell $(-\frac{3}{4}\,\phi^{-3/2}F^2_{\alpha\beta\gamma})$actions, plus a kind of Noether-type couplings $\bar{\psi}_\mu\,\not{\partial}\phi\,\Gamma^\mu\lambda$, $\bar{\psi}\psi F$ and $\bar{\psi}\psi\lambda$. Most of the transformation laws are of familiar form: $\delta\phi \sim \bar{\eta}\lambda$, $\delta\lambda \sim \not{\partial}\phi\eta + \Gamma\cdot\hat{F}\,\eta$, $\delta A_{\mu\nu} \sim \bar{\eta}\psi + \bar{\eta}\lambda$, $\delta e_\mu{}^m = \frac{1}{2}\,\bar{\eta}\Gamma^m\psi_\mu$. In $\delta\lambda$ no $\not{\lambda}\lambda$ terms are present, but in $\delta\psi_\mu$ one finds $D_\mu(\hat{\omega})\eta + \not{\lambda}\lambda$ terms, complicated $\bar{\psi}\lambda$ terms and $\Gamma.\hat{F}$ terms.

The $\bar{\chi}\lambda\epsilon$ terms in $\delta\lambda$ (and $\lambda^3\psi$ couplings in the action) should be absent since they would violate the integrability conditions of a certain fermionic differential equation (namely that the $\lambda$ field equation be supercovariant).

Next one couples the d=10 N=1 gauge action to d=10 N=1 Yang-Mills action by means of the Noether procedure. Since there are scalar fields $\phi$, one adds wherever possible arbitrary functions of $\kappa\phi$. Hence one starts from

$$\mathcal{L}^{(0)} = -\frac{e}{4} F_{\mu\nu}^2 f^2(\phi) - \frac{e}{2} \bar{\chi} \Gamma^\mu D_\mu(\omega(e)) \chi$$

$$\delta A_\mu = \tfrac{1}{2}\bar{\epsilon}\Gamma_\mu \chi\, g(\phi) \quad \text{and} \quad \delta\chi = -\tfrac{1}{4}\Gamma^{mn} F_{mn} \epsilon\, h(\phi) \tag{1}$$

(A possible $k(\phi)$ in the Dirac action can be removed by rescaling $\chi$). The (Q,Q) commutator produces the correct translation provided gh=1, and the order $\kappa^0$ variations are cancelled if the Noether coupling is

$$\mathcal{L}^N = -\frac{e}{4}\left(\bar{\psi}_\mu \Gamma^{mn} F_{mn} \Gamma^\mu \chi\right) h \;, \quad h = f^2 g \tag{2}$$

Much of the Noether analysis goes like in d=4, but there are some interesting differences. Unlike in d=4, $\bar{\epsilon}\Gamma_{[\alpha}\,{}_{\beta}\psi_{\gamma]}$ is not a field equation in d=10, but the theory provides another way to cancel such terms: the curl of the 2-index photon $A_{\beta\gamma}$ has the same variation.

An interesting feature is the appearance of $A_\mu F_{\alpha\beta}\ F^{\mu\alpha\beta}$ terms ($F_{\alpha\beta}$ being the curl of $A_\alpha$ ) in the action which seem to break the Maxwell invariance $\delta A_\mu = \partial_\mu\Lambda$. However, in the (Q,Q) commutator one finds the expected two Maxwell transformations: $\delta_M^2$ acting only on $A_{\mu\nu}$, and $\delta_M^1$ acting both on $A_\mu$ and on $A_{\mu\nu}$

$$\begin{aligned}
[\delta_Q(\epsilon_1), \delta_Q(\epsilon_2)] &= \delta_G(\xi^\mu) + \delta_Q(-\xi\cdot\psi + \lambda \text{ terms}) \\
&+ \delta_L\left(\xi^\mu \hat{\omega}_{\mu mn} + \text{terms with } \hat{F}_{\alpha\beta\gamma} \text{ and } \bar{\chi}\lambda \text{ and } \bar{\chi}\chi\right) \\
&+ \delta_M^{(1)}(-\xi^\mu A_\mu) + \delta_M^{(2)}\left(-\xi^\nu A_{\nu\mu} - \tfrac{1}{\sqrt{2}}\phi^{3/4}\xi_\mu\right)
\end{aligned} \tag{3}$$

where $\delta_M^{(1)}(\Lambda)A_\mu = \partial_\mu\Lambda$ (and idem $\delta_M^{(2)}$) but also $\delta_M^{(1)}(\Lambda)A_{\mu\nu} = \frac{1}{\sqrt{2}}\kappa\Lambda\ F_{\mu\nu}$.

Even more interesting is that supersymmetry and Maxwell invariance no longer commute: The Maxwell transformation are no longer central charges, but all charges form one large irreducible group.(16)

$$[\delta_M^{(1)}(\Lambda), \delta_M^{(2)}(\Lambda_\mu)] = 0, \quad [\delta_Q(\epsilon), \delta_M^{(2)}(\Lambda_\mu)] = 0$$

$$[\delta_Q(\epsilon), \delta_M^{(1)}(\Lambda)] = \delta_M^{(2)}\left[\tfrac{1}{\sqrt{2}}\Lambda\phi^{3/8}\bar{\epsilon}\Gamma_\mu\chi\right] \tag{4}$$

This leads one to introduce a Maxwell covariant curl $F'_{\mu\nu\rho} = F_{\mu\nu\rho} - \frac{1}{\sqrt{2}} A_{[\mu}F_{\nu\rho]}$. If one replaces in the Maxwell-Einstein action $F_{\mu\nu\rho}$ by $F'_{\mu\nu\rho}$, all bare $A_\mu$ terms disappear, and a great simplification occurs. One can go further and introduce a $\hat{F}'_{\mu\nu\rho}$ which is equal to $F'_{\mu\nu\rho}$ plus those supercovariantizations of $F'_{\mu\nu\rho}$ which are only due to the pure gauge (but not the matter) fields. Although $\hat{F}'_{\mu\nu\rho}$ differs from $\hat{F}_{\mu\nu\rho}$ both are supercovariant, but only $F'_{\mu\nu\rho}$ is Maxwell covariant as well. In terms of $\hat{F}'_{\mu\nu\rho}$ the action simplifies again considerably.

A most interesting property of the coupling of the Maxwell system to supergravity in d=10 is its Weyl invariance. Usually one says that Weyl invariance of a theory with photons is only possible at d=4 (because photons must have vanishing Weyl weights to avoid $\partial_\nu\Lambda$ terms in the varied action). Supergravity improves this situation, because of the scalar function multiplying the Maxwell action. By taking the Weyl weight of $\phi$ appropriately and subsequently the weight of all other fields as well, one arrives at a Weyl invariant coupling. There is a coupling $F'_{\mu\nu\rho}\bar{\chi}\Gamma^{\mu\nu\rho}\chi$ which is only Weyl invariant if $F_{\mu\nu\rho}$ transforms homogeneously. This can be achieved by introducing an auxiliary field $\tau_{\mu\nu\rho}$. As explained in section 4, $\tau_{\mu\nu\rho}$ becomes equal to $F'_{\mu\nu\rho}$ only on-shell, but it is the first order formulation with $\tau_{\mu\nu\rho}$ which is Weyl invariant. All transformation rules are covariant under constant scale transformations (with $\epsilon$ having weight $-\frac{1}{2}$ as follows from the (Q,Q) commutator), but only $\delta A_\mu$ and $\delta\chi$ are Weyl covariant. It looks as if there were a Weyl invariant N=1 d=10 supergravity, and one had coupled to a compensating matter multiplet.(17) In N=1 and N=2 supergravity in d=4, one finds similar features: for example, in the N=2 d=4 superconformal model there is an auxiliary field $T_{mn}$ which couples in the ordinary model to the matter photon as

$$T_{mn}^2 + T_{mn}\,\partial_m B_n + (\partial_m B_n - \partial_n B_n)^2 \,. \tag{5}$$

and the couplings (the last two terms) are Weyl invariant.

## 4. CURRENTS AND AUXILIARY FIELDS IN d=10

There is no royal road to auxiliary fields, but rather several approaches may combine to a solution. Historically the first method was to simply look at some coupling of matter to supergravity and to "see" that certain bilinears in fields always appeared. In the d=4 Einstein-Maxwell system one found all over the place $\bar{\chi}\gamma_5\gamma_\mu\chi$ which strongly suggested an axial auxiliary field $A_\mu$.[18] (Without matter coupling its field equation is $A_\mu = 0$ and one could not see it. In general auxiliary fields only become excited in the presence of matter). Another method is to start with a larger symmetry group (usually conformal supergravity), again couple it to matter (in a conformally invariant way) and then fix the extra (conformal) symmetries by putting some of the matter and gauge fields equal to zero or one. In that gauge, gauge plus matter fields form an irreducible representation,[19] and if one originally knew the auxiliary fields (of conformal supergravity and of the conformal supersymmetric matter) then closure is not lost. Coupling the Wess- Zumino model (A, B, $\chi$, F, G) to the N=1 conformal supergravity fields ($e_\mu{}^m$, $\psi_\mu$, $A_\mu$, $b_\mu$), one fixes A=1, B=0, $\chi$=0, $b_\mu$ drops from the coupling, while (F, G, $A_\mu$ ) become the ordinary supergravity auxiliary fields (S, P, $A_\mu$).

One approach which is nowadays popular, is to use currents from a globally supersymmetric matter system, and associate to each current a field. The matter fields are on-shell, but the currents are not. The fields coupling to the currents are then the off-shell gauge fields with closed gauge algebra of global supersymmetry.[20] In d=11 dimensions only pure supergravity but no supersymmetric matter exists. However, in d=10 dimensions an N=1 globally supersymmetric Yang-Mills theory exists, and this system we will use to extract information about the auxiliary fields of the corresponding supergravity theory, namely N=1 d=10 supergravity. The matter fields are (we consider photons, not Yang-Mills bosons) the photon with 8 (transversal) states, and a spinor $\lambda$. As discussed in chapter 1, spinors in d=10 can be real and chiral, and thus have 32 x ½ x ½ = 8 states. Indeed, the sum of the Maxwell and Dirac actions is supersymmetric, the transformation laws being

$$\delta A_\mu = \tfrac{1}{2}\bar{\epsilon}\Gamma_\mu\chi \quad , \quad \delta\chi = -\tfrac{1}{4}\Gamma^{mn}F_{mn}\epsilon \qquad (1)$$

To explain the general method in a more familiar example, consider the Wess-Zumino model in d=4

$$\mathcal{L} = -\tfrac{1}{2}(\partial_\mu A)^2 - \tfrac{1}{2}(\partial_\mu B)^2 - \tfrac{1}{2}\bar{\chi}\partial\!\!\!/\chi \qquad (2)$$

$$\delta A = \tfrac{1}{2}\bar{\epsilon}\chi \ , \ \delta B = -\tfrac{i}{2}\bar{\epsilon}\gamma_5 B \ , \ \delta\chi = \tfrac{1}{2}\partial\!\!\!/(A - i\gamma_5 B)\epsilon$$

Note that all matter fields are on-shell, hence $\Box A = \Box B = \partial\!\!\!/\chi = 0$.

The currents we consider include the energy momentum tensor (since we want to couple to the graviton), obtained in the usual way by putting $\mathcal{L}$ into curved spacetime.

$$T_{\mu\nu}(A) = \partial_\mu A \partial_\nu A - \tfrac{1}{2}\delta_{\mu\nu}(\partial_\lambda A)^2 + \text{idem } B + \tfrac{1}{4}\bar{\chi}(\gamma_\mu\partial_\nu + \gamma_\nu\partial_\mu)\chi \qquad (3)$$

Using such on-shell identities as $(\partial\!\!\!/ A)\chi = \partial\!\!\!/(A\chi)$ one can factor out one derivative in $\delta T_{\mu\nu}$ and finds the Noether current $J^S_\mu$

$$\delta T_{\mu\nu} = -\tfrac{1}{8}\bar{\epsilon}\left(\gamma_{\mu\lambda}\partial_\lambda J^S_\nu + \gamma_{\nu\lambda}\partial_\lambda J^S_\mu\right) , \ J^S_\mu = \partial\!\!\!/(A + i\gamma_5 B)\gamma^\mu\chi \qquad (4)$$

Varying $J^S_\mu$ , the terms quadratic in A and B yield $T_{\mu\nu}$ while the AB terms are proportional to $\epsilon^{\mu\nu\rho\sigma}\ \partial_\rho A\partial_\sigma B$ which is the dual of the curl of an axial current $k^5_\sigma = A\ \overleftrightarrow{\partial_\sigma} B$. The variation of the $\bar{\chi}\partial\gamma\chi$ terms completes $T_{\mu\nu}$ , while the rest is again a total derivative, namely of the usual axial current $j^5_\mu = \bar{\chi}\gamma_5\gamma_\mu\chi$ .

$$\delta J^S_\mu = T_{\mu\nu}(\gamma_\nu\epsilon) - \tfrac{1}{8}\epsilon_{\mu\nu\rho\sigma}\left(\partial_\rho j^5_\sigma + 4i\partial_\rho k^5_\sigma\right)(\gamma_\nu\epsilon) + \tfrac{1}{4}\partial_\nu j^5_\mu(\gamma_5\gamma_\nu\epsilon) \qquad (5)$$

The current $j^5_\mu$ rotates back into $J^S_\mu$ as $\delta j^5_\mu = \bar{\epsilon}\gamma_5 J^S_\mu$. But the variation of the axial current $k^5_\sigma$

$$\delta k^5_\sigma = -\tfrac{i}{2}\bar{\epsilon}\gamma_5(A - i\gamma_5 B)\overleftrightarrow{\partial_\sigma}\chi \qquad (6)$$

would yield a new current. However, since only the curl of $k^5_\mu$ appears in $\delta J^S_\mu$ let us see whether its variation leads us back to known currents

$$\delta\left(\epsilon^{\mu\nu\rho\sigma}\partial_\rho k^5_\sigma\right) = -i\epsilon^{\mu\nu\rho\sigma}\bar{\epsilon}\gamma_5\left(\partial_\rho(A - i\gamma_5 B)\right)\partial_\sigma\chi \qquad (7)$$

Since we have at this point equal numbers of bosonic and fermionic currents (10-4 for $T_{\mu\nu}$, 4-1 for $j^5_\mu$ and 6-3 for (7), versus 16-4 for $J^S_\mu$ ), we expect that (7) can be expressed into $J^S_\mu$ alone. The most general possibility is

$$(7) = \bar{\epsilon}\left[A\epsilon_{\mu\nu\rho\sigma}\gamma_5 + B\delta_{\mu\rho}\delta_{\nu\sigma} + C\gamma_\mu\gamma_\rho\delta_{\nu\sigma} + D\gamma_\mu\gamma_\sigma\delta_{\nu\rho} - \mu\leftrightarrow\nu\right]\partial_\rho J^S_\sigma \tag{8}$$

Using the following new identiy

$$\gamma_{\mu\nu}(\partial_\lambda A)(\partial_\lambda\chi) = -\epsilon_{\mu\nu\rho\sigma}\gamma_5(\partial_\rho A)(\partial_\sigma\chi) + \left[\gamma_{\mu\sigma}(\partial_\sigma A)(\partial_\nu\chi) - \mu\leftrightarrow\nu\right] \tag{9}$$

which follows from $(\gamma_\mu\gamma_\sigma\gamma_\nu\gamma_\rho)\,\partial_\sigma\chi = \gamma_\mu[\gamma_\sigma,\gamma_\nu\gamma_\rho]\,\partial_\sigma\chi$ one finds a unique solution for (8).

Hence, defining $t^{\mu\nu} = \epsilon^{\mu\nu\rho\sigma}\,\partial_\rho k^5_\sigma$ , the currents close

$$\delta T_{\mu\nu} \text{ as in (4)}, \qquad \delta j^5_\mu = \bar{\epsilon}\gamma_5 J^S_\mu$$

$$\delta J^S_\mu = T_{\mu\nu}\gamma_\nu\epsilon - \frac{i}{2}t_{\mu\nu}\gamma_\nu\epsilon + \frac{1}{4}\partial_\nu j^5_\mu\gamma_5\gamma_\nu\epsilon - \frac{1}{8}\epsilon_{\mu\nu\rho\sigma}\partial_\rho J^5_\sigma\gamma_\nu\epsilon$$

$$\delta t_{\mu\nu} = \frac{i}{4}\left[\epsilon_{\mu\nu\rho\sigma}\gamma_5 + \gamma_{\mu\rho}\delta_{\nu\sigma} - \gamma_{\nu\rho}\delta_{\mu\sigma}\right]\partial_\rho J^S_\sigma \tag{10}$$

Instead of the canonical energy momentum tensor, we could have started with the improved $T^{imp}_{\mu\nu} = T_{\mu\nu} + \frac{1}{6}(\delta_{\mu\nu}\Box - \partial_\mu\partial_\nu)(A^2 + B^2)$, which follows from $\mathcal{L} = -\frac{1}{2}\sqrt{g}((\partial_\mu A)^2 - \frac{1}{6}RA^2)$ in the usual way. On shell $T^{imp}_{\mu\nu}$ is not only conserved but also traceless. By variation one finds the same result as in (4)

$$\delta T^{imp}_{\mu\nu} = -\frac{1}{8}\bar{\epsilon}\left(\gamma_{\mu\lambda}\partial_\lambda J^{S,imp}_\nu + \gamma_{\nu\lambda}\partial_\lambda J^{S,imp}_\mu\right) \tag{11}$$

where $J^{S,imp}_\mu = J^S_\mu + \frac{2}{3}\Gamma_{\mu\nu}\,\partial_\nu\,((A - i\gamma_5 B)\,\chi)$ is conserved and $\gamma\cdot J^{S,imp} = 0$ on shell. Its variation introduces only one new current(!)

$$\delta J^{S,imp}_\mu = T^{imp}_{\mu\nu}\gamma_\nu\epsilon + \frac{1}{4}\partial_\nu J^5_\mu\gamma_5\gamma_\nu\epsilon + \frac{1}{8}\epsilon_{\mu\nu\rho\sigma}\partial_\rho J^5_\sigma\gamma_\nu\epsilon \tag{12}$$

$$J^5_\mu = \frac{1}{3}\left(j^5_\mu - 4ik^5_\mu\right) \tag{13}$$

(Note the different sign in (13) and (5)). Already at this point the currents close

$$\delta J^5_\mu = \bar{\epsilon}\gamma_5 J^{S,imp}_\mu \tag{14}$$

which follows easily if one decomposes $\Gamma_{\mu\nu} = \Gamma_\mu\Gamma_\nu - \delta_{\mu\nu}$ in the improvement part of $J_\mu^{S,imp}$ and uses again $\Gamma_\mu \not\partial(A\chi) = \Gamma_\mu(\not\partial A)\chi$ . On the left hand side of (14), one finds (6) which is rewritten as

$$A\overleftrightarrow{\partial}_\mu \chi = \alpha(\not\partial A)\gamma^\mu \chi + \beta \tfrac{2}{3}\gamma_{\mu\nu}\partial_\nu(A\chi),\ \beta = \alpha - \tfrac{1}{2} \text{ and } \alpha = -1 \qquad (15)$$

Let us now first couple the multiplet of improved currents to fields

$$\mathcal{L} = \tfrac{1}{2} h_{\mu\nu} T_{\mu\nu}^{imp} + \tfrac{1}{2}\bar{\psi}_\mu J_\mu^{S,imp} + \tfrac{3i}{8} A_\mu J_\mu^5 \qquad (16)$$

Requiring the action to be invariant one finds from $\tfrac{1}{2}\,\delta h_{\mu\nu}\, T_{\mu\nu}^{imp} + \tfrac{1}{2}\ \bar{\psi}_\mu\gamma_\nu\varepsilon\, T_{\mu\nu}^{imp} = 0$

$$\delta h_{\mu\nu} = \tfrac{1}{2}\left(\bar{\epsilon}\gamma_\mu\psi_\nu + \bar{\epsilon}\gamma_\nu\psi_\mu\right) \qquad (17)$$

which is the usual linearized result if $g_{\mu\nu} = \delta_{\mu\nu} + h_{\mu\nu}$ . The terms with $J_\mu^{S,imp}$ yield

$$\delta\psi_\mu = (\partial_n h_{m\nu})\left(\tfrac{1}{4}\gamma^{mn}\epsilon\right) - \tfrac{3i}{4} A_\mu \gamma_5 \epsilon \qquad (18)$$

which is the linearized form of

$$\delta\psi_\mu = \left(D_\mu - \tfrac{3i}{4} A_\mu \gamma_5\right)\epsilon \qquad (19)$$

Finally, the terms with $J_\mu^5$ in the varied action yield

$$\delta A_\mu = -\tfrac{i}{3}\bar{\epsilon}\gamma_5 \not\partial\psi_\mu - \tfrac{i}{6}\epsilon_{\mu\nu\rho\sigma}\bar{\epsilon}\gamma_\nu\partial_\rho\psi_\sigma \qquad (20)$$

Since $J_\mu^5$ is conserved, $\not\partial\psi_\mu$ may be replaced by $\not\partial\psi_\mu - \partial_\mu\gamma\cdot\psi$ in the first term, and one finds that $\delta A_\mu$ is proportional to the gravitino field equation $R^\mu = \varepsilon^{\mu\nu\rho\sigma}\ \gamma_5\gamma_\nu\partial_\rho\psi_\sigma$

$$\delta A_\mu = -\frac{i}{2}\bar{\epsilon}\gamma_5\left(R_\mu - \frac{1}{3}\gamma_\mu \gamma\cdot R\right) \tag{21}$$

Thus we have found the linearized global parts of the Q-supersymmetry variations of the gauge fields of conformal supergravity.[1] The nonlinear extensions can be found by the order by order in $\kappa$ Noether method. In a similar way one may derive the global linearized parts of the other superconformal symmetries. In addition there are the local invariances $\delta h_{\mu\nu} = \partial_\mu \xi_\nu + \partial_\nu \xi_\mu$, $\delta h_{\mu\nu} = \lambda_{\mu\nu} = -\lambda_{\nu\mu}$, $\delta h_{\mu\nu} = \delta_{\mu\nu}\Lambda_D$, $\delta\psi_\mu = \partial_\mu \varepsilon$, $\delta\psi_\mu = \gamma_\mu \varepsilon$, $\delta A_\mu = \partial_\mu \Lambda_A$ corresponding to the conservation laws of the currents. These local invariances are the local field-independent parts of the gauge invariances of conformal supergravity.

One might wonder why no dilaton current and no accompanying dilaton field was found - after all, the dilaton field $b_\mu$ plays a role in conformal supergravity.[1] (Since conformal boosts only act on $b_\mu$, namely as $\delta b_\mu = \xi_\mu^K$, their absence in the list of local field-independent symmetries is understood). The reason is that in the complete Q-conformal law $\delta\psi_\mu = D_\mu\varepsilon - \frac{3i}{4} A_\mu\gamma_5 + \frac{1}{2} b_\mu\varepsilon$, the spin connection contains $b_\mu$, too, in such a way that the total dilaton content is $\delta\psi_\mu = \frac{1}{2}\gamma_\mu \not{b}\varepsilon$. Hence, in $\delta\bar{\psi}_\mu J_\mu^{S,imp}$ the $b_\mu$ terms cancel. In fact, also in the full nonlinear coupling of the scalar multiplet to conformal supergravity, $b_\mu$ cancels. This full coupling is given by the tensor calculus as

$$\mathcal{L} = F + \frac{1}{2}\bar{\psi}\cdot\gamma\chi + \frac{1}{2}\bar{\psi}_\mu \sigma^{\mu\nu}(A - i\gamma_5 B)\psi_\nu \tag{22}$$

for a conformal multiplet (A, B, $\chi$, F, G) which in our case is $\Sigma \times T(\Sigma)$ with $\Sigma = (A, B, \chi, F, G)$ and $T(\Sigma) = (F, -G, \not{D}^c\chi, \Box^c A, -\Box^c B)$. As one may show, $b_\mu$ drops out of $T(\Sigma)$ altogether. (Use that R contains a factor $6\,\partial.b$. Note: In eq.(16) page 305 of ref (1) one should replace $A_\mu^c$ by $A_\mu^{aux}$.) Thus the analysis of currents has indeed produced the fields and linearized (global and local field-independent) transformation rules.

If one would instead have started with a nonimproved $T_{\mu\nu}$, one would have found as gauge fields ($h_{\mu\nu}$, $\psi_\mu$, $A_\mu$, $C_{\mu\nu}$). This is not the usual set ($h_{\mu\nu}$, $\psi_\mu$, $A_\mu$, S, P), but a reshuffling: since $\partial_\mu \tau^{\mu\nu}=0$, the field $C_{\mu\nu}$ has a gauge invariance $\delta C_{\mu\nu} = \partial_\mu\lambda_\nu - \partial_\nu\lambda_\mu$ and has thus $6 - 3 = 3$ components. Similarly, $A_\mu$ is invariant under $\delta A_\mu = \partial_\mu \lambda$ and has also 3 states. Their sum is the same as the six fields ($A_\mu$,S,P). The set with $C_{\mu\nu}$ is further discussed by Sohnius and West, while their connection to currents is discussed by Townsend.

[We would like to add here that the trace $T_{\lambda\lambda} = -\frac{1}{2}\Box(A^2 + B^2)$ defines yet another multiplet ($T_{\lambda\lambda}$, $\frac{1}{2}\partial\!\!\!/\gamma\cdot J^S$, $-\frac{1}{8}\varepsilon^{\mu\nu\rho\sigma}\partial_\rho \ell_\sigma^5$). This is an antisymmetric tensor multiplet (A, $\chi$, $A_{\mu\nu}$) defined by

$$\delta A = \tfrac{1}{2}\bar{\epsilon}\chi\ , \ \delta\chi = \tfrac{1}{2}\partial\!\!\!/ A\epsilon + \tfrac{1}{2}\epsilon_{\mu\nu\rho\sigma}\partial_\nu A_{\rho\sigma}\gamma_5\gamma_\mu\epsilon\ , \ \delta A_{\mu\nu} = \tfrac{1}{2}\bar{\epsilon}\sigma_{\mu\nu}\chi$$

$$[\delta(\epsilon_1), \delta(\epsilon_2)]A_{\mu\nu} = \xi^\tau\partial_\tau A_{\mu\nu} + \partial_\mu\Lambda_\nu - \partial_\nu\Lambda_\mu\ , \ \Lambda_\mu = A_{\mu\tau}\xi^\tau \tag{23}$$

Indeed $\delta T_{\lambda\lambda} = \frac{1}{4}\bar{\varepsilon}\partial\!\!\!/\gamma\cdot J^S$ and the rest follows easily from $\delta(\gamma\cdot J^S) = T_{\lambda\lambda}\varepsilon + \frac{1}{2}(\gamma^{\mu\nu}\gamma_5\varepsilon)(\partial_\mu j_\nu^5)$ where $\ell_\mu^5 = j_\mu^5 + 2ik_\mu^5$ ].

Having seen how currents can lead one to auxiliary fields, let us now leave d=4 and return to d=10.

In d=10, we define $T_{\mu\nu} = F_{\mu\alpha}F_{\nu\alpha} - \frac{1}{4}(F)^2\delta_{\mu\nu} + \frac{1}{4}\bar{\chi}(\gamma_\mu\partial_\nu + \gamma_\nu\partial_\mu)\chi$ and $J_\mu^S = (\sigma\cdot F)\Gamma_\mu\chi$. The matter fields are on-shell: $\partial\!\!\!/\chi = \partial_\mu F^{\mu\nu} = 0$, and these two currents are conserved but not traceless. (Hence: non-improved currents). The variation of $T_{\mu\nu}$ is the same as in (4) or (11), as expected since it yields the $\delta\psi_\mu = D_\mu\varepsilon$ part. In

$$\delta J_\mu^S = T_{\mu\nu}\Gamma_\nu\epsilon + \tfrac{1}{4}\Gamma_{\mu\alpha\beta\gamma\delta}\epsilon\,\partial_\alpha V_{\beta\gamma\delta} - \tfrac{1}{48}\Gamma_{\mu\nu}\Gamma_{\alpha\beta\gamma}\epsilon\,\partial_\nu X_{\alpha\beta\gamma} \tag{24}$$

the first term is again universal as it yields $\delta e_\mu{}^m$, but two new currents are found

$$V_{\alpha\beta\gamma} = A_{[\alpha}F_{\beta\gamma]} + \tfrac{1}{12}\bar{\chi}\Gamma_{\alpha\beta\gamma}\chi$$
$$X_{\alpha\beta\gamma} = \bar{\chi}\Gamma_{\alpha\beta\gamma}\chi \tag{25}$$

There is always some ambiguity in what to choose as the new abstract current, but in this case the choice is strongly suggested by the fact that in the nonlinear action it is $X_{\alpha\beta\gamma}$ which couples to $\partial_{[\alpha}A_{\beta\gamma]}$, while $V_{\alpha\beta\gamma}$ does not rotate into new currents. For $X_{\alpha\beta\gamma}$ we took the only gauge independent current.

Clearly $X_{\alpha\beta\gamma}$ is the d=10 generalization of the axial current $j^5_\mu$ while $V_{\alpha\beta\gamma}$ resembles $k^5_\mu$ in the sense that only its curl is present in $J^S_\mu$. In d=4 we chose the curl $\tau^{\mu\nu} = \varepsilon^{\mu\nu\rho\sigma}\partial_\rho k^5_\sigma$ as an abstract new current which was then conserved. In d=10, however, we take $V_{\alpha\beta\gamma}$ as the new current which then has the gauge invariance $\delta V_{\alpha\beta\gamma} = \partial_{[\alpha}\Lambda_{\beta\gamma]}$. (We could also take $W_{\alpha\beta\gamma\delta} = \partial_{[\alpha}V_{\beta\gamma\delta]}$ as new current, and would then again constrain it such that the dual of W is conserved). The current $V_{\alpha\beta\gamma}$ rotates back into $J^S_\mu$ just as $j^5_\mu$ did in d=4. However, the current $X_{\alpha\beta\gamma}$ rotates into $X_{\mu\nu} \equiv F_{\mu\nu}\chi$. With $\Gamma_\mu\hat{\chi}_{\mu\nu} = 0$, $\hat{\chi}_{\mu\nu} = \chi_{\mu\nu} - \frac{1}{4}\Gamma_{[\mu}J^S_{\nu]} - \frac{7}{9x24}\Gamma_{\mu\nu}\Gamma\cdot J^S$ does not contain $J^S_\mu$ and is a new current. The on-shell condition $\partial\cdot J^S = 0$ is equivalent to $\Gamma_{\mu\nu}\not\partial\hat{\chi}_{\mu\nu} = 0$ so this is a constraint on $\chi_{\mu\nu}$( or $\hat{\chi}_{\mu\nu}$). If one varies $\chi_{\mu\nu}$ one gets a tremendous number of new currents (see ref(15)). It is an open question after how many steps the multiplet will close (it will always close eventually). In d=5 N=4, Howe and Lindström only found one new current in $\delta J^5_\mu$ , and after dimensional reduction to d=4, they found the multiplet of N=4 <u>conformal</u> supergravity.[21] This is interesting and disappointing because in d=5 no conformal supergravity exists, whereas one really would prefer to end up with the auxiliary fields of N=4 ordinary supergravity.

The currents we have obtained are the lowest dimensional ones, so the fields they couple to are the highest dimensional ones. The only fields which can occur quadratically in the action (like $S,P,A_m$ in d=4) are the fields which couple to $X_{\alpha\beta\gamma}$ and $V_{\alpha\beta\gamma}$ . Fields which couple to higher dimensional currents have too low dimension to appear quadratically (physical fields have also lower dimensions, but they can appear quadratically since they carry derivatives). So at this point, we should be able to say something about the action.

Coupling as follows:

$$\mathcal{L}=\tfrac{1}{2}h'_{\mu\nu}T_{\mu\nu}+\tfrac{1}{2}\bar{\psi}'_{\mu}J^{S}_{\mu}+\frac{3\sqrt{2}}{4}t_{\alpha\beta\gamma}V^{\alpha\beta\gamma}+u_{\alpha\beta\gamma}X^{\alpha\beta\gamma}+\cdots \tag{26}$$

we see that there are no currents for $\lambda,\phi$ . Dimensionally possible sources are $(F_{\mu\nu})^2$ and $\Gamma.F\chi$ but these are already contained in $T_{\lambda\lambda}$ and $\Gamma\cdot J^S$. Are also $\lambda,\phi$ perhaps contained in $\psi'_\mu$ and $h'_{\mu\nu}$ ? Expanding the d=10 action we indeed find

$$\mathcal{L}=\tfrac{1}{2}(h_{\mu\nu}-\tfrac{1}{4}\delta_{\mu\nu}\ln\phi)T_{\mu\nu}+\tfrac{1}{2}(\bar{\psi}_{\mu}-\frac{\sqrt{2}}{12}\bar{\lambda}\Gamma_{\mu})J^{S}_{\mu}+\frac{3\sqrt{2}}{4}\partial_{\alpha}A_{\beta\gamma}V^{\alpha\beta\gamma} \tag{27}$$

Hence, the primed fields are not the same as the unprimed fields. Moreover, $F_{\alpha\beta\gamma}$ is related to $t_{\alpha\beta\gamma}$, but whereas the the gauge invariance of the abstract current $V_{\alpha\beta\gamma}$ implies $\partial_\alpha t^{\alpha\beta\gamma}=0$ <u>off shell</u>, the 2-photon curl $F_{\alpha\beta\gamma}$ is only conserved on-shell.

Such a situation, however, is not new. If one introduces a Lagrange multiplier $A_{\mu\nu}$ by

$$\mathcal{L}=\tfrac{3}{4}t^{2}_{\alpha\beta\gamma}+\tfrac{3}{2}A_{\nu\rho}\partial_{\mu}t_{\mu\nu\rho}+\frac{3\sqrt{2}}{4}t_{\alpha\beta\gamma}V^{\alpha\beta\gamma} \tag{28}$$

then integration over $A_{\nu\rho}$ says that $t_{\mu\nu\rho}$ is conserved off-shell, whereas the t-field equation equates $t_{\mu\nu\rho}=(F_{\mu\nu\rho}-\frac{1}{\sqrt{2}}V_{\mu\nu\rho})=(F'_{\mu\nu\rho}+\bar{\chi}\chi\text{ term})$, whereas $F_{\mu\nu\rho}$ is only conserved on-shell. <u>Thus</u> $A_{\mu\nu}$ <u>is a Lagrange multiplier</u>. Are $\lambda,\phi$ further Lagrange multipliers? In terms of $F_{\mu\nu\rho}+V_{\mu\nu\rho}$ the nonlinear Maxwell-Einstein action simplifies considerably: one "sees" $t_{\mu\nu\rho}$ .

As far as $u_{\alpha\beta\gamma}$ is concerned, one can add for free a coupling $u^2_{\alpha\beta\gamma}$ since $(\bar{\chi}\Gamma_{\alpha\beta\gamma}\chi)^2$ vanishes identically, due to typical d=10 Majorana-Weyl identities. However, the transformation law of $u_{\alpha\beta\gamma}$ requires the knowledge of how other currents, not present in $\mathcal{L}$, transform, and that we did not work out.

The conclusion seems to be that there are two "axial" auxiliary fields ($t_{\alpha\beta\gamma}$ and $u_{\alpha\beta\gamma}$) and a large but finite number of lower-dimensional ones. It is questionable whether an invariant action exists for them but they certainly form a closed multiplet. Whether this set yields, upon dimensional reduction to d=4, a set of auxiliary fields for conformal or ordinary N=4 supergravity is not known either. In Howe and Lindström's case of d=5 they fell back onto the d=4 conformal set, but in our d=10 case we found more currents than correspond to theirs. One should do the reduction.

## REFERENCES


1. For an introduction (and many of the details used below) to supergravity, see P. van Nieuwenhuizen, Phys. Rep. 68, 189-398(1981)

2. E. Cremmer, B.Julia and J. Scherk, Phys. Lett. 76B, 409(1978).

3. H. Nicolai, P. K. Townsend and P. van Nieuwenhuizen, Nuovo Cimento Lett. 30, 315(1981).

4. see the "Lectures at the 1981 Trieste School of Supergravity" by P. van Nieuwenhuizen (S. Ferrara and J. G. Taylor, editors).

5. F. Gliozzi, J. Scherk and D. Olive, Nucl. Phys. B122, 253(1977). L. Brink, J. Jcherk and J. H. Schwarz, Nucl. Phys. B121, 77 (1977).

6. M. Kaku, P. K. Townsend and P. van Nieuwenhuizen, Phys. Rev. D17, 3179(1978).

7. P. K. Townsend and P. van Nieuwenhuizen, Phys. Lett. B67, 439(1977).

8. For a review, see L. Castellani, P. Fre and P.van Nieuwenhuizen Ann. of Physics.

9. The BRS transformations of $\phi^i$ and $C^\alpha$ are nilpotent, but those of $C^{*\alpha}$ only if one introduces an auxiliary field into the quantum action. See ref(1), page 239.

10. We follow here R. Kallosh, Nucl. Phys B141, 141(1978). Similar results were obtained by E. Fradkin and M. A.Vassilev Phys. Lett. 72B, 70(1977) and by G. Sterman, P. K. Townsend and P. van Nieuwenhuizen, Phys Rev. D17, 1501(1978).

11. From here on we follow B. de Wit and J. W. van Holten, Phys. Lett. 79B, 389(1978).

12. In detail: $\delta(\xi^\mu\psi_\mu)$ contains $\xi^\mu\eta^\alpha\partial_\alpha\psi_\mu$, while the f-f term contains $-\eta^\alpha\partial_\alpha(\xi^\mu\psi_\mu)$.

13. N. K. Nielsen, to be published

14. R. Ore and P. van Nieuwenhuizen, to be published.

15. E. Bergshoeff, M. de Roo, B. de Wit and P. van Nieuwenhuizen, to be published.

16. This phenomenon was already found in N=4 d=4 supergravity in H.Nicolai and P. K. Townsend, Phys. Lett. 98B, 257(1981).

17. In N=1 d=4, the coupling of the Maxwell system to ordinary supergravity is Weyl invariant (in fact, superconformally invariant), whereas the transformation laws of e, $\psi$ are covariant only under constant scale transformation. The ordinary gauge action itself can be viewed as due to fixing the superconformal local symmetries in the coupling of the scalar multiplet to the fields of conformal N=1 supergravity. After fixing fields, say, A=1, the global weights will differ from the local weights.

18. S. Ferrara, F. Gliozzi, J. Scherk and P. van Nieuwenhuizen, Nucl. Phys. B117, 333(1976).

19. Because the gauge $\chi = 0$ is only consistent (i.e., $\partial\chi = 0$ as well) if the gauge parameters depend on matter fields. Without fixing the conformal symmetries, one has a reducible (but not completely reducible) representation.

20. This was first done by S. Ferrara and B. Zumino for the case of the spin (1/2, 1) multiplet, Nucl. Phys. B87, 207(1975). See also V. Ogievetski and E. Sokatchev, Nucl. Phys. B124, 309(1977).

21. P. Howe and U. Lindstöm, Phys. Lett. 103B, 422(1981).

# BUILDING LINEARISED EXTENDED SUPERGRAVITIES

J.G. Taylor
Department of Mathematics, King's College, London

Abstract. We describe how candidates for auxiliary field multiplets may be discovered for unextended and extended supergravities using field redefinition rules. These may then be supplemented by component transformation rules to select non-linearisable supergravities. The existence of a barrier at N = 3 is demonstrated and candidate auxiliary multiplets with central charges presented for N = 4 supergravity. The extension of this analysis to N = 8 and N = 4 Yang-Mills is finally discussed.

## 1 INTRODUCTION

Much effort has been expended recently in attempting to obtain an off-shell version of maximally extended N = 4 Yang-Mills supersymmetry and N = 8 supergravity, in which the gauge algebra is closed without the use of field equations. ( van Nieuwenhuizen (1981); Hawking and Rocek (1981); Ferrara and Taylor (1982)). This requires the discovery of a suitable set of auxiliary fields which must be added to the known physical fields. Such a discovery could be expected to lead to a superfield formulation of those theories. This latter could then allow an incorporation of the superfield perturbation rules which have proved so effective in including the known bose-fermi ultra-violet divergence cancellation mechanism. Furthermore attempts to construct the 63 composite gauge vector bosons of the on-shell local SU(8) symmetry, discovered recently (Cremmer and Julia 1979) depend on this symmetry being a dynamical one. Only if the off-shell theory with closed gauge algebra has such a symmetry should we even hope that such composite particles could exist. Thus the auxiliary fields of N = 4 Yang-Mills and N = 8 supergravity are of considerable interest.

Initial success for N = 1 and 2 Yang-Mills and supergravity (Stelle and West, 1978; Ferrara and van Nieuwenhuizen, 1978; de Wit and van Holten, 1979; Fradkin and Vasiliev, 1979 ) has not continued to higher N. For N = 4 Yang-Mills an irreducible multiplet of the N = 4 global supersymmetry algebra $S_4$ has been proposed as a candidate (Sohnius, Stelle and West (1980); J.G. Taylor (1980). This multiplet arises in the presence of a spin-reducing central charge, so is the Maxwell multiplet with maximum spin 1 and is classified by the internal symmetry group USp(4). It has the spin and USp(4) content $\phi^z_o = (1_P, 1_A, \frac{1}{2}^4, 0^5_P, 0^5_A)$ where $j^m$ denotes a component field of spin-j and USp(4) dimensional representation m; the subscripts P and A denote fields of dimension 1 and 0 (modulo 2) respectively. Both $1_P$ and $1_A$ are constrained, being described by vectors $A_\mu$, $B_\mu$ satisfying $\partial_\mu A_\mu = \partial_\mu B_\mu = 0$ (at the linearised level). In order to obtain (1 + 5) scalars on-shell the field $B_\mu$ must be used to describe the USp(4) singlet scalar. For various purposes, such as quantisation, it is necessary to solve the constraint on $B_\mu$, and whilst this can be done satisfactorily in the Abelian case this is not so in the non-Abelian case.

A similar unsatisfactory state of affairs occurs for N ≥ 3 supergravity. No-go theorems were proved recently for all such extended supergravities, showing that no such off-shell formulations could exist which had the correct spectrum unless central charges were present (Rivelles and Taylor (1981a); Taylor (1981a)). More recently (Taylor (1981b)) central charges had been used to obviate the above-mentioned theorems and candidates for auxiliary fields for linearised supergravity were given for N ≥ 3. Whilst the physical fields did not need to have central charges for N = 3 or 4, these being carried only on the auxiliary supersymmetric multiplets, this was not the case for N ≥ 5. For these latter values of N all multiplets possessed central charges, including the Weyl multiplet containing the graviton. This multiplet had superspin 0 and we denote it by $\bar{\Phi}^z_o$, with spin and USp(8) content $(2_P, 2_A, \frac{3}{2}^8, 1^{27}_P, 1^{27}_A, \frac{1}{2}^{48}, 0^{42}_P, 0^{42}_A)$.

The use of $\bar{\Phi}^z_o$ alone clearly leads to constrained component fields, the latter constraints not even being soluble at the linearised level. (Cremmer, et al. (1980). It was suggested recently (Taylor (1981b)) that three copies of $\bar{\Phi}^z_o$, one with negative kinetic energy (which we denote by $2\bar{\Phi}^z_o - \bar{\Phi}^z_o$) remove the constraints in a supersymmetric fashion. This is because we can use field redefinition rules (Rivelles and Taylor (1981b)) to allow the physical scalars to be represented purely by scalar fields with no differential constraints.

To describe these developments we will commence by reviewing the techniques of building linearised N = 1 and 2 supergravities from irreducible multiplets by use of the field redefinition rules. These latter rules are described in the next section and their use in selecting suitable multiplets for N = 1 and 2 supergravities analysed in the following section. We then discuss the no-go theorems in section 4 and the way central charges may be included to construct candidate multiplets for N = 4 supergravity. In a concluding section we discuss the extension of our programme to N = 8 and N = 4 Yang-Mills.

## 2 FIELD REDEFINITION RULES

Field redefinition rules (Rivelles and Taylor (1981a)) allow the differentially constrained component fields in different irreducible supersymmetric multiplets to be combined together to give fields without such non-local constraints. These rules are therefore at the basis of constructing 'local' representations of supersymmetry algebras from non-local ones. More specifically the rules are of two sorts, 'annihilation' or 'creation'. The former combine fields so that the resulting redefined field is purely auxiliary (with equation of motion setting the field to vanish); the latter allow the graviton to be constructed from its pure spin 2 and 0 components.

Let us denote by $j_P$ or $j_A$ a field of spin j and dimension - 1 or - 2 respectively, as well as its quadratic kinetic energy Lagrangian. Thus $0_P$ denotes a physical scalar $\phi$ as well as denoting the kinetic energy term $\frac{1}{2}(\partial_\mu\phi)^2$. The simplest field redefinition rule is the annihilation rule

$$1_A - 0_P \approx 0 \qquad (2.1)$$

where $\approx 0$ denotes vanishes by virtue of the field equations. The proof of (2.1) is by the (non-local) decomposition of a 4-vector $A_\mu$ into its longitudinal and transverse parts:

$$A_\mu = \partial_\mu A^L + A^T_\mu, \quad \partial_\mu A^T_\mu = 0 \qquad (2.2)$$

Then from (2.2), $A^L_\mu = \square^{-1}\partial_\mu A_\mu$. But from (2.2)

$$\frac{1}{2} A_\mu^2 = \frac{1}{2} (\partial_\mu A^L)^2 + (A_\mu^T)^2 \tag{2.3}$$

Since the l.h.s. of (2.3) vanishes by virtue of the equation of motion $A_\mu = 0$, whilst the r.h.s. is the difference of the k.e. terms for $0_P$ and $1_A$ then (2.1) is valid. Another simple annihilation rule is

$$0_P - 0_P \approx 0 \tag{2.4}$$

proved from the re-arrangement

$$\phi_1 p^2 \phi_1 - \phi_2 p^2 \phi_2 = S_1 S_2 \tag{2.5}$$

with $S_1 = (\phi_1 - \phi_2)$, $S_2 = p^2(\phi_1 + \phi_2)$. Since $S_1$ and $S_2$ are independent scalars, with equation of motion $S_1 = S_2 = 0$, rule (2.4) results.

By multiplying the rules (2.1) and (2.4) by spin j (and reducing) many new annihilation rules may be obtained; care must be taken in ascribing the correct dimensions to the resulting fields. Thus the fermion rule

$$\frac{1}{2} - \frac{1}{2} \approx 0 \tag{2.6}$$

and the vector rule

$$1_P - 1_P \approx 0 \tag{2.7}$$

are important cases.

The creation rules allow the spin - $\frac{1}{2}$ mode of the gravitino and the dilation mode of the graviton to be included in the pure spin $\frac{3}{2}$ and 2 kinetic energy to give the linearised Rarita-Schwinger and Einstein Lagrangians. The rules in these cases may be obtained, after some algebraic manipulation, to be

$$2_P - 0_P = L_{Einst} (f_{\mu\nu}) \tag{2.8}$$

$$\frac{3}{2} - \frac{1}{2} = L_{R.S.} (\psi_\mu) \tag{2.9}$$

where

$$f_{\mu\nu} = e_{\mu\nu} + \frac{1}{3}\eta_{\mu\nu} S \qquad (2.10)$$

$$\psi_\mu = \tau_\mu + \frac{1}{3}\gamma_\mu \tau \qquad (2.11)$$

and S, $\tau$, $\tau_\mu$ and $e_{\mu\nu}$ are the component fields of spins 0, $\frac{1}{2}$, $\frac{3}{2}$ and 2 respectively which enter on the l.h.s. of (2.8) and (2.9).

We may use the field redefinition rules in two ways: firstly schematically to discover sets of multiplets which marry together to give the correct off-shell linearised supergravity Lagrangian plus auxiliary fields, and secondly by more detailed analysis to determine which of those sets of multiplets have purely local supersymmetry transformation rules when expressed in terms of the redefined fields. We will demonstrate these two levels of analysis in the next section for N = 1 supergravity.

## 3 LINEARIZED N = 1 AND 2 SUPERGRAVITIES

The basic multiplet of N = 1 global supersymmetry is the chiral multiplet or its related real counterpart with 4 boson and 4 fermion degrees of freedom (A, B, F, G, $\psi$) which can be denoted as $\Phi_o = (0^+_P, 0^-_P, \frac{1}{2}, 0^+_A, 0^-_A)$, where the ± superscripts denote the parity of the field. The Weyl multiplet containing the graviton may be constructed from this by multiplying by spin $\frac{3}{2}$, and has the content $\Phi_{\frac{3}{2}} = (1^+_A, \frac{3}{2}, 2^+_P)$. We may then take the difference of the two Lagrangians $L_{\frac{3}{2}} - L_o$, and recombine it by means of the field redefinition rules of the previous section as

$$L_{\frac{3}{2}} - L_o = (2^+_P - 0^+_P) + (\tfrac{3}{2} - \tfrac{1}{2}) + (1^+_A - 0^-_P) - 0^+_A - 0^-_A \qquad (3.1)$$

This gives the minimal form of Poincaré supergravity with auxiliary fields $A_m$ (unconstrained vector), F and G. Another form (Sohnius and West 1981) uses the alternate auxiliary multiplet $\Phi_{\frac{1}{2}+} = (0^+_P, \frac{1}{2}, 1^+_A)$, to give

$$L_{\frac{3}{2}} - L_{\frac{1}{2}} = (2^+_P - 0^+_P) + (\tfrac{3}{2} - \tfrac{1}{2}) + (1^+_A - 1^+_A) \qquad (3.2)$$

where the 6 auxiliary field degrees of freedom in $1^+_A - 1^+_A$ are described by a gauge-invariant 4-vector and an antisymmetric tensor with appropriate gauge invariance.

There are also possible solutions with three auxiliary multiplets. Thus the non-minimal set involves the superspin $\frac{1}{2}$ multiplets $\Phi_{\frac{1}{2}+} = (0^+_P, \frac{1}{2}, 1^+_A)$ and $\Phi_{\frac{1}{2}-} = (0^-_P, \frac{1}{2}, 1^-_A)$ (constructed from $\Phi_0$ by multiplying by a spin $-\frac{1}{2}$) as well as $\Phi_0$ and $\Phi_{\frac{3}{2}}$. The resulting expression is

$$\Phi_{\frac{3}{2}} + \Phi_0 - \Phi_{\frac{1}{2}+} - \Phi_{\frac{1}{2}-} = (2^+_P - 0^+_P) + (\tfrac{3}{2} - \tfrac{1}{2}) + (\tfrac{1}{2} - \tfrac{1}{2})$$

$$+ (0^+_P - 1^-_A) + (0^-_P - 1^+_A) + (1^+_A - 0^-_P) + 0^+_A + 0^-_A \tag{3.3}$$

and involves three unconstrained vector auxiliary fields as well as two spinor, one scalar and one pseudo-scalar auxiliaries.

Another solution with three auxiliary multiplets (Rivelles and Taylor (1981c))is obtained by taking $\Phi_{\frac{3}{2}}$ , $\Phi_{\frac{1}{2}+}$ and $\Phi_0$ (twice), and is

$$\Phi_{\frac{3}{2}} + \Phi_0 - \Phi_0 - \Phi_{\frac{1}{2}+} = (2^+_P - 0^+_P) + (\tfrac{3}{2} - \tfrac{1}{2}) + (\tfrac{1}{2} - \tfrac{1}{2})$$

$$+ (0^+_P - 0^+_P) + (1^+_A - 0^-_P) + (0^-_P - 1^+_A)$$

$$+ 0^+_A + 0^-_A - 0^+_A - 0^-_A \, . \tag{3.4}$$

This is clearly distinct from the non-minimal solution (3.3) since it has scalar auxiliary fields of dimension 1 and 3.

It is reasonable to determine if further candidates for local representations exist with more than three auxiliary multiplets. This is possible using multiplets of higher super-spin. Thus if $\Phi_Y$ denotes the multiplet of superspin Y (constructed by attaching spin Y to each exponent field of $\Phi_0$), then there is the (36 + 36) solution:

$$\Phi_{\frac{3}{2}} + \Phi'_{\frac{3}{2}} + \Phi_0 - \Phi_1 - \Phi_0 = (2_P^+ - 0_P^+) + (\tfrac{3}{2} - \tfrac{1}{2}) + (\tfrac{3}{2} - \tfrac{3}{2})$$

$$+ (2_A^+ - 1_P^- + 0_A^+) + (1_P^+ - 1_P^+) + (1_A^+ - 0_P^-) + (0_P^- - 1_A^+)$$

$$- (1_A^- - 0_P^+) \qquad (3.5)$$

and $\Phi'_{\frac{3}{2}}$ is constructed from $\Phi_{\frac{3}{2}}$ by changing the dimension of all boson fields by one. There are also similar solutions, but with more auxiliary multiplets. For example one exists with (60 + 60) components, involving Y = 0 (twice), 1, $\frac{3}{2}$, 2 and $\frac{5}{2}$. Another with (92 + 92) components can be constructed. Higher maximum values of superspin are also allowed.

To analyse these further we must determine if the transformation rules for the auxiliary and physical fields are local. This can be checked directly once the transformation laws of the component fields in the separate multiplets are known, since each of the above solutions (3.1), (3.2), (3.3) and (3.4) allows the auxiliary and physical fields to be written down directly using the rules of the previous section.

Thus we can immediately derive the known transformation laws in the cases (3.1), (3.2) and (3.3), whilst we obtain a new set of local laws in the case of (3.4). In the process both this and (3.3) require a certain amount of rotation of the scalars, spinors and vectors from different multiplets amongst each other, as occurs for N = 2.

It seems that, on requiring local transformation laws, the solution (3.5) and further similar ones can all be reduced to having a subset of multiplets being removeable from the remainder. This subset can be shown to include the Weyl multiplet with superspin $\frac{3}{2}$ and only multiplets with superspins 0 or $\frac{1}{2}$. It is further very likely that no more than three such multiplets can be taken, without loss of generality, so reducing to the known solutions (3.1) - (3.4).

For N = 2 there is expected to be a similar situation. A minimal solution can be written in terms of the superspin 0 and 1 multiplets $\Phi_0 = (1_P, \frac{1}{2}^2, 0_P^{1^2}, 0_A^{1^2})$ and $\Phi_1 = (2_P, \frac{3}{2}^2, 1_P^{1^2}, 1_A^{3+1}, \frac{1}{2}^2, 0_P^1)$, where $j^{m^n}$ denotes a component field of spin j belonging to the m-dimensional representation of SU(2) with n-fold degeneracy (and we have dropped parity assignments for simplicity). The minimal solution for N = 2 is then

$$\Phi_1 - \Phi_0 - \Phi_0' =$$

$$(2_P - 0_P) + (\tfrac{3}{2} - \tfrac{1}{2})^2 + 1_P + (1_A - 0_P)^3 + (1_A - 0_P)$$

$$+ (1_P - 1_P) + (0_P - 1_A) + (\tfrac{1}{2} - \tfrac{1}{2})^2 - 0_A^3 + 0_A^{1^2} \qquad (3.6)$$

where $\Phi_0'$ is constructed from $\Phi_0$ by changing the dimension of all boson fields in $\Phi_0$ by one. It may be seen that, again after suitable mixings, the field transformation laws in (3.6) are local.

No auxiliary fields for higher N have been discovered without the presence of central charges and we turn to that problem now.

## 4 NO-GO THEOREMS

We will consider the problem of constructing linearised N = 3 Poincaré supergravity using the basic annihilation rule (2.6) and its higher spin companions

$$(\tfrac{3}{2} + \tfrac{1}{2}^2) - (\tfrac{3}{2} + \tfrac{1}{2}^2) \approx 0 \qquad (4.1)$$

$$(\tfrac{5}{2} + \tfrac{3}{2}^2 + \tfrac{1}{2}^2) - (\tfrac{5}{2} + \tfrac{3}{2}^2 + \tfrac{1}{2}^2) \approx 0 \qquad (4.2)$$

We will use the rules (2.6), (4.1) and (4.2) and their higher analogues in such a way that the lower spin partners in (4.1) and (4.2) give no further contribution, so the general fermion annihilation rule becomes

$$(j + \tfrac{1}{2}) - (j + \tfrac{1}{2}) \approx 0 \qquad (4.3)$$

for any non-negative integer j. The rules (4.3) appear to exhaust all known annihilation rules for fermions; one way to circumvent our no-go theorems is to develop new rules which do not require fermions to be monogamous, marrying off against each other.

The set of irreps which we will consider will be specified by choosing only those contained in the extended superfields $E_A{}^M$ and $\Omega_{AB}{}^C$. From the representation theory of $S_N$ (Taylor, 1982) we conclude that these have superspin Y and I-spin I values (we use I-spin instead of SU(3) irreps of $S_3$ since decomposition of products of SO(3) irreps is even simpler than for SU(3)): $Y = \frac{7}{2}$, $I = 0$; $Y = 3$, $I \leq 1$; $Y = \frac{5}{2}$, $I \leq 2$; $Y = 2$, $I \leq 3$; $Y \leq \frac{3}{2}$, $I \leq 4$.

The basic irrep of $S_3$ with $Y = I = 0$ has fermionic content $\frac{3}{2}$, $\frac{1}{2}^{5+3+3+3}$ (where $j^m$ denotes a fermion m-dimensional irrep of SO(3) of spin j). All other irreps are obtained by multiplying this by irreps of the Lorentz and SO(3) group and reducing suitably. We can thus construct explicitly the irreps with the above limitations.

We may write the general linearised Lagrangian constructed from these irreps as

$$L = \sum_{Y,I} (a_{YI} - b_{YI})\, L_{YI} \tag{4.4}$$

where $a_{YI}$, $b_{YI}$ are non-negative integers denoting the multiplicity and sign of the kinetic energy term of the irrep (Y,I) in L and $L_{YI}$ is the corresponding kinetic energy written in terms of constrained component fields. Thus a fermion of spin $(n + \frac{1}{2})$ will give contribution $\bar{\psi}_n \not\partial \psi_n$, where $\psi_n$ is a symmetric tensor-spinor of rank n totally symmetric on its vector indices, traceless in any pair, and with divergence and $\gamma$-trace on any index zero. Any other field representation of spin $(n + \frac{1}{2})$ can always be written in this form by field redefinition. Since we then use all possible field redefinitions to achieve our goal of obtaining an off-shell formulation such a component field choice is not restrictive.

We now rewrite (4.4) as a sum of fields of a given spin. Since we are interested specifically in the fermion contributions we write

$$L_f = \sum_{j,I} d_{j,I}\, \bar{\psi}_{j,I} \not\partial \psi_{j,I} \tag{4.5}$$

where $\psi_{jI}$ is the fermion field of spin j and I-spin I, and is a linear combination of the $c_{YI} = (a_{YI} - b_{YI})$. In order that $L_f$ correctly describe linearised off-shell N = 3 supergravity we require

$$d_{j;I} = 0 \text{ for } j > \tfrac{3}{2}\text{, or } j = \tfrac{3}{2},\ I \neq 1 \text{ or } j = \tfrac{1}{2},\ I > 1 \tag{4.6}$$

and

$$d_{\frac{3}{2},1} = d_{\frac{1}{2},0} = 1, \quad d_{\frac{1}{2},1} = -1 \tag{4.7}$$

The value of $d_{\frac{1}{2},1}$ is obtained from the 'creation rule' $(\frac{3}{2} - \frac{1}{2}) = L_{R.S.}$, where $L_{R.S.}$ is the Rarita-Schwinger Lagrangian for a purely massless spin $\frac{3}{2}$ field. We write the values of $d_{j,I}$ in terms of $C_{Y,I}$, and the solution of the 23 equations (4.6) for the $C_{Y,I}$ gives (Taylor 1981a) after some algebra,

$$d_{\frac{1}{2},0} = 2[c_{\frac{1}{2},0} - 5c_{3,0} + 2c_{\frac{5}{2},1}] \tag{4.8}$$

Since this is always an even number then equation (4.7) can never be satisfied. In other words the requirements (4.6) of vanishing on-shell of the unwanted higher spin and I-spin fields contributions of the various chosen irreps of $S_3$ in the linearised Lagrangian (4.4) also destroys the possibility that the desired physical spin $\frac{1}{2}$ and $\frac{3}{2}$ fields for N = 3 supergravity propagate in the requisite fashion. This was to be expected by a cursory glance at the irreps since it does indeed appear difficult to remove the unwanted modes by the annihilation rules (4.3); the analysis through (4.6) and (4.7) confirms this analytically by the result (4.8).

There are various ways around our no-go theorem for off-shell extension of N = 3 supergravity:

(i) Super-differential geometry is too restrictive and a larger class of irreps of $S_3$ must be chosen. This destroys the elegance of the geometric approach which since the time of Einstein has been important as a general framework within which to construct theories of gravity and matter. More importantly such as extension would not seem to help us since irreps of $S_3$ with higher Y and I values than used above bring along many further annihilation conditions like (4.6). A cursory examination indicates that what is needed is an infinite set of auxiliary irreps with ever increasing Y and I values. This is clearly a very unsatisfactory solution.

(ii) The annihilation rules (4.3) can be modified so that fermions no longer have to be annihilated only in pairs but an odd number can be removed by suitable field redefinition rules. Such alternate rules are unknown to the author, so cannot be discussed further.

(iii) Central charges are present in the algebra $S_3$, so that the irreps no longer have SO(3) (and SU(3)) as symmetry group for their classification. This possibility will be investigated in the next section.

We now extend our no-go theorem to N greater than 3. If there were an off-shell formulation of N > 3 supergravity without central charges its linearised version could trivially be truncated to N = 3 to give a contradiction to our result. A similar situation would arise if an off-shell version of 11-dimensional supergravity were obtained, for again it could be reduced to 4 dimensions by assuming triviality in the other dimensions and then truncated to N = 3 at the linearised off-shell level.

Our result indicates the need for the construction of extended supergravities in the presence of central charges; the corresponding destruction of the on-shell SO(8) symmetry on going off-shell augurs badly for the binding of the 'hidden symmetry' SU(8) gauge vector bosons required in phenomenological applications of N = 8 supergravity.

## 5 CENTRAL CHARGES AND N = 4 SUPERGRAVITY

The off-shell no-go theorems presented above indicate that a satisfactory N-extended supergravity Lagrangian must have non-trivial central charges. These are charges which commute with the generators of momentum, angular momentum and supersymmetry transformations, though they may not commute with generators of internal symmetries. This lack of commutativity turns out to be essential in order that unphysical representations of high dimensions of the internal symmetry, which otherwise cannot be made auxiliary, are destroyed. This was suspected to be the case for N = 8, (Taylor, 1981b) where our construction of linearised auxiliary fields depended crucially on the absence of any internal symmetry whatsoever of the supersymmetry algebra $S_N$ off-shell. We present here the central charge structure which destroys such internal symmetries.

The central charges $X_{ij}$, $Y_{ij}$ ($1 \leq i, j \leq N$) arise from the anti-commutation brackets for the supersymmetry generators $S_{\alpha i}$ as

$$[S_{\alpha i}, S_{\beta j}]_+ = (\not{p}\eta)_{\alpha\beta}\, \delta_{ij} + i\eta_{\alpha\beta}\, X_{ij} + i(\gamma_5\eta)_{\alpha\beta}\, Y_{ij} \qquad (5.1)$$

where $\eta = -C^{-1}$, C being the charge conjugation matrix, $\gamma_5 = \gamma_0\gamma_1\gamma_2\gamma_3$

and metric (+ - - -). From the antisymmetry of $\eta$ and $\gamma_5\eta$ follows the antisymmetry of the central charge matrices X and Y. The representation of $S_{\alpha i}$ by differential operators on superfields uses the representation $X_{ij} = i\partial/\partial x^{ij}$, $Y_{ij} = i\partial/\partial y^{ij}$ on further bose co-ordinates.

Massive irreducible representations (irreps) of the algebra $S_N$ of (5.1) have at least Poincaré spin $\frac{1}{2}$ N, so have at least spin 4 for N = 8. Spin reduction to maximal Poincaré spin $\frac{1}{4}$ N will occur if only half of the $S_{\alpha i}$ are independent. Such spin reduction will thus lead to maximal spin 2 for N = 8. Without spin reduction, auxiliary fields of spin $\frac{5}{2}$, 3, $\frac{7}{2}$ and 4 must occur in N = 8 supergravity, leading to possible problems of consistency of the associated 2- and 4-f ghost interactions. To avoid such difficulties we will require spin reduction for $S_8$.

We turn to analyse in detail how such spin reduction occurs and what are the resulting symmetries. The r.h.s. of (5.1) may be written as $(\not{p} + iZ)\eta$, with $Z = X + \gamma_5 Y$. One method of spin reduction has been to consider (5.1) in the rest frame of $p^2 = 0$, (Sohnius 1978; Derendinger et al 1981), but this is only valid for on-shell analysis when X and Y are numerical matrices. Instead we follow the off-shell analysis given earlier (Taylor, 1980) where a Dirac equation associated to $(\not{p} + iZ)$ is to be constructed. The relevant associated operator is $(\not{p} - i\bar{Z})$ $= (\not{p} - iX + i\gamma_5 Y)$, the product of these 2 operators being $Q = p^2 + X^2 + Y^2 - \gamma_5[X,Y]_-$. The decomposition $S = (2\not{p})^{-1}\,[(\not{p} - i\bar{Z})S + (\not{p} + i\bar{Z})S]$ defines $U = (2\not{p})^{-1}\,(\not{p} - i\bar{Z})S$, $V = (2\not{p})^{-1}\,(\not{p} + i\bar{Z})S$. From (5.1), U will be nilpotent on irreps with Q = 0. However this condition is only sensible for a scalar superfield when Q has no internal symmetry indices. We thus need $[X,Y]_- = 0$ and $X^2 + Y^2$ proportional to the identity. In that case $(\not{p} - iZ)V = 0$ on the irrep considered, so we expect reduction of the independent degrees of freedom in V by a factor of 2.

If X and Y are expanded in a complete basis of real anti-symmetric matrices $\Gamma_\ell$ then the conditions on X and Y require $[\Gamma_\ell,\Gamma_m] \propto \delta_{\ell m}$. The matrices are the basis of the SO(N - 1) Clifford algebra, with $N = 2^k$ for some integer k; for N = 8 we may choose $[\Gamma_\ell,\Gamma_m]_+ = -2\delta_{\ell m}$ with $1 \leq \ell, m \leq 7$. We also need X = 0 or Y = 0, so that there are at most seven non-zero central charges on any spin reduced irrep of $S_8$. If we write $X = \sum_\ell X_\ell \Gamma_\ell$, Y = 0, with $X_\ell = i\partial/\partial x^\ell$, the associated superfield $\Phi$ satisfies the 11-dimensional massless wave equation (with only one time direction)

$$(p^2 + \sum_{\ell=1} \partial^2/\partial x^{\ell^2})\Phi = 0$$

The similar case of $X = 0$, $Y = \sum \Gamma_\ell \; i\partial/\partial y^\ell$ also leads to an 11-dimensional theory, with V satisfying the 11-dimensional massless Dirac equation.

The algebra (5.1) with $X = \sum_\ell X_\ell \Gamma_\ell$, $Y = 0$ (or that with X and Y interchanged) has internal symmetries depending on the number of non-zero $X_\ell$'s. To determine these symmetries we take the real $8 \times 8$ representation of the $\Gamma_\ell$ as $(\underline{\alpha}\,\tau_3, \underline{\beta}\,\tau_1, i\tau_2)$, where $\underline{\alpha}$ and $\underline{\beta}$ are the $4 \times 4$ real antisymmetric mutually commuting generators of the two SO(3)'s in SO(4) with $[\alpha_i,\alpha_j] = 2\delta_{ij}$, $[\alpha_i,\alpha_j]_- = 2\varepsilon_{ijk}\,\partial_k$, $[\alpha_i,\beta_j]_- = 0$, and similar commutation relations for $\beta_i$ and $\beta_j$. The maximal internal symmetry of $S_8$ is given by the unitary matrices U for which $UXU^T = X$; in terms of anti-hermitian infinitesimal generators A this latter condition is $AX + XA^T = 0$.

When $X = 0$, this set of 63 generators of SU(8) are the set of 7 $\Gamma_\ell$, 21 $\Gamma_{[\ell m]} = (\underline{\alpha}\,\underline{\beta}\,\tau_2, \underline{\alpha}, \underline{\beta}, \underline{\beta}\,\tau_3, \underline{\alpha}\,\tau_1)$ and 35 $\Gamma_{[\ell mn]} = i(\underline{\alpha}\,\underline{\beta}\,\tau_1, \underline{\alpha}\,\underline{\beta}\,\tau_3, \underline{\alpha}\,\underline{\beta}, \underline{\alpha}\,\tau_2, \underline{\beta}\,\tau_2, \tau_1, \tau_3)$. We may consider an increasing number of $X_\ell$'s, starting with $X_1$ with $\Gamma_1 = \alpha_1\,\tau_3$. Then the matrices from SU(8) split into antisymmetric ones A commuting with $\alpha_1\,\tau_3$ and anti-symmetric ones S anti-commuting with $\alpha_1\,\tau_3$. The first set comprises the 16 matrices $(\alpha_1, \alpha_1\,\tau_3, \underline{\beta}\,\tau_3, \underline{\beta}, \alpha_2\,\tau_2\,\underline{\beta}, \alpha_3\,\tau_2\,\underline{\beta}, \alpha_2\,\tau_1, \alpha_3\,\tau_1)$ and the second the 20 matrices $i(\tau_1, \alpha_1\underline{\beta}\,\tau_1, \alpha_2\underline{\beta}\,\tau_3, \alpha_3\,\underline{\beta}\,\tau_3, \alpha_2\,\underline{\beta}, \alpha_3\,\underline{\beta}, \alpha_1\,\tau_2, \underline{\beta}\,\tau_2)$; they generate USp(8), as is well known. The same method used for $N = 4$ reduces the initial symmetry of SU(4) to USp(4). With increasing numbers of central charges the symmetries are $SU(8) \supset USp(8) \supset USp(4) \times USp(4) \supset U(2) \supset U(1) \supset \phi$ for $N = 8$ and $SU(4) \supset USp(4) \supset USp(2) \times USp(2) \supset USp(2) \supset U(1) \supset \phi$ for $N = 4$.

Let us consider explicitly the case of $N = 4$ supergravity. The $Y = 0$ multiplet $\Phi_0^W$ (with no central charge) has the Lorentz spin and USp(4) content

$$\Phi_0^W = (2_P, \tfrac{3}{2}^4, 1_P^{1^2+5^2}, 1_A^{5+10}, \tfrac{1}{2}^{4^2+16}, 0_P^{1^3+5+14}, 0_A^{10^2}) \qquad (5.2)$$

whilst the lowest central charge multiplet was given in the introduction as

$$\Phi_o^z = (1_P, 1_A, \tfrac{1}{2}^4, 0_P^5, 0_A^5) \tag{5.3}$$

We can also construct

$$\Phi_{0,5}^z = (1_P^5, 1_A^5, \tfrac{1}{2}^{4+16}\ 0_P^{1+10+14}, 0_A^{1+10+14}) \tag{5.4}$$

where $\Phi_{0,5}^z$ is obtained from $\Phi_o^z$ by addition of an extra $\underline{5}$ of USp(4). We may then write down a linearised N = 4 supergravity Lagrangian

$$\Phi_o^W - \Phi_o^z - \Phi_{0,5}^z = (2_P - 0_P) + 1_P^{1+5} + 0_P^{1^2} + (\tfrac{3}{2} - \tfrac{1}{2})^4 + \tfrac{1}{2}^4$$

$$+ (1_A - 0_P)^{10} + (0_P - 1_A)^{1+5} + (1_A - 0_P)^5 + (0_P - 0_P)^{14}$$

$$+ (1_P - 1_P)^{1+5} + 0_A^{10^2} - 0_A^5 - 0_A^{1+10+14} + (\tfrac{1}{2} - \tfrac{1}{2})^{4+16} \tag{5.5}$$

We note that this Lagrangian has the Weyl multiplet entering with no central charge, and furthermore the on-shell physical fields belong to the correct SU(4) representation. A similar solution can be given for N = 2, viz $(\Phi_1 - \Phi_o^z - \Phi_{\frac{1}{2}}^z)$, which appears to be the N = 2 analogue of (5.5).

As we have discussed in section 3 we must analyse the supersymmetry transformation rules for the fields on the r.h.s. of (5.5) to see if they are local or not. It does not seem possible to obtain locality without further breaking of USp(4) to at least USp(2) × USp(2). We are presently investigating if the representations (5.2), (5.4) when classified by USp(2) × USp(2) have local transformation laws. This corresponds to requiring at least two extra central charges or dimensions.

## 6 HIGHER N

We have presented linearised off-shell versions of N = 5, 6 and 8 supergravities elsewhere (Taylor 1981b), using only multiplets carrying non-trivial off-shell central charges. A similar solution was given for N = 4 super-Yang-Mills theory. This set of multiplets had the advantages (a) it involved no component fields of spin greater than two (so avoiding possible coupling inconsistencies well known for higher spin fields) (b) it was minimal in the sense that no breaking of the internal symmetry to the lowest possible could remove the constraints by using fewer multiplets. However it had 2 serious disadvantages (i) the graviton multiplet had to possess a central charge. This would seem to be in contradiction to the folklore that no massless particle can have a charge (ii) the physical scalars had to arise from different multiplets so apparently breaking supersymmetry on-shell, when all the auxiliary fields vanish. Because of these latter problems it was felt necessary to see if the Weyl multiplet for N = 8 supergravity and N = 4 Yang-Mills could have no central charge and simultaneously all the physical fields arise from the same multiplet. In the process we will have to learn how to deal with higher spin auxiliary fields but that may be a small penalty to pay to avoid (i) and (ii).

Solutions to the above criteria (i) and (ii) are now available both for N = 8 supergravity and N = 4 super Yang-Mills theory. There are various solutions, with differing internal symmetries and numbers of central charges. As in the cases of lower N these are still only candidates for the corresponding non-linear theories, and will only be acceptable provided the corresponding auxiliary and physical fields have local transformation laws. This is presently being analysed for such solutions.

We conclude that through field redefinition rules we may discover the set of all possible sets of multiplets (irreps) of $S_N$ which may be combined to give a linearised version of N-extended supergravity. By direct analysis we can then check if the corresponding redefined fields have local transformation laws. Those solutions which satisfy this criterion are then expected to be able to be non-linearised fairly directly; the others never.

REFERENCES

Cremmer, E, Ferrara , S., Stelle, S. and West P. (1980) Phys. Lett. 94B, 349.
Cremmer, E. and Julia, B. (1979). Nucl. Phys. B129 141
Derendinger, J.P., Ferrara, S. and Savoy C.A.P. (1981) Phys. Lett. 100B 393.
de Wit, B. and van Holten, J.W. (1979). Nucl. Phys. B. 155, 530.
Ferrara, S. and Taylor, J.G. (1982) (ed.) "Supergravity 1981", Camb. Univ. Press.
Ferrara, S. and van Nieuwenhuizen, P. (1978) Phys. Lett. 74B, 333.

Fradkin, E. and Vasiliev, M. (1979). Lett. N. Cim. 25, 79.
Hawking, S. and Rocek, M. (ed.) (1981) "Superspace and Supergravity", Camb. Univ. Press.
van Nieuwenhuizen, P. (1981) Phys. Rep. 68, 192.
Rivelles, V. and Taylor, J.G. (1981a) Phys. Lett. 104B 131.
Rivelles, V. and Taylor, J.G. (1981b) J. Phys. A (to appear)
Rivelles, V. and Taylor, J.G. (1981c) "N = 1 Supergravities", King's College preprint.
Sohnius, M. (1978), Nucl. Phys. B138 100.
Sohnius, M., Stelle, K. and West, P.C. (1980) Phys. Lett. 92B, 123.
Sohnius, M. and West, P.C. (1981), King's College preprint 1981.
Stelle, K. and West, P.C. (1978) Phys. Lett. 74B, 330.
Taylor, J.G. (1980) Phys. Lett. 94B, 174.
Taylor, J.G. (1981a), J. Phys. A. (to appear).
Taylor, J.G. (1981b), Phys. Lett. 105B, 429, 434.
Taylor, J.G. (1982) in "Supergravity '81", ed. S. Ferrara and J.G. Taylor, Camb. Univ. Press.

† Talk given at the Nuffield Quantum Gravity Conference, Imperial College, August, 1981.

# (SUPER)GRAVITY IN THE COMPLEX ANGULAR MOMENTUM PLANE

M. T. Grisaru
Department of Physics, Brandeis University
Waltham, Massachusetts 02254, USA

## INTRODUCTION

I wish to describe some features of gravity and supergravity theories from a rather unusual point of view, by studying the properties of the corresponding scattering amplitudes in the complex angular momentum plane. In doing so not only will I deal with a procedure which is unfamiliar to most physicists working in quantum gravity (indeed to most physicists of the younger generation) but also I will apply it in a context different from that for which it was originally devised; viz. unrenormalizable weakly interacting massless gravity rather than renormalizable, strongly interacting massive hadron theory. My motivation for doing this is the following: By working exclusively with on-shell quantities one can avoid the off-shell divergences of the theory and by using unitarity and analyticity properties of the S-matrix one can say something about higher orders of perturbation theory. Furthermore one can say something about the on-shell divergences of the theory. I will describe some old results--absence of two-loop divergences in supergravity and some new results--possible bound states in N=8 supergravity.

I will divide my discussion into three parts: a) Life in the complex angular momentum plane, b) Relation between Kronecker delta singularities and counterterms and c) Regge poles in N=8 supergravity. Most of the arguments I shall present are quite nonrigorous. They are often based on intuition developed by working with renormalizable theories and even there the degree of rigour is less than desired. It may be that some of the conclusions I reach may have to be modified. On the other hand, to convince you that correct results may be inferred, I will quote from a 1975 paper (Grisaru et al. 1975b) where, on the basis of complex angular momentum properties of gravity we concluded "if helicity conservation is true...the two loop amplitudes should be renormalizable on shell". This antedated by two years a different proof that supergravity, where helicity conservation is indeed true and the same methods could be applied, is two-loop finite (Grisaru 1977).

## THE COMPLEX ANGULAR MOMENTUM PLANE

Consider a two-particle scattering amplitude $f(s,z)$, $s$=(center of mass energy)$^2$, $z=\cos\theta$ where $\theta$ is the scattering angle. We define a partial wave amplitude (ignoring spin for the moment)

$$f_J(s) = \int_{-1}^{1} f(s,z)\ P_J(z)dz \qquad (1)$$

where $P_J$ is a Legendre polynomial. Typically the scattering amplitude enjoys analyticity properties which allow the representation (a fixed energy dispersion relation)

$$f(s,z) = a_o(s) + a_1(s)z + \ldots + a_N(s)z^N + \frac{z^{N+1}}{\pi} \int \frac{dz'}{(z'-z)z'^{N+1}} A(s,z') \qquad (2)$$

This implies

$$f_J(s) = b_o(s)\ \delta_{Jo} + b_1(s)\delta_{J1} + \ldots + b_N(s)\delta_{JN}$$

$$+ \int dz P_J(z)\ \frac{z^{N+1}}{\pi} \int \frac{dz'}{(z'-z)z'^{N+1}} A(s,z')$$

$$= \int Q_J(z')\ A(s,z')dz' \quad \text{for} \quad \mathrm{Re}J>N \qquad (3)$$

because one can then interchange the orders of integration and use standard properties of the Legendre functions. This expression can be continued to complex values of $J$ since $Q_J$ is defined there and the integral converges and defines an analytic function $F(s,J)$ regular in $\mathrm{Re}J>N$. (For a discussion of this and other topics to be mentioned below a standard reference is Collins & Squires (1968)).

In general, for fixed $s$, $F(s,J)$ can be continued to the left of $\mathrm{Re}J=N$ and one will encounter singularities (for simplicity we shall only discuss poles). In particular we can compare $F(s,j)$, where $j<N$ is a real integer to $f_j(s)$ which is also defined there:

$$f_j(s) = b_j(s) + \int dz\ P_j(z)\ \frac{z^{N+1}}{\pi} \int \frac{dz'}{(z'-z)z'^{N+1}} A(s,z') \qquad (4)$$

and

$$F(s,j) = \text{analytic continuation of } \int Q_j(z') A(s,z')\, dz' \;. \quad (5)$$

A priori F(s,j) and $f_j(s)$ need not agree at j in which case F does not describe the physics there (of course F=f for integral values of J>N). If they do not agree we say that the physical amplitude has a Kronecker delta singularity at j. If they do agree we say that the physical amplitude "Reggeizes" at j. In that case the value of $f_J(s)$ at J=j is determined from a knowledge of the function F(s,J) defined at large J. As we will see the discussion of counterterms in (super) gravity is based on this.

Let me now review Regge poles. In the neighborhood of a pole of F(s,J) we can write

$$F(s,J) = \frac{\beta(s)}{J - \alpha(s)} + F_{reg}\,(s,J) \quad (6)$$

In general both α, the Regge trajectory, and β, the Regge residue, are functions of s. Suppose that for $s=m_o^2$ $\alpha(s)=j_o$, a physical value where we have reasons to believe F is equal to the physical amplitude. Then

$$F(s,\ j_o) \sim \frac{\beta(s)}{j_o - \alpha(s)} = \frac{\beta(m_o^2)}{(s-m_o^2)\alpha'(m_o^2)} + \ldots \quad (7)$$

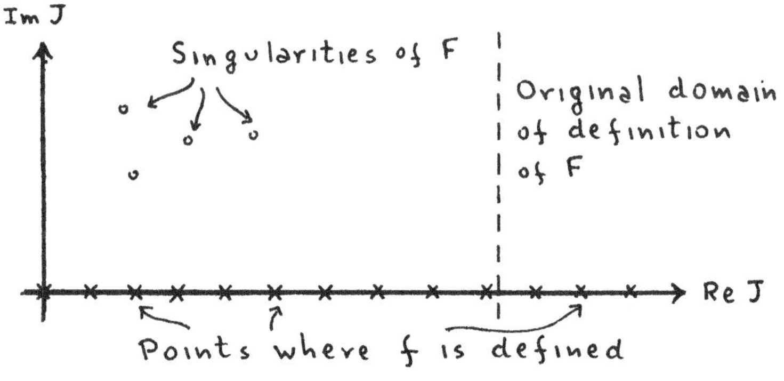


Figure 1

and we conclude that a particle of mass $m_0$ and spin $j_0$ (a "bound state" of the initial or final particles) is present

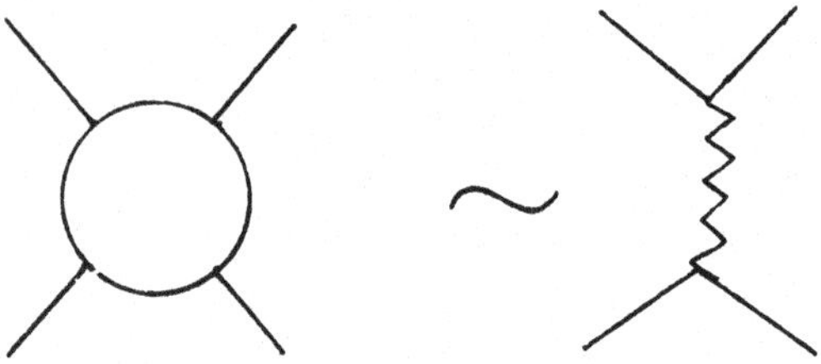

Fig. 2

In general the Regge trajectory $\alpha(s)$ may give rise to several particle poles, one for each value of s such that $\alpha(s)$ is an integer (see Fig. 3 where two trajectories are plotted). To be precise only every other point (J=0, 2, 4... or J=1, 3, ...) gives rise to a pole because of a "signature" factor ($1 \pm \exp i\pi J$) which is actually present. These are called "right signature points". An exception occurs if $\beta(m_0{}^2)=0$ at a right signature point. Then we say that the trajectory "chooses nonsense" and again no particle pole appears. There are however occasions when $\beta=0$ for kinematical reasons. The particle exists but one has to look at a different amplitude to see the pole. For example the graviton-graviton tree approximation amplitudes have the form

$$f(s,z) = \frac{\kappa^2 s}{1-z^2} \qquad (8)$$

and there is no graviton 1/s pole in $f_J(s)$ at J=2. This is simply for kinematical reasons: The $\kappa^2$ requires a factor of s in the numerator. In Regge pole language it corresponds to $\alpha(0)=2$ but $\beta(0)=0$ for dimensional reasons since $\beta$ must be a function of $\kappa^2 s$, vanishing as $\kappa \to 0$. Of course the pole could be seen in more complicated, multiparticle amplitudes. As

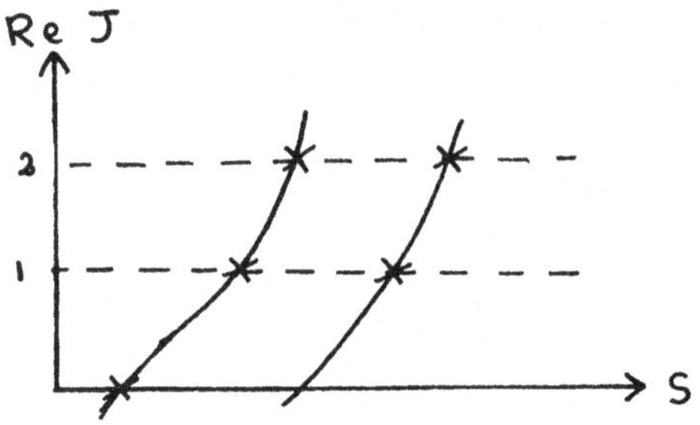


Fig. 3

we will discuss later, all the Regge poles we will find in N=8 supergravity have this property. We find trajectories but cannot be sure without looking at more complicated processes whether they choose nonsense (i.e. do not correspond to particles) or their $\beta$ vanishes for kinematical reasons.

Finally let me mention the relation between Regge poles and asymptotic behavior. It can be shown that a Regge pole in F(s,J) corresponds to a contribution

$$f(s,z) \underset{z\to\infty}{\sim} \beta(s)\, z^{\alpha(s)} \tag{9}$$

Therefore one way of finding Regge poles is to look at the asymptotic behavior for large (unphysical) $z=\cos\theta$.

KRONECKER DELTA SINGULARITIES AND COUNTERTERMS

As we have already mentioned we deal with two kinds of amplitude, $f_J(s)$ and F(s,J), defined in Eqs. (4) and (5) respectively. Are there conditions under which they can be equal for small J? For example, suppose that in some theory the scattering amplitude has a spin-zero particle pole

$f(s,z) \sim$ 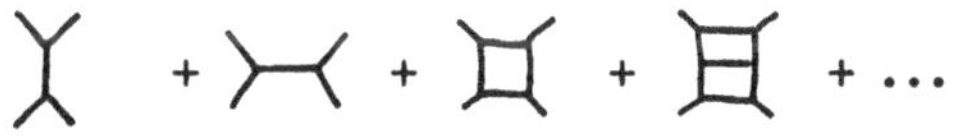

$$= \frac{g^2}{s-m^2} + \tilde{f}(s,z) \tag{10}$$

Then

$$f_J(s) = \frac{g^2}{s-m^2}\,\delta_{J0} + \ldots$$

The question is: How can this equal an analytic function F(s,J)? The answer is this: We recognize that

$$\frac{g^2}{s-m^2}\,\delta_{J0} \simeq \lim_{g^2\to 0} f_J(s) \tag{11}$$

i.e. it is the small g limit of a more complicated expression. Now suppose

$$F(s,J) \sim \frac{\beta(g^2,s)}{J-(s-m^2)\gamma(g^2,s)} \tag{12}$$

such that, as $g^2 \to 0$

$$\beta(g^2,s) \to -g^4 \quad , \quad \gamma(g^2,s) \to g^2 \tag{13}$$

Then

$$\lim_{g^2 \to 0} F(s,J) \sim \frac{g^2}{s-m^2}\,\delta_{J0} \tag{14}$$

and therefore it is not impossible for $F(s,J)$ to equal $f_J(s)$. In this case we see an example of a particle which is "elementary" at the Lagrangian level and yet can be described by a Regge pole as a bound state. One says that the elementary particle "Reggeizes".

Obviously this is not guaranteed. In order for $F(s,J)$, defined at large J, to equal at low J $f_J(s)$ the two quantities must be severely constrained. In the mid 60's Mandelstam (1965) pointed out that for certain field theories constraints do exist. He noted that both $f_J(s)$ and $F(s,J)$ satisfy the same (nonlinear) integral equations and if they differ it is because the solutions are not unique. They can differ by the values of a certain number P of arbitrary parameters. At the same time they both satisfy a certain number C of identical constraints (threshold behavior and conditions at $s=0$), which restrict the freedom in choosing the P parameters, and if $C>P$ they must be identical. Therefore by counting C and P at each j one can check whether or not the two amplitudes must be equal. This is known as "Mandelstam counting" and is a sufficient condition for Reggeization (not necessary; cases are known where Reggeization occurs although $C<P$).

Mandelstam counting and this whole subject were set up to discuss the S-matrix for strongly interacting particles. However in 1975 we decided to apply the procedure to graviton-graviton scattering (Grisaru et al. 1975b) and discuss the divergences of the scattering amplitude in perturbation theory. In doing so we were aware that our conclusions could at best be plausible rather than definitive and the remarks which follow should also be considered the same way. Let us consider for example the graviton-graviton amplitude up to some loop order, calculated in dimensional regularization

$$f(s,z) = f^{finite}(s,z) + \frac{1}{\varepsilon}\sum_{\ell=0}^{L} a_\ell z^\ell \qquad (15)$$

The important point is that the (local) divergent part is a polynomial in the fields and a finite number of their derivatives and hence the corresponding amplitude is a polynomial in z (and s). (We will not discuss here a possible nonlocal divergent part.) Clearly F(s,J) calculated for large J is finite since it does not see the polynomial. Now suppose one continues it to low J and suppose Mandelstam counting criteria require that F(s,J) must equal $f_J(s)$. One would then conclude that the original amplitude must have been finite to begin with.

In gravity theories the number P of arbitrary parameters increases with loop order (number of subtraction constants in dispersion relations increases because of the dimensionality of the coupling constant) and therefore one must make a loop-by-loop counting. We did this and in particular concluded that if helicity is conserved then the two-loop amplitudes should be finite. While it is not known if pure gravity conserves helicity, supergravity does and the same arguments apply. Thus complex angular momentum considerations provide an alternative proof of the two loop finiteness of supergravity.

Unfortunately the same methods do not require the $f_J(s)$ = F(s,J) for three or more loops and give no more information than conventional methods. However, as mentioned earlier the Mandelstam counting conditions are sufficient but not necessary and it is conceivable that the two amplitudes are equal. In any event the attitude that I would like to take is the following: To any finite order N of perturbation theory one can find a large enough ReJ so that $f_J(s)$ defines an analytic function F(s,J). This function contains information about the dynamics of the theory and when continued to low J it may make useful predictions about properties of the theory. For example it appears that in the neighborhood of J=2 the amplitude for graviton-graviton scattering has the form (Grisaru <u>et al.</u> 1975b)

$$f_J(s) \sim b_2(s)\ \delta_{J2} + \frac{\beta(s)}{J-\alpha(s)} \qquad (16)$$

with $\alpha(o)=2$. (In fact up to two loops at least $\beta_2(s)=0$; See also remarks in the concluding section.) This would indicate that a Regge trajectory corresponding to the graviton is present i.e. that one can think of the graviton as a "bound state" the same way that, in this language, a non-Abelian vector meson is a bound state (Grisaru et al. 1973; Grisaru & Schnitzer 1979). In general a possible picture of the J-plane structure is given in Fig. 4.

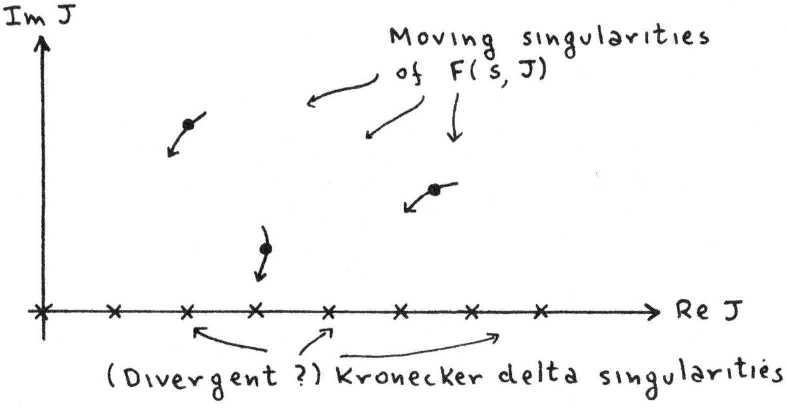


Figure 4

## HOW TO FIND REGGE POLES (=BOUND STATES)

The attitude expressed in the previous section makes it possible to attack the problem of bound states in supergravity. The conventional way of finding bound states in field theory is to solve some kind of Bethe-Salpeter equation, or to find a (static) potential to be used in conjunction with the Schrodinger equation. In gravity the Bethe-Salpeter equation leads immediately to divergences (especially since some of the external lines are off-shell) while the potential cannot be defined for massless particles.

An alternative method is to recognize (or hope) that bound states lie on Regge trajectories and to look for Regge poles in F(s,J). In principle this can be done in two ways: Sum diagrams in perturbation theory and look for $z^{\alpha}$ behavior, or use analyticity and unitarity (solving partial wave dispersion relations) to find directly F(s,J). The first procedure is out of the question because we simply do not have the techniques to compute loop diagrams in gravity. The second one is possible and leads rather quickly to results which, in theories where comparison is possible, agree with those obtained by summing diagrams (for example, see Grisaru 1976). Roughly speaking it extracts the Regge poles that one would obtain from the high energy behavior of the ladder diagrams of Fig. 5.

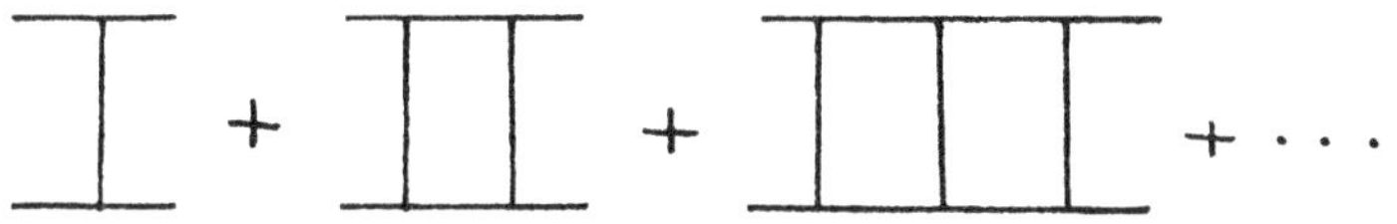

Figure 5

by solving the integral equation satisfied by helicity amplitudes

$$F_{\lambda_3\lambda_4,\lambda_1\lambda_2}(s,J) = V_{\lambda_3\lambda_4,\lambda_1\lambda_2}(s,J) + \int_0^\infty \frac{ds'}{s'-s}\, \rho(s') \sum_{\mu,\nu} F^*_{\lambda_3\lambda_4,\mu_1\mu_2}(s',J)\, F_{\mu_1\mu_2,\lambda_1\lambda_2}(s',J) \tag{17}$$

where V is the partial wave projection of the Born approximation continued to complex J, $\rho$ is a phase space factor and the integral uses elastic unitarity in a partial wave dispersion relation. (This is an approximation to an exact equation where V includes contributions from all left-hand and inelastic cuts.)

For example in graviton-graviton scattering the analytically continued helicity amplitude $F_{2-2,2-2}(s,J)$ satisfies, near J=2 (this is a slight oversimplification)

$$F_{2-2,2-2}(s,J) \sim \frac{\kappa^2 s}{J-2} + \int \frac{ds'}{s'-s}\, \rho(s')\, |F|^2 \tag{18}$$

whose solution (K(s) is a certain integral)

$$F_{2-2,2-2}(\sigma,J) \sim \frac{\kappa^2 s^2 K(s)}{J-2-\kappa^2 sK(s)} \tag{19}$$

exhibits a Regge pole near J=2 which we can identify with the graviton itself. We remark that at J=2 $F_{2-2,2-2}$ is not a physical amplitude (physical values of the helicities must satisfy $|\lambda_1-\lambda_2|<J$, $|\lambda_3-\lambda_4|<J$) but a so-called "nonsense-nonsense" amplitude. (It is of course perfectly well defined by analytic continuation from larger J>4.) In general however the same Regge pole will appear in physical amplitudes coupled by unitarity to it. The reason for looking at this kind of amplitude is that only there the "potential" V in Eq. (17) has a $1/J-2$ singularity which gets turned into a moving Regge pole by higher order corrections. Other, nonsingular terms in the potential are present but only the singular term is "strong" enough to produce binding.

The general procedure (Grisaru et al. 1973) is to look at Eq. (17) as a matrix equation for all the amplitudes coupled by unitarity, but to keep only the "nonsense-nonsense" potential that is, near the integral (or half-integral) value $J_o$ of interest, those elements $V_{\lambda 3\lambda 4,\lambda 1\lambda 2}$ such that $|\lambda_3-\lambda_4|>J_o$, $|\lambda_1-\lambda_2|>J_o$. They are obtained by projecting the Born approximation helicity amplitudes and have the form $v_{\nu',\nu}/J-J_o$ where $v_{\nu',\nu} = v_{\lambda 3\lambda 4,\lambda 1\lambda 2}(s)$ are polynomials in s. The general solution is

$$F_{\lambda'\lambda}(s,J) = v_{\lambda'\nu'}\left[\frac{K(s)}{J-J_o-vK(s)}\right]_{\nu'\nu} v_{\nu\lambda} \qquad (20)$$

and exhibits Regge pole behavior with trajectories $\alpha(s)=J_o$ + eigenvalues of $v_{\nu'\nu}K$. The number of trajectories equals the rank of the matrix $v_{\nu'\nu}$, and they pass through $J_o$ (correspond to particles of spin $J_o$ if the corresponding residue does not vanish) for $(\text{mass})^2$ values equal to the zeroes of $v_{\nu'\nu}(s)$. We also remark that the requirement that for a given $J_o$ "nonsense" values of the helicities be present restricts the spin s ($=J_o$) of the bound state by $s<\min(s_1+s_2-1,\ s_3+s_4-1)$. (This generalizes to multiparticle states to read $s-1<\Sigma(s_i-1)$. Thus in renormalizable theories where $s_i\leq 1$ one always has $s<1$ whereas in gravity multiparticle states can lead to bound states of arbitrary high spin. I shall return to this point later.)

## BOUND STATES IN N=8 SUPERGRAVITY

The possible existence and phenomenological implications of such bound states have been raised recently (Ellis et al. 1980). We have carried out an investigation in the complex angular momentum plane (Grisaru & Schnitzer 1981) with the following conclusions: A priori the two-body scattering amplitudes in N=8 supergravity have the form (omitting helicity labels)

$$f_J(s) = \frac{a_0}{\varepsilon}\,\delta_{J0} + \frac{a_{1/2}}{\varepsilon}\,\delta_{J1/2} + \ldots + \frac{a_n}{\varepsilon}\,\delta_{Jn} + \ldots$$

$$+ \frac{b_0}{s}\,\delta_{J0} + \frac{b_{1/2}}{s}\,\delta_{J1/2} + \ldots + \frac{b_2}{s}\,\delta_{J2} \qquad (21)$$

$$+ \int A(s,z)\,Q_J(z)dz$$

where the first line corresponds to possible divergences while the second line contains contributions from elementary particle (preon) poles. The J plane situation is depicted in two different ways in Fig. 6a or 6b.

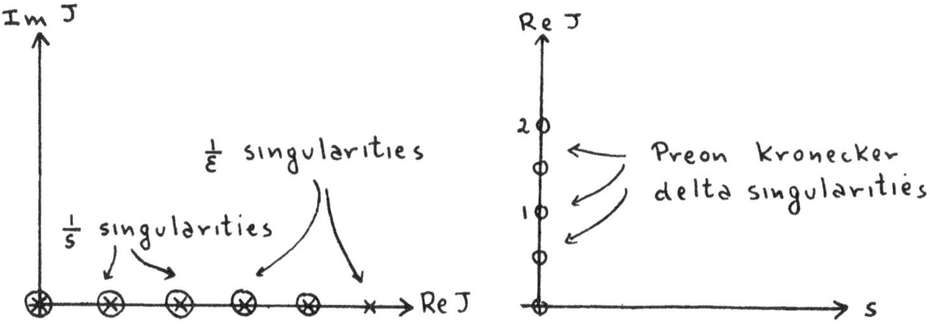


(a) (b)

Figure 6

What we find in fact is an analytic continuation of the form (this is oversimplified)

$$f(s,J) = \frac{\beta_0(s)}{J-s\gamma_0} + \frac{\beta_{1/2}(s)}{J-\frac{1}{2}-s\gamma_{1/2}} + \ldots + \frac{\beta_2(s)}{J-2-s\gamma_2}$$

$$+ \frac{\tilde\beta_{1/2}(s)}{J-\frac{1}{2}-s\tilde\gamma_{1/2}} + \ldots + \frac{\tilde\beta_{5/2}}{J-\frac{5}{2}-s\tilde\gamma_{5/2}} + \ldots \qquad (22)$$

corresponding to the picture of Fig. 7a or 7b.

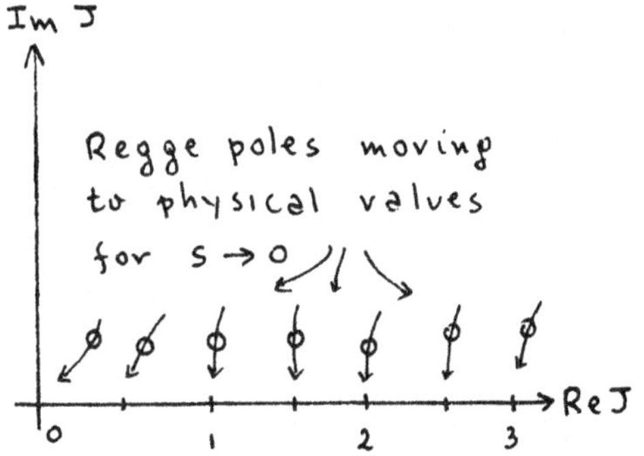


(a)

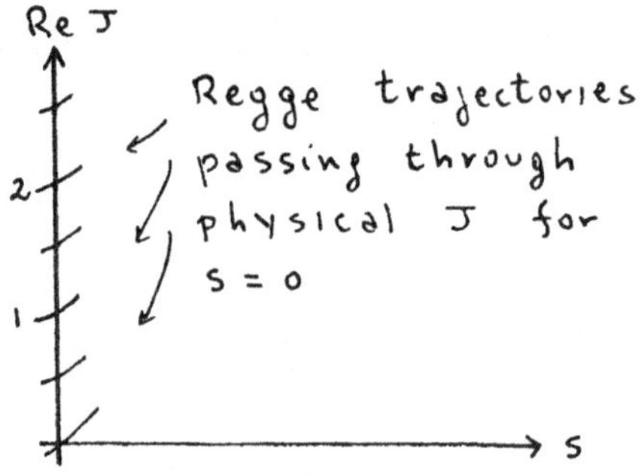


(b)

Figure 7

We find Regge poles which correspond to the preons (they "Reggeize") and additional poles which correspond to various multiplets of bound states (although the residues vanish in our approximation and it is rather difficult to settle this issue).

The actual procedure and results are as follows: In two particle scattering, where the maximum spin is 2 we can have nonsense states for $J<3$. Near each $J=0,1/2,1,\ldots.3$ we must find partial wave projections of the Born helicity amplitudes and construct the nonsense-nonsense matrix $V_{\nu'\nu}$ which gives the information about Regge trajectories. The first step is to find the helicity amplitudes. We classify the preons of N=8 supergravity by a global SU(8) label corresponding to O(8) and helicity i.e. $-2$, $-3/2^A$, $-1^{AB}\ldots.3/2_A$, 2. We start with the known (e.g. Grisaru et al. 1975a) graviton-graviton helicity amplitude

$$F(2,-2;2,-2) = \frac{\kappa^2 s^3}{tu} \tag{23}$$

and generate the others by supersymmetry transformation (Grisaru & Pendleton 1977) (this is actually the most time consuming part of the work). For example

$$f(1^{AB}, -1/2_{CDH};\ 3/2_E\ -1^{FG}) = i\kappa^2 u^2 \sqrt{\frac{-st}{u^2}}\ \cdot \tag{24}$$

$$\cdot \left[ \frac{\delta^{AB}_{DE}\delta^{FG}_{HC} + \delta^{AB}_{CE}\delta^{FG}_{DH} + \delta^{AB}_{HE}\delta^{FG}_{CD}}{t} + \frac{\delta^{AB}_{HC}\delta^{FG}_{DE} + \delta^{AB}_{DH}\delta^{FG}_{CE} + \delta^{AB}_{CD}\delta^{FG}_{HE}}{u} \right]$$

where $\delta^{AB}_{CD} = \delta^A_C\delta^B_D - \delta^A_C\delta^B_D$. We then project onto irreducible SU(8) representations and take partial wave projections which give us matrices v and sets of Regge trajectories corresponding to such irreducible representations all passing through various physical J's for s=0.

In the figures below we indicate the helicities of the corresponding massless "bound states" plotted against their SU(8) content except that we group together states belonging to different representations by absorbing traces. In Fig. 8 we present right signature trajectories, i.e. trajectories for which the corresponding values of angular momentum are right signature points, while in Fig. 9 we present wrong signature trajectories which do not correspond to bound states unless the theory has "exchange degeneracy" (see Collins & Squires 1968).

Here x : $(-\lambda)_{AB\ldots}$, ■ : $(-\lambda)^E_{AB\ldots}$, ● : $(-\lambda)^{EF}_{AB\ldots}$. The open circles and squares correspond to trajectories we expect are present but which cannot be found by looking only at two-body scattering. The states are left-handed but they have TCP conjugate right-handed partners. For technical reasons we have not investigated the situation at J=0.

The ranks and SU(8) representations of the nonsense matrices are just right to give us a pattern of multiplets which are those required by N=8 supersymmetry if they correspond to actual bound states. In Fig. 8 we find a set corresponding to the preons and a second set corresponding to the so-called "current multiplet" (see Ellis et al. 1981). Both have vanishing residues at s=0 (as do the preons themselves in the original partial wave projections) but we see no obvious reasons why the preons should form genuine bound states while the other set do not.

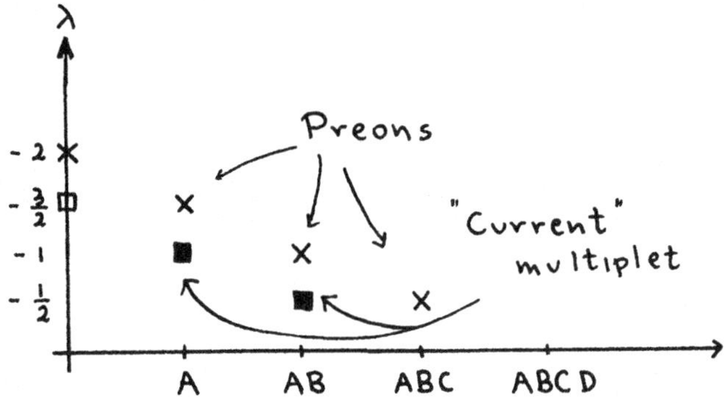


Figure 8

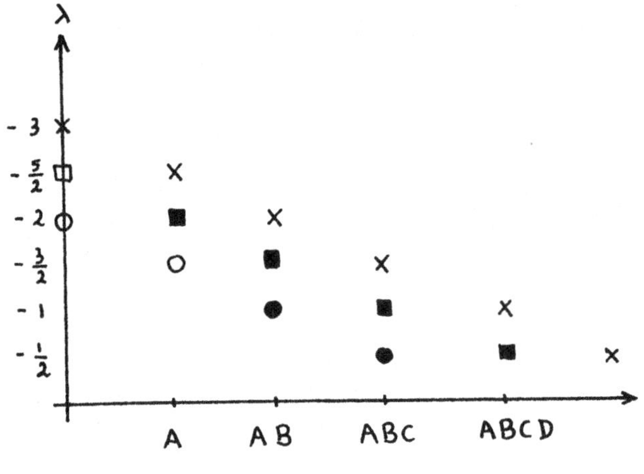


Figure 9

The trajectories in Fig. 9 do not correspond to particles unless, as already mentioned, they are exchange degenerate with another set of trajectories of exactly the same nature but for which the corresponding J's are right signature points. Such trajectories would have to appear as multiparticle bound states. In any event even if such additional bound states are present they do not improve the phenomenological situation.

Besides, an equally important role might be played by multi-particle Regge poles. For example in three-graviton scattering it is possible to have a nonsense state at J=4 and therefore a bound state of spin 4 could exist. With four gravitons one can produce spin 5, and various preons can produce a large variety of high spins corresponding to various SU(8) representations. Methods for handling the situation are not well developed but it appears that it could be reduced to a quasi-two-body situation by considering the scattering of preons and "Reggeons" predicted by two body scattering, then of the "Reggeons" etc. We plan to look at this.

At this point I would like to return briefly to the nonrenormalizability issue and the extent to which one can trust our results if on-shell infinities are present in supergravity beyond two loops. To begin with let me point out that if one were able to compute the (finite) one- and two-loop four-particle amplitudes in N=8 supergravity and examine their large angle behavior one would already be able to fit it uniquely with the Regge poles we have found if our experience with Yang-Mills theory is any guide. At three loops the fit should still work but there might also be a (infinite $1/\varepsilon$) term which goes like a fixed power of the angle (corresponding to the possible three-loop divergence ~ $(R_{\mu\nu\alpha\beta})^4$). Our analytically continued amplitude does not see this.

To conclude let me describe the situation in N=4 supersymmetric Yang-Mills theory. We find right signature trajectories as shown in Fig. 10a, b, classified by global SU(4) and also by the internal symmetry group (e.g. SU(2)) (I=2, as well as wrong signature trajectories also exist).

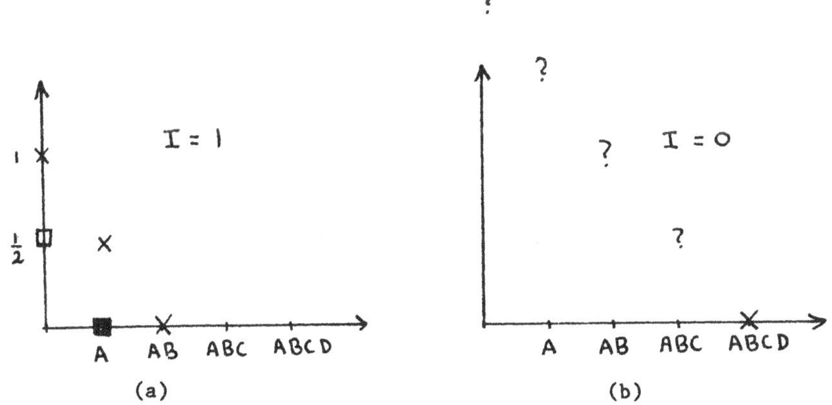


Figure 10

For I=1 we recognize the "preons" again, and another set. In addition at I=0 we find one trajectory passing through J=0 at s=0 which is an SU(4) singlet (indicated as $X_{ABCD}$ in Fig. 10b). If this singlet becomes a bound state N=4 supersymmetry requires the existence of a massless multiplet to which it belongs and it is easy to see that it must be 2, $3/2_A$, $1_{AB}$, $1/2_{ABC}$, $0_{ABCD}$. The theory produces a graviton! This must be a highly non-perturbative effect because here, with maximum preon spin of one, states with any finite number of particles are never nonsense for J>1. N=4 Yang-Mills is a rather remarkable theory and perhaps it manages to evade the Weinberg-Witten argument (1980) in which case this theory, in which the graviton is a bound state provides a viable alternative to conventional quantum (super)gravity.

ACKNOWLEDGMENTS

Much of the research described here is the result of work done in collaboration with H. J. Schnitzer. It was supported in part by NSF grant number PHY 79-20801.

REFERENCES

Collins, P. D. B. & Squires, E. J. (1968). Regge poles in particle physics, New York: Springer Verlag.

Ellis, J., Gaillard, M. K. & Zumino, B. (1980). A grand unified theory obtained from broken supergravity, Phys. Lett. 94B, pp. 343-348.

Grisaru, M. T., Schnitzer, H. J. & Tsao, H.-S. (1973). Reggeization of elementary particles in renormalizable gauge theories: vectors and spinors, Phys. Rev. D8, pp. 4498-4509.

Grisaru, M. T., van Nieuwenhuizen, P. & Wu, C. C. (1975a). Gravitational Born amplitudes and kinematical constraints, Phys. Rev. D12, pp. 397-405.

Grisaru, M. T., van Nieuwenhuizen, P. & Wu, C. C. (1975b). Reggeization of the graviton and the question of higher loop renormalizability of gravity, Phys. Rev. D12, pp. 1563-72.

Grisaru, M. T. (1976). High energy behavior of scattering amplitudes in Yang-Mills theory, Phys. Rev. D13, pp. 2916-18.

Grisaru, M. T. (1977). Two loop renormalizability of supergravity, Phys. Lett. 66B, pp. 75-77.

Grisaru, M. T. & Pendleton, H. N. (1977). Some properties of scattering amplitudes in supersymmetric theories, Nucl. Phys. B124, pp. 81-92.

Grisaru, M. T. & Schnitzer, H. J. (1979). Reggeization of vector mesons and unified theories, Phys. Rev. D20, pp. 784-793.

Grisaru, M. T. & Schnitzer, H. J. (1981). Dynamical calculation of bound state multiplets in N=8 supergravity, Phys. Lett. B (to be published).

Mandelstam, S. (1965). Non-Regge terms in the vector-spinor theory, Phys. Rev. 137B, pp. 949-954.

Weinberg, S. & Witten, E. (1980). Limits on massless particles, Phys. Lett. 96B, pp. 59-62.

# THE MULTIPLET STRUCTURE OF SOLITONS IN THE O(2) SUPERGRAVITY THEORY

G.W. Gibbons
D.A.M.T.P.
University of Cambridge
Silver Street
Cambridge
CB3 9EW

## INTRODUCTION

The notion of a black hole soliton in Quantum Gravity has recently been clarified by Hajicek (1981a) and the relevance of these ideas to supergravity theories has been discussed by Gibbons (1981). In the present report I wish to expand on these ideas a little and to describe in more detail the fermion zero mode structure.

## ZERO MODES AND GAUGE FIXING

Translation zero modes for the Schwarzschild solution were discussed by Gibbons and Perry (1978a). In that case the functional method was used. One could have adopted a different procedure and used quantum field theory about a fixed background, regarding the metric fluctuations $h_{\alpha\beta}$ as an operator. In either case one must employ a gauge condition which, for simplicity, we take to be De Donder gauge $h_{\alpha\beta}{}^{;\beta} - \frac{1}{2} h^{\sigma}_{\sigma;\alpha} = 0$. The covariant derivatives being that obtained from the background metric. To what extent does this really fix the guage? That is to what extent does it eliminate fluctuations of the form

$$h_{\alpha\beta} = \nabla_\alpha V_\beta + \nabla_\beta V_\alpha \tag{1}$$

corresponding to dragging the metric along the integral curves of the vector field $V^\alpha$ where

$$\nabla^\sigma \nabla_\sigma V^\alpha = 0 \tag{2}$$

On Euclidean Schwarzschild there are no $L^2$ solutions of (2) but there exist four vectors giving rise to $L^2$ $h_{\alpha\beta}$'s. These gauge transformations do not become the identity at infinity. One corresponds to time translations, $V^\alpha = \delta^\alpha_0$, the other three correspond to spatial translations of the form

$$V_i^{\alpha} = g^{\alpha\beta} \partial_\beta X_i , \quad i = 1, 2, 3 \tag{3}$$

$$X_i = (r - M) (\sin\theta \cos\phi, \sin\theta \sin\phi, \cos\theta) \tag{4}$$

The three functions $X^i$ may be thought of as cartesian co-ordinates for the hole. Since $V_i^{\alpha} V_{i\alpha} = 0$ at $r = 2M$ the translation moves the hole with respect to infinity. Usually we say that two metrics related by a diffeomorphism represent the same spacetime but for many purposes this is too large an equivalence class - we want to distinguish a black hole at rest from one moving with a non zero velocity and so we restrict our diffeomorphisms to being the identity at infinity. In this way we are left with a "global" finite dimensional Poincare group as the remnant of the infinite dimensional group of "local" gauge transformations. One can go further and introduce collective co-ordinates. Hajicek (1981b) has shown how the momentum conjugate to the translations contribute to the Hamiltonian in the expected way (although he has not made explicit use of the Gibbons - Perry modes). In particular one can construct quantum operations $\exp(iP_\mu x^\mu)$ which translate the hole.

In supergravity things are similar, one must eliminate gravitino fluctuations of the form

$$\psi_\mu = \hat{D}_\mu \varepsilon \tag{5}$$

where $\hat{D}_\mu$ is the covariant derivative in O(1) and the supercovariant derivative in the O(2) theory. If one imposes the gauge condition $\gamma^\mu \psi_\mu$ the analogue of (2) is the Dirac equation

$$\gamma^\mu \nabla_\mu \varepsilon = 0 \tag{6}$$

Solutions of (6) which are asymptotically constant give rise to the super translations. If $\hat{D}_\mu \varepsilon = 0$ we get the analogue of Killing vectors - these leave the background invariant. Corresponding to $\psi_\mu$ of the form (5) with zero energy we have creation operators $a_i^+$, and multiplets of states $|0\rangle$, $a_i^+ |0\rangle$, $a_i^+ a_j^+ |10\rangle$... which degenerate in energy with the ground state $|0\rangle$.

## FERMION MODES IN THE O(1) THEORY

Any static spherically symmetric metric has the form:

$$ds^2 = -A^2(\underline{x})\, dt^2 + B^2(\underline{x})\, d\underline{x}^2 \qquad (6)$$

The Dirac equation has solutions of the form

$$\psi = \psi_0 \; A^{-\frac{1}{2}} B^{-1} \qquad (7)$$

where $\psi_0$ is a time independent solution of the flat space Dirac equation. The current vector for $\psi$, if $\gamma^0\psi_0 = \psi_0$ is

$$\bar{\psi}\, \gamma^\mu\, \psi = (AB)^{-2}\, K^\mu = (AB)^{-2}\, \delta^\mu_0 \qquad (8)$$

where $K^\mu$ is the timelike Killing vector. If $\psi$ is constant at infinity $\psi_0$ must be constant everywhere. Then $\psi$ is singular if $AB^2 = 0$. In the Schwarzschild case, in isotropic co-ordinates, $A = (1 - \frac{M}{2r})\,(1 + \frac{M}{2r})^{-1}$ $B = (1 + \frac{M}{2r})^2$ and so there are no non-singular fermion zero modes. This result is disagreement with that of Yoneya (1978) who seems to have made an error in his calculation. The singularity in $\psi$ can be understood from the point of view of the periodicity of the metric in imaginary time. Any well behaved spinor field on the Euclidean section must be antiperiodic in imaginary time (Gibbons and Perry 1978b). For the same reason the candidate gravitino field proposed by Cordero & Teitelboim (1978) is in fact singular.

There are no zero modes in the Kerr background either as is shown by explicit calculations invoking separating variables. This is true even in the extreme, zero temperature limit. A similar result applies to non gauge solutions of the Rarita-Schwinger equation which are independent of time and which fall off at infinity. Güven (1980) has shown that neutral black holes have no "superhair".

## FERMION ZERO MODES IN THE O(2) THEORY

Aichelberg & Güven (1981) have extended the "No super hair" theorem to the Kerr-Newman solutions except in the non rotating extreme Reisner-Nordstrom case for which $M = G^{-\frac{1}{2}}(Q^2+P^2)^{\frac{1}{2}}$, M the mass, Q electric and P magnetic charge, G is Newtons constant.

An analogous - and not unrelated - result holds for the Dirac equation. In the extreme limit the metric has the form (6) with $AB = 1$ and so we get 4 zero modes. A further calculation shows that 2 of these are supercovariantly constant,

$$\nabla_\mu \psi - \tfrac{1}{4} F_{\alpha\beta}\gamma^\alpha\gamma^\beta\gamma_\mu \psi = 0 \tag{9}$$

This situation also applies in the Papapetrou-Majundar multi black hole metrics (Hartle and Hawking 1972) for which $AB = 1$

$$\nabla^2 B = 0 \tag{10}$$

with a Maxwell field (in pure electric complexion)

$$A_\mu = A\,\delta_\mu{}^0 \tag{11}$$

Recent work with C. Hull strongly **suggests** that these solutions are the <u>only</u> ones admitting supercovariantly constant spinors with a timelike current vector. A null current vector is also possible - this corresponds to the case of p - p waves which is of less interest from the point of view of solitons. C. Hull and I have also shown that only asymptotically flat solution of the Einstein - Maxwell equations must satisfy the Bogomolny inequality

$$M \geq G^{-\frac{1}{2}} (Q^2 + P^2)^{\frac{1}{2}} \tag{12}$$

equality only being possible if there exists a supercovariantly constant spinor.

Given the zero modes one can construct the soliton multiplets as indicated above. The procedure is entirely analogous to the Yang-Mills monopole case discussed by Osborn (1979). One gets the states $|0\rangle$, $a^+|0\rangle$, $a_2^+|0\rangle$, $a_1^+a_2^+|0\rangle$, where $|0\rangle$ is the Boulware-Hartle-Hawking state and $a_1^+$ and $a_2^+$ create the zero energy fermionic states. One thus gets 4-states not 16 as would be the case for general massive supermultiplet. The reason is that Q is a central charge (**Teitelboim** 1977) and if (12) is saturated the multiplets are 4-fold not 16-fold (cf. Olive & Witten 1978).

REFERENCES

Aichelberg, P.C. & Güven, R. (1981). "Can Charged Black Holes have a Super Hair?" Vienna preprint UW Th Ph-81-10.

Cordero, P. & Teitelboim, C. (1978). Phys. Letts 78B 80.

Gibbons, G.W. (1981). "Soliton States and Central Charges in Extended Supergravity Theories" to appear in Proceeding of the Heisenberg Memorial Symposium ed. P. Breitenlohner & H.P. Durr in the Springer Lecture Notes in Physics series.

Gibbons, G.W. & Perry, M.J. (1978a). Proc. Roy. Soc. A 358 467.

Gibbons, G.W. & Perry, M.J. (1978b). Nucl. Phys. B146 90

Güven, R. (1980) Phys. Rev. D22 2327.

Hajicek, P. (1981a). Nucl. Phys. B185 254.

Hajicek, P. (1981b). "Quantum Theory of Wormholes" to appear in "Proceedings of the Tenth Conference on Differential Geometric Methods in Theoretical Physics, Trieste 1981".

Hartle, J.B. & Hawking, S.W. (1972). Commun. Math. Phys. 26 87.

Olive, D. and Witten, E. (1978). Phys. Letts. 78B 99.

Osborn, H. (1979). Phys. Letts. 83B 321.

Teitelboim, C. (1977). Phys. Letts. 69B 240.

Yoneya, T. (1978). Phys. Rev. D17 2567.

# ULTRA-VIOLET PROPERTIES OF SUPERSYMMETRIC GAUGE THEORIES

S. Ferrara
Theory Division, CERN, 1211 Geneva 23, Switzerland

## I HIERARCHIES OF SUPERSYMMETRIES

Supersymmetric gauge theories of the fundamental interactions at present offer the possibility of consistently describing our low energy world as well as the physics at the Planck scale (Fayet & Ferrara 1977; Van Nieuwenhuizen 1981).

In grand unified theories of electroweak and strong interactions it is well known that one is faced with the so-called hierarchy problem of interactions (Gildener & Weinberg 1976; Weinberg 1979). This problem is related to the fact that non-supersymmetric renormalizable theories suffer from quadratic mass renormalization for Higgs particles thus preventing the theory from a natural explanation of a big mass scale $M_X \geq 10^{15}$ GeV and a low mass scale $M_W \geq 100$ GeV. This naturalness problem can be overcome (Maiani 1979; Witten 1981) in $N = 1$ supersymmetric Yang-Mills theories where, under some restrictions, quadratic divergences cannot occur (Wess & Zumino 1974 a,b; Iliopoulos & Zumino 1974; Ferrara & Piguet 1975).

At a more ambitious level, supersymmetric gauge theories of gravity, called supergravities (Van Nieuwenhuizen 1981) have the chance of describing a consistent theory of gravity and at the same time they offer the scheme for a superunification of the gravitational interactions with the non gravitational strong and electroweak forces. Although different options remain open, the hidden local SU(N) symmetry discovered in N-extended supergravity (Cremmer & Julia 1978) suggests that our low energy physics be described in terms of states belonging to composite supermultiplets of the elementary preonic fields of the basic supergravity Lagrangian (Ellis, Gaillard, Maiani & Zumino 1980).

In the dynamical framework offered by extended supergravity theories in which quarks and leptons are composite different scenarios can be envisaged. For example, one may consider the situation in which all supersymmetries are broken at the Planck scale. This is the situation which has been considered by some authors (Ellis, Gaillard & Zumino 1980; Derendinger, Ferrara & Savoy 1981; Frampton 1981; Curtright & Freund 1979). One can give, in this case, some criteria for extracting from a certain set of composite N-extended supermultiplets an effective low energy. It has been argued that the only possible superGUT which can be obtained in this way is the minimal SU(5) GUT with precisely three families in N = 8 supergravity (Ellis, Gaillard & Zumino 1980).

It has recently been pointed out by Barbieri, Ferrara & Nanopoulos(1981 a,b) (Nanopoulos 1981) and also by Ellis, Gaillard & Zumino (1981) that the picture of composites in N-extended supergravity and that of a supersymmetric N = 1 grand unified theory may be consistent, provided not all supersymmetries are broken at the Planck scale. In particular, if we want to reconcile N-extended supersymmetry with the standard model of strong and electroweak interactions based on the gauge group SU(3) × SU(2) × U(1) or with the minimal GUT SU(5) with one supersymmetry surviving at low energies then we are led with $N \geq 6$. The reason is that any N-extended supermultiplet can be decomposed in k-extended supermultiplets with a residual symmetry SU(N-k) commuting with the $k < N$ supersymmetry generators. For k = 1 one gets SU(N-1) as a maximal internal symmetry group. For N = 6 one would get precisely SU(5) while for N = 8 one can get at most SU(7).

In this connection it is interesting to notice that Fayet (1980) has pointed out that if one wants to get an acceptable mass splitting between quarks, leptons and their scalar superpartners one must have an extra chiral U(1) gauge group beyond the usual weak hypercharge which survives at moderate (1 TeV) energies. An independent argument for the existence of an extra low energy gauge symmetry beyond SU(3) × SU(2) × U(1) has been recently given by Weinberg (1981) to provide a natural explanation for the absence of certain couplings giving rise to a too fast proton decay rate in certain supersymmetric GUT models (Dimopoulos & Georgi 1981). It has been pointed out (Barbieri, Ferrara & Nanopoulos 1981b) that the Weinberg argument for a too fast proton decay rate is not very strong. For example, it is sufficient to change by an order of magnitude the product $M_X M_W$ to produce an acceptable rate of the order of $10^{30} \sim 10^{31}$ years. What is very interesting is that if this is the case the dominant modes would be $p \to \nu + \ldots$ rather than the usual modes $p \to e^+ + \ldots$ which are suppressed by a factor $10^4$ - $10^8$ due to the increasing unification point $M_X$. The above authors concluded that this would be evidence for a supersymmetric GUT in proton decay searches and it would rule out the standard SU(5) model.

It has been observed (Barbieri, Nanopoulos & Ferrara 1981b) that if this extra U(1) gauge group is really needed to make a realistic N = 1 supersymmetric low energy theory, then the only supergravity which may provide a low energy gauge theory based on a group SU(3) × SU(2) × U(1) × $\tilde{G}$ is just N = 8 supergravity where G could be a subgroup of SU(2) × U(1) contained in the decomposition SU(8) → (N=1 SUSY) × SU(7) → (N=1 SUSY) × × U(1) × SU(2) × [$SU_C(3)$ × $SU_L(2)$ × U(1)]. It is not without interest to remark that the hierarchy of supersymmetries from N = 8 down to N = 1 would naturally allow the additional gauge group SU(2) × U(1) commuting with SU(3) × SU(2) × U(1) and even with SU(5).

## II ULTRA-VIOLET PROPERTIES IN N = 1 SUPERSYMMETRY YANG-MILLS THEORIES

The great interest raised by supersymmetric gauge theories for the unification of fundamental interactions, some aspects of which we tried to stress in the first part of this report, is mainly due to their exceptional ultra-violet properties. The ultra-violet behaviour is improved with respect to conventional renormalizable theories because of mutual cancellations of boson and fermion loops. These cancellations which occur in quantum corrections are due to the restriction on the couplings dictated by supersymmetry (Fayet & Ferrara 1977).

The absence of quadratic mass renormalization in unbroken (or very softly broken) supersymmetric super Yukawa theories, where only chiral multiplets are involved, was proven a long time ago (Wess & Zumino 1974a; Iliopoulos & Zumino 1974)

The persistence of this phenomenon in Abelian supersymmetric gauge theories was noticed, at the one-loop level, soon after (Wess & Zumino 1974b). The general proof for the absence of quadratic mass renormalization in the physically interesting situation of non-Abelian gauge theories, which is directly related to the hierarchy problem of GUTs, was obtained later using supergraph techniques (Ferrara & Piguet 1975). The very point in supersymmetric gauge theories is that the usual Weinberg power counting formula for the superficial degree of divergence is inadequate if applied to Feynman diagrams for individual component fields. Power counting must be established in superspace, where cancellations between boson and fermion loops are automatically taken into account and this reveals the lowering of independent renormalization constants needed in a general supersymmetric gauge theory. Unfortunately this technique is not yet understood for extended superspace where more spectacular cancellations seem to occur for higher N. In particular, it is generally believed that the maximally extended supersymmetric theories have additional convergence properties due to the PCT self conjugate nature of their elementary supermultiplets. This is confirmed by the fact that in N = 4 Yang-Mills theory there is no infinite charge renormalization in the first three-loop (Tarasov & Vladimirov 1980; Grisaru, Rocek & Siegel 1980; Caswell & Zanon 1980) orders and it is hoped that similar properties are shared by the maximally extended N = 8 supergravity which may eventually lead to a finite quantum theory of gravity.

We would now like to confine our attention to some relations which are responsible for some ultra-violet cancellations in N = 1 supersymmetric gauge theories. Algebraic identities responsible for ultra-violet suppressions in extended supergravities will be considered in the next section.

Let us first consider again N = 1 globally supersymmetric theories with a gauge group G commuting with supersymmetry. A general property of these theories is the absence of quadratic mass renormalization and of independent renormalization for the Yukawa and scalar couplings among chiral multiplets (non-renormalization theorem) (Wess & Zumino 1974a; Iliopoulos & Zumino 1974; Ferrara, Iliopoulos & Zumino 1974).

The first result can be intuitively understood from the fact that scalars are supersymmetric partners of chiral fermions and fermions can at most have a logarithmically divergent self-mass. In other words, scalar masses, through supersymmetry, are protected by chiral invariance. This argument however is only heuristic because it does not explain why even an independent fermionic mass renormalization does not occur, i.e., why the mass and coupling constant renormalization for matter chiral multiplets are only induced by a common wave function renormalization (Wess & Zumino 1974a; Iliopoulos & Zumino 1974; Ferrara, Iliopoulos & Zumino 1974).

For example one has

$$\frac{m_{\text{ren}}}{m_{\text{bare}}} = \left(\frac{g_{\text{ren}}}{g_{\text{bare}}}\right)^{\frac{2}{3}} = Z(\text{wave function renormalization}) \tag{1}$$

This result can be understood from superspace counting arguments. An even stronger result has been obtained by Grisaru, Rocek & Siegel (1979); Grisaru (1981). This result states that in a supersymmetric renormalizable theory no term which can only be written as the last (F or G) component of a chiral multiplet can be generated to any finite order of perturbation theory. In fact this means that some allowed supersymmetric effective couplings and mass terms are never generated in perturbation theory if they are not present in the original bare Lagrangian (supernaturalness).

Let us now consider in more detail the important case of quadratic mass renormalization.

One can derive the general condition under which quadratic divergences are absent by observing that in an arbitrary supersymmetric N = 1 Yang-Mills theory with a gauge group $G = G \times U(1)^P$ where G is semi-simple and with a set of chiral multiplets $S^i$ which transform according to some complex representations $R^i$ of G and with (chiral) charges $q_k^i$ (k=1...p) with respect to $U(1)^P$, the following functional relation is true (Ferrara, Girardello & Palumbo 1979; Ferrara & Palumbo 1981; Girardello & Grisaru 1981)

$$\sum_{J=0,1/2,1} (-)^{2J}(2J+1)m_J^2(Z^i,Z^{*i})d_J = \sum_k \mathrm{Tr}\, Q_k D^k(Z^i,Z^{*i}) \qquad (2)$$

In the left-hand side is the supertrace of the square mass $m_J^2(Z,Z^*)$ of the particle fields of the theory and on the right-hand side is the sum of the D-auxiliary fields of the U(1) factors of G, expressed in terms of the physical complex scalar fields $Z^i$ contained in the chiral multiplets $S^i$ weighted with a factor which is the trace of the corresponding charge matrix $Q_k$: $\mathrm{Tr}\, Q_k = \Sigma_i\, q_k^i$.

The important point is that Eq. (2) is a functional relation and does not depend on the particular value of the fields $Z^i$. In fact, Eq. (2) is a simple consequence of the general form of the scalar, fermion and vector mass matrix of an arbitrary N = 1 Yang-Mills theory, with arbitrary self-interactions among the chiral multiplets (Ferrara, Girardello & Palumbo 1979). As a consequence of this, Eq. (2) is also true when supersymmetry is spontaneously broken, in which case the right-hand side may not vanish.

In the absence of U(1) factors it has been shown that Eq. (2) (no contribution to the right-hand side) is only modified by finite corrections in perturbation theory (Iliopoulos & Girardello 1979). It is reasonable to assume that the same is true in the more general case given by Eq. (2). Equation (2) gives a rather strong constraint on the physical theory based on a spontaneously broken supersymmetry if quadratic divergences must be avoided. This requires that

$$\mathrm{Tr}\, Q_k = \Sigma_i\, q_k^i = 0 \qquad k = 1...p \qquad (3)$$

If Eq. (3) is valid, quadratic divergences are avoided irrespectively of the vacuum expectation value of $D_k$. Incidentally, we observe that Eq. (3) is automatically fulfilled if the gauge group $G \times U(1)^P$ is embedded in a larger semi-simple group G'. This would be the most attractive situation although we cannot exclude that unification in a susy theory requires a fundamental U(1) group which may eventually be related to some space-time property of a more general theory which also includes gravity. Equation (3) was independently obtained using superspace arguments as a necessary and sufficient condition for the absence of quadratic divergences (Fischler et al. 1981). However, we notice the important fact that Eq. (3) does indeed put rather stringent constraints on the mass spectrum of the theory because of Eq. (2).

It is worth remarking that the models proposed by Fayet, which include an extra U(1) gauge factor, do in general suffer from quadratic divergences in addition to dangerous U(1) anomalies which spoil the renormalizability of the theory. Because of the previous difficulties it appears clear that it is non-trivial to build up a realistic GUT model which contains SU(3) × SU(2) × U(1) and which exhibits conventional supersymmetry breaking, the correct hierarchy of gauge symmetries and a realistic mass spectrum (Fayet 1977).

Recently Weinberg has proposed (Weinberg 1981) a systematic approach for selecting a phenomenologically acceptable gauge theory which at low energy reduces to $SU(3) \times SU(2) \times U(1) \times \tilde{G}$ by analyzing the constraints which arise by the requirement that the low energy theory does not break $SU_C(3) \times U(1)_{em}$ and gives a realistic mass spectrum for quarks, leptons and their superpartners.

An alternative possibility to the picture discussed above is that supersymmetry is softly but explicitly broken (Dimopoulos & Georgi 1981) or that supersymmetry is broken by non-perturbative effects (Witten 1981). The first alternative we find unsatisfactory although we must admit the nice feature that the ultra-violet properties of a softly broken supersymmetric theory are not deteriorated by explicit breaking terms if carefully selected (Girardello & Grisaru 1981). A much more appealing possibility is a non-perturbative supersymmetry breaking as emphasized by Witten (1981) and others (Dimopoulos & Raby 1981; Dine, Fischler & Szednicki 1981).

Unfortunately, no example of non-perturbative effects which break supersymmetry in four dimensions are known.

## III ULTRA-VIOLET PROPERTIES OF EXTENDED SUPERGRAVITIES

We now turn to extended supergravity theories where the ultra-violet regime is still poorly known. We know that all extended supergravities from $N = 1$ up to $N = 8$ are one and two-loop finite (Van Nieuwenhuizen 1981) however we do not know what happens at three-loops and beyond. In spite of this fact we may argue that, in analogy to globally supersymmetric gauge theories, extended supergravities with higher N have better quantum properties and in particular, the maximally extended $N = 8$ supergravity, with a PCT self-conjugate multiplet, will have unexpected convergence properties (Townsend 1981). In the $N = 4$ maximally extended supersymmetric Yang-Mills theory it has been pointed out (Ferrara & Zumino 1979b; Sohnius & West 1981) that the global SU(4) [rather than U(4)] invariance may be responsible for the vanishing of the Callan-Symansik function $\beta(g)$ and may eventually give rise to a finite field theory. In analogy we may hope that the SU(8) [rather than U(8)] invariance of the maximally extended $N = 8$ supergravity may be responsible for finiteness of this non-renormalizable theory.

At a less speculative level we know at present that in N-extended supergravities there are some ultra-violet cancellations which cannot be explained with simple invariance arguments but which have an explanation in terms of sum rules which arise from the Clifford algebra structure of the rest frame supersymmetry algebra. One of these properties is the vanishing of the one-loop $\beta$ function (Christensen et al. 1980; Duff 1981) in $N > 4$ extended supergravities with a gauged SO(N) group (Freedman & Das 1977; Fradkin & Vasiliev 1977; de Wit & Nicolai 1981 a,b). Incidentally because of the relation (Freedman & Das 1977; Fradkin & Vasiliev 1977), due to supersymmetry, between the gauge coupling g and the cosmological constant $\Lambda$

$$\Lambda \sim g^2/K^4 \tag{4}$$

this also implies the absence of renormalization of the cosmological terms in these theories. The above property is due to the following sum rules for the helicity spectrum $\lambda$ of extended particle supermultiplets (Curtright 1981)

$$\Sigma_\lambda (-)^{2\lambda} d(\lambda)\lambda^p = 0 \qquad p < N \tag{5}$$

$$\Sigma_\lambda (-)^{2\lambda} d(\lambda)\lambda^p C_2(\lambda) = 0 \qquad p + 2 < N \tag{6}$$

$d(\lambda)$ is the multiplicity of the state of helicity $\lambda$ and $C_2(\lambda)$ is the quadratic SO(N) [or SU(N)] Casimir for the corresponding SO(N) [or SU(N)] representation. Equation (5) also explains the vanishing at the one-loop of the $\beta$ function in N = 4 Yang-Mills theory by observing that the one-loop $\beta$ function for a particle of helicity $\lambda$ can be parametrized as follows (Curtright 1981)

$$\beta_\lambda \sim (-)^{2\lambda} C_2(\lambda)(a+b\lambda^2) \qquad (7)$$

where a, b are $\lambda$-independent coefficients and $C_2(\lambda)$ is the quadratic Casimir of the corresponding representation of the gauge group. For N = 4 Yang-Mills $C_2(\lambda)$ is $\lambda$-independent and Eq. (7) reduces to Eq. (5). This is not the case in extended supergravity with gauged SO(N) where the more restrictive formula (6) must be used. Equation (5) is used in this case to prove, as an alternative proof, the absence of the renormalization of $\Lambda$. As pointed out by Christensen et al. (1980), Eq. (4) gives an amusing relation between apparently unrelated Feynman graphs because, for instance, the graviton loop contributes to the one-loop computation of $\Lambda$ but not to the one-loop computation of $\beta$ since the graviton is an SO(N) singlet.

Another ultra-violet cancellation occurring in extended supergravity, which at first sight seems to be unrelated to the one we have been discussing till now, is connected to the mass formulae of spontaneously broken supergravity through dimensional reduction (Scherk & Schwarz 1979). These mass relations, which are a generalization of Eq. (3), imply that the one-loop cosmological term is finite (Zumino 1975) in N $\geq$ 6 supergravities and are crucial to prove that these theories are one-loop renormalizable (Sezgin & Van Nieuwenhuizen 1981).

In an N-extended supergravity, spontaneously broken through dimensional reduction from five-dimensions, one has (Scherk & Schwarz 1979; Cremmer, Scherk & Schwarz 1979; Ferrara & Zumino 1979a)

$$\Sigma_J (-)^{2J}(2J+1)\, m_J^{2k}(m_i) = 0 \qquad 2k < N \qquad (8)$$

where $m_J$ is the mass of the particle of spin-J of the graviton supermultiplet and $m_i$ are the parameters (i=1...4 for N=8) introduced through dimensional reduction which set up the scale of the supersymmetry breaking.

As anticipated above it is remarkable that the spin and mass sum rules given by Eqs. (5), (6) and (8) have the same origin. They are in fact a consequence of the Clifford algebra structure of the rest-frame supersymmetry algebra for massive and massless supermultiplets (Ferrara, Savoy & Girardello 1981).

These sum rules can be summarized as follows: the supertrace of the p-power of any operator O which is bilinear in the rest-frame supersymmetry generators, belonging to the SO(4N) algebra [SO(2N) in the massless case] satisfies the following property (Ferrara, Savoy & Girardello 1981)

$$\text{Supertrace } O^p = 0 \quad p < 2N \quad (p < N \text{ in the massless case}) \tag{9}$$

If we consider the helicity $\Lambda$ in the massless case we get

$$\text{Supertrace } \Lambda^p = 0 \quad p < N \tag{10}$$

On one particle states this gives

$$\Sigma_{k=0}^{N} \binom{N}{k} (\lambda_{MAX} - \tfrac{k}{2})^p (-)^k = 0 \qquad p < N \tag{11}$$

this is nothing but formula (5). In the massless case we can combine the helicity $\Lambda$ with an arbitrary SU(N) generator [contained in O(2N)] and we get

$$\text{Supertrace } \Lambda^h T^{i_1}_{j_1} \ldots T^{i_n}_{j_n} = 0 \qquad h + n < N \tag{12}$$

For n = 2, by contraction, we obtain

$$\text{Supertrace } \Lambda^h C_2 = 0 \qquad h + 2 < N \tag{13}$$

which is Eq. (6).

Another interesting sum rule which involves the group theoretical factor of the SU(N) axial anomaly is

$$\text{Supertrace } \Lambda^h A = 0 \qquad h + 3 < N \tag{14}$$

where

$$A(R) = \frac{1}{2} \underset{(R)}{\text{Tr}}\, T^i_j \{T^j_k, T^k_i\} \tag{15}$$

Finally, in order to obtain (8) and its possible generalization we must apply (9) to the massive supersymmetry algebra with central charges and with maximal spin reduction. In this case we have still an SO(2N) algebra in the rest-frame but we apply Eq. (9) to the generators of $SU(2)_{spin} \times$ $\times\, USp(N) \subset SO(2N)$. In spontaneously broken supergravity through dimensional reduction, the mass operator is an element of the Cartan subalgebra of USp(N) (Scherk & Schwarz 1979; Cremmer, Scherk & Schwarz 1979) (which has rank N/2) so the particle masses depend on N/2 parameters.

Since $\mathcal{M}$ belongs to USp(N) from (9) we get

$$\text{Supertrace } (S^2)^h \mathcal{M}^{2k} = 0 \quad \text{for} \quad 2h + 2k < N \quad \text{or} \quad h + k < [N/2] \tag{16}$$

$S^2$ in the spin Casimir.

For h = 0 we obtain the mass sum rules given by Eq. (8). For h > 0 Eq. (16) gives rise to new sum rules which combine spin factors and the masses.

As previously mentioned, the mass formulae given by (8) or by (16) for h = 0 and k = 1, 2 imply that the one-loop induced cosmological constant is finite. Sum rules with higher values of h and k have not yet an analogous physical interpretation.

However, it might be possible that these sum rules turn out to be relevant for higher loop calculations and eventually help to find hidden ultra-violet suppressions which could not be deduced in the absence of invariance arguments.

REFERENCES


Barbieri, R., Ferrara, S. & Nanopoulos, D.V. (1981a). CERN preprint TH.3159.

Barbieri, R., Ferrara, S. & Nanopoulos, D.V. (1981b), to be published.

Caswell, W.E. & Zanon, D. (1980). Phys. Lett. B100, 152.

Christensen, S.M., Duff, M.J., Gibbons, G.W. & Rocek, M. (1980). Phys. Rev. Lett. 45, 161.

Cremmer, E. & Julia, B. (1978). Phys. Lett. 80B, 48; Nucl. Phys. B159, 41 (1979).

Cremmer, E., Scherk, J. and Schwarz, J. (1979). Phys. Lett. 84B, 83.

For an earlier attempt see also:

Curtright, T.L. & Freund, P.G.O. (1979). In Proceedings of the Supergravity workshop at Stony Brook (Sept. 1979), ed. by P. van Nieuwenhuizen and D.Z. Freedman (North Holland, Amsterdam), p. 197.

Curtright, T.L. (1981). Phys. Lett. 102B, 17.

Derendinger, J.P., Ferrara, S. & Savoy, C.A. (1981). Nucl. Phys. B188, 77.

de Wit, B. & Nicolai, H. (1981a). Nucl. Phys. B188, 98.

de Wit, B. & Nicolai, H. (1981b). CERN preprint TH.3183.

Dimopoulos, S. & Georgi, H. (1981). Nucl. Phys. B193, 150.

Dimopoulos, S. & Raby, S. (1981). Nucl. Phys. B192, 513.

Dine, M., Fischler, W. & Szednicki, M. (1981). Princeton preprint.

Duff, M. (1981). To appear in the proceedings of the Supergravity School held at Trieste, Cambridge University Press, ed. by S. Ferrara and J.C. Taylor.

Ellis, J., Gaillard, M.K., Maiani, L. & Zumino, B. (1980). In "Unification of the Fundamental Particle Interactions", ed. by S. Ferrara, J. Ellis and P. van Nieuwenhuizen (Plenum Press, New York, 1980).

Ellis, J., Gaillard, M.K. & Zumino, B. (1980). Phys. Lett. 94B, 343.

Ellis, J., Gaillard, M.K. & Zumino, B. (1981). LAPP-TH-44/CERN-TH.3152, to be published in Acta Physica Polonica.

Fayet, P. (1977). Phys. Lett. 69B, 489; 70B, 461. New Frontiers in High Energy Physics, ed. by A. Perlmutter and L.F. Scott (Plenum Press, New York, 1978).

See, for instance

Fayet, P. (1980). In "Unification of the Fundamental Particle Interactions", ed. by S. Ferrara, J. Ellis and P. van Nieuwenhuizen (Plenum Press, New York, 1980).

For a review on supersymmetry see:

Fayet, P. & Ferrara, S. (1977). Phys. Rep. 32C, No.5, 249.

Ferrara, S., Iliopoulos, J. & Zumino, B. (1974). Nucl. Phys. B77, 413.

Ferrara, S., Girardello, L. & Palumbo, F. (1979). Phys. Rev. D20, 403.

The mass relation with the inclusion of a chiral U(1) factor where the right-hand side may not vanish was obtained by:

Ferrara, S. & Palumbo, F. (1981) (unpublished).

Ferrara, S. & Piguet, O. (1975). Nucl. Phys. B93, 261.

Ferrara, S., Savoy, C.A. & Girardello, L. (1981). Phys. Lett. 105B, 363.

Ferrara, S. & Zumino, B. (1979a). Phys. Lett. 86B, 279.

Ferrara, S. & Zumino, B. (1979b). Unpublished.

Fradkin, E.S. & Vasiliev, M.A. (1977). Lebedev Institute preprint.

Frampton, P. (1981). Phys. Rev. Lett. 46, 381.

Freedman, D.Z. & Das, A. (1977). Nucl. Phys. B170, 221.

Fischler, W., Nilles, H.P., Polchinski, J., Raby, S. & Susskind, L. (1981). SLAC preprint PUB-2760.

Gildener, E. & Weinberg, S. (1976). Phys. Rev. D13, 3333.
Girardello, L. & Grisaru, M.T. (1981). Harvard preprint.
Grisaru, M.T. (1981). Brandeis preprint, to appear in the Proceedings of the Supergravity School, Trieste, 1981.
Grisaru, M.T., Rocek, M. & Siegel, W. (1979). Nucl. Phys. B159, 429.
Grisaru, M.T., Rocek, M. & Siegel, W. (1980). Phys. Rev. Lett. 45, 1063.
Iliopoulos, J. & Zumino, B. (1974). Nucl. Phys. B76, 310.
Iliopoulos, J. & Girardello, L. (1979). Phys. Lett. B88, 85.
Maiani, L. (1979). In Proc. of the Summer School at Gif-sur-Yvette, p. 3.
Nanopoulos, D.V. (1981). To appear in the Proceedings of the EPS Conference on "Unification of the Fundamental Particle Interactions", ed. by J. Ellis and S. Ferrara, Erice (1981).
Scherk, J. & Schwarz, J. (1979). Nucl. Phys. B153, 61.
Sezgin, E. & van Nieuwenhuizen, P. (1981). Stony Brook preprint NSF-ITP-81-47.
Sohnius, M.F. & West, P.C. (1981). Phys. Lett. B100, 245.
Tarasov, O.V. & Vladimirov, A.A. (1980). Dubna preprint.
Townsend, P. (1981). CERN preprint TH.3066. Lecture given at the 18th Winter School of Theoretical Physics, Karpacz, Poland, March 1981.
For a review on supergravity see:
van Nieuwenhuizen, P. (1981). Phys. Rep. 68, No.4, 189.
Weinberg, S. (1979). Phys. Lett. 82B, 387.
Weinberg, S. (1981). Harvard preprint HUTP-81/A047.
Wess, J. & Zumino, B. (1974a). Phys. Lett. 49B, 52.
Wess, J. & Zumino, B. (1974b). Nucl. Phys. B78, 1.
Witten, E. (1981). Nucl. Phys. B188, 513.
Zumino, B. (1975). Nucl. Phys. B89, 535.

# EXTENDED SUPERCURRENTS AND THE ULTRAVIOLET FINITENESS OF N = 4 SUPERSYMMETRIC YANG-MILLS THEORY

K.S. Stelle
Laboratoire de Physique Théorique de l'Ecole Normale Supérieure
24 rue Lhomond, F-75231 Paris cedex 05, France

## INTRODUCTION

Without a doubt the most striking formal property of supersymmetric theories is the cancellation of ultraviolet infinities between bosonic and fermionic contributions to the quantum amplitudes. Such cancellations have generally been discovered in explicit calculations, with a fundamental understanding of their origin following later from a formal analysis that strives to keep the supersymmetry manifest. Indeed, the desire to understand the deeper reasons for the ultraviolet properties of supersymmetric theories has been the spring for much of the large body of work on the representations of supersymmetry on fields.

Two basic situations arise in connection with the cancellations of infinities : either the cancellations are forced by the non-existence of an appropriate supersymmetric invariant that could serve as a counterterm, or such an invariant does exist but nonetheless cancellation takes place due to more involved details of the quantum calculations. The first situation is encountered in the quantum corrections to supergravity theories at the one- and two-loop orders where, for example, there is no supersymmetric invariant to generalize the counterterm $\int e\, R_{\mu\nu}{}^{\alpha\beta} R_{\alpha\beta}{}^{\rho\sigma} R_{\rho\sigma}{}^{\mu\nu}$ expected on the basis of general covariance alone. The second situation is encountered in the absence of mass and coupling constant renormalizations in the interacting Wess-Zumino model of a single chiral supermultiplet. In this case, although the existence of an appropriate counterterm is guaranteed since the model is renormalizable, the cancellations may be understood by use of the manifestly supersymmetric superfield Feynman rules, which do not give rise to chiral type invariants in the quantum corrections to the effective action.

As we press the investigation of infinities on to the extended supersymmetric theories, we are hampered by the absence of manifestly covariant formulations of the Feynman rules such as we have for N = 1 supersymmetry. This is true even for N = 2 Yang-Mills theory and supergravity, where despite our knowing the auxiliary fields needed to complete the physical fields to a representation of a closed algebra of supersymmetry plus gauge transformations, the fixing of the gauge transformations cannot yet be done while preserving the supersymmetry. Nonetheless, it is reasonable to assume that the general form of the extended supersymmetry transformations and associated invariants will be preserved by the quantum corrections, with appropriate wave function and coupling constant renormalizations.

In supersymmetric theories the full invariance of the effective action implies not only relations between infinities, but also between the finite parts left after renormalization, and in particular between anomalies. There is thus a complementarity between the study of supersymmetric renormalization and the study of the quantum supermultiplet containing the stress tensor, the supersymmetry currents and the U(N) currents.

This article examines the questions of renormalization and stress tensor multiplet structure for N = 1 to N = 4 supersymmetries. The most interesting question to be faced is why the N = 4 supersymmetric Yang-Mills theory appears to have a vanishing β-function, as has been discovered by explicit calculations carried out to the three loop order (Tarasov et al. 1980 a, b ; Grisaru et al. 1980, 1981; Caswell & Zanon 1981). Since a manifestly N = 4 supersymmetric set of Feynman rules for this theory is not known, any attempt to explain the vanishing of the β-function from fundamental principles must rely upon a set of assumptions concerning the preservation of the extended supersymmetries or of the internal symmetries of the model. Arguments directed toward a demonstration that the β-function vanishes have been given by Ferrara & Zumino (1978) and by Sohnius & West (1981 a). These arguments rely on properties of the stress tensor multiplet in the presence of anomalies in N = 1 and N = 2 supersymmetry. After a discussion of the multiplet structure for anomalies in N = 1 and N = 2 supersymmetries, we re-examine the arguments given concerning N = 4 Yang-Mills theory in the light of the structure of the full N = 4 stress tensor supermultiplet. In the end, the arguments based purely on symmetries of abstract stress tensor multiplets are not sufficient, as can be seen from the existence of an abstract multiplet containing a spin two component that is not traceless plus nine spin one components that are not conserved and six that are conserved. If this abstract multiplet could describe the anomalous stress tensor multiplet under N = 4 supersymmetry, with anomalies for the nine chiral currents of SU(4) while leaving conserved the six non-chiral currents of SO(4), then there would seem to be no reason why the β-function and the associated trace anomaly should vanish.

In order to explain the β-function's vanishing, it is necessary to take into account more specific details of the quantum calculations. A direct examination of the constraints imposed upon the renormalization constants by gauge invariance, extended supersymmetry and a non-renormalization theorem for the scalar potential term in the theory reveals the underlying reason for the theory's finiteness. The implications of gauge invariance can be assured by use of the background field method, while the non-renormalization of the scalar potential can be clearly seen if the calculation is performed using N = 1 superfield Feynman rules. The essential assumption in this approach is that all four supersymmetries are preserved in the quantum corrections, or alternatively, that the SO(4) non-chiral internal symmetry is preserved. This type of argument should also be generalizable to explain the vanishing β-function in $N \geq 5$ gauged supergravities with a scalar field potential and a cosmological constant.

## 1 SUPERCURRENTS AND ANOMALIES IN N = 1 SUPERSYMMETRY

The structure of the supercurrent multiplet in N = 1 supersymmetry was derived by Ferrara & Zumino (1975). For conformally invariant theories this multiplet contains the stress tensor $T_{\mu\nu}$, the supersymmetry current $J_\mu{}^\alpha$ and an axial current $J_\mu{}^5$, satisfying two sets of constraints : conservation of the stress tensor and the supersymmetry current,

$$\partial^\mu T_{\mu\nu} = \partial^\mu J_\mu{}^\alpha = 0 \qquad (1.1)$$

and three conditions related to the conformal invariance of the theory

$$T^\mu{}_\mu = \sigma^\mu_{\alpha\dot\alpha} J_\mu{}^{\dot\alpha} = \partial^\mu J_\mu{}^5 = 0 \, . \qquad (1.2)$$

The constraints (1.1) and (1.2) leave an irreducible supermultiplet, with an equal number of Bose and Fermi components, technically classified by the value of its superspin, which is 3/2. To summarize the multiplet, we may make a little table,

Table 1
N=1 SS=3/2

| Spin | SL(2, C) | dim | constraints |
|---|---|---|---|
| 2 | (1, 1) | 4 | $\partial^\mu T_{\mu\nu} = 0$ |
| 3/2 | (1/2, 1) + (1, 1/2) | 7/2 | $\partial^\mu J_\mu = 0$ |
| 1 | (1/2, 1/2) | 3 | $\partial^\mu J^5_\mu = 0$ |

Superfield : $V^{\alpha\dot\alpha}(x, \theta, \bar\theta)$ real, with $\bar D_{\dot\alpha} V^{\alpha\dot\alpha} = 0$ (2 component notation)

Since the stress tensor is required to be traceless in this irreducible representation, its explicit form must include the correct improvement terms for scalar fields. Similar terms must be included in the supersymmetry current to ensure "σ-tracelessness".

In the case of a non-conformally-invariant theory, the trace of the stress tensor cannot be zero. Since complete supermultiplets must have equal numbers of Bose and Fermi components, some of the other constraints on the supercurrent multiplet must be relaxed as well. Just as, in the absence of supersymmetry, a $T_{\mu\nu}$ which is not traceless describes a reducible representation of the Lorentz group, (1, 1) + (0, 0), so in supersymmetric theories the supercurrent multiplet is reducible for non-conformal theories. The various possible forms of non-conformal supercurrents are determined by the extra lower-superspin representations that can be present. There are two possibilities, corresponding to the representations with superspin 0 and 1/2, both of which contain a component scalar that can become the trace of $T_{\mu\nu}$.

The superspin 0, or chiral multiplet is

Table 2
N=1 SS=0

| Spin | SL(2, C) | dim | identification |
|---|---|---|---|
| 1/2 | (1/2, 0) + (0, 1/2) | 7/2 | $\sigma^\mu J_\mu$ |
| 0 | (0, 0) x 2 | 4 | $T^\mu{}_\mu$, $\partial^\mu J_\mu{}^5$ |
|  | (0, 0) x 2 | 3 | $\hat M$, $\hat N$ |

Superfield : $C(x, \theta, \bar\theta)$ complex, with $\bar D_{\dot\alpha} C = 0$ .

If the "trace supermultiplet" is a chiral multiplet, then not only are the Lorentz constraints on the lower spin parts of $T_{\mu\nu}$ and $J_\mu$ relaxed, but also the axial current $J_\mu{}^5$ is no longer conserved. In addition, there is now a new pair of scalar and pseudoscalar components denoted $\hat{M}$ and $\hat{N}$ that do not have any direct connection with currents, but which are required to balance the numbers of Bose and Fermi fields in the multiplet. The superfield equation that expresses the reducibility of the supercurrent multiplet in this case is

$$\bar{D}_{\dot{\alpha}} V^{\alpha\dot{\alpha}} = D^{\alpha} C \, . \qquad (1.3)$$

The other basic possibility for a trace supermultiplet is the superspin 1/2, or linear multiplet,

Table 3
N=1 SS=1/2

| Spin | SL(2, C) | dim | constraint | identification |
|---|---|---|---|---|
| 1 | (1/2, 1/2) | 3 | $\partial^{\mu} H_{\mu} = 0$ | |
| 1/2 | (1/2, 0) + (0, 1/2) | 5/2 | | $(\not\partial)^{-1} \sigma^{\mu} J_{\mu}$ |
| 0 | (0, 0) | 2 | | $(\Box)^{-1} T^{\mu}{}_{\mu}$ |

Superfield : $L(x, \theta, \bar{\theta})$ real, with $D^2 L = \bar{D}^2 L = 0$.

In order to balance the number of Bose and Fermi components in this multiplet, the vector component must be required to be transverse. The superspin 1/2 multiplet does not appear in the supercurrent directly in the non-local form given above. The reducibility of the supercurrent is given by

$$\bar{D}_{\dot{\alpha}} V^{\alpha\dot{\alpha}} = \bar{D}^2 D^{\alpha} T \qquad (1.4)$$

where T is a general real scalar superfield. It is not necessary to impose the superspace differential constraints to make T contain only the superspin 1/2 part, because the operator acting on T in (1.4) projects onto this irreducible representation only. Equivalently, one may say that the right-hand side of (1.4) is a spinor superfield $\lambda^{\alpha}(x, \theta, \bar{\theta})$ which is chiral, $\bar{D}_{\dot{\beta}} \lambda^{\alpha} = 0$. This is another way of representing the superspin 1/2 multiplet, in which the vector appears only through its curl, and $T^{\mu}{}_{\mu}$ and $\sigma^{\mu} J_{\mu}$ appear directly as components of dimension 4 and 7/2.

The most salient feature of a non-conformal supercurrent in which the linear multiplet appears is that the axial current remains conserved. This can be seen from the fact that there is only one scalar component in the linear multiplet, and this must be $T^{\mu}{}_{\mu}$. It can also be seen directly from

(1.4), for

$$\partial_{\alpha\dot{\alpha}} V^{\alpha\dot{\alpha}} = \frac{1}{2} \mathrm{Im}(D_\alpha \bar{D}^2 D^\alpha T) = 0 . \tag{1.5}$$

Both types of non-conformal supercurrents occur in field theory examples. In a model containing scalars, even if the theory is conformally invariant, a stress tensor may be constructed without the improvement terms, or with the wrong amount of improvement. Depending upon the coefficients, the trace supermultiplet will be chiral, linear, or will contain terms of both types. This is discussed in detail in the article by M. Duff and P.K. Townsend in this volume. Another simple example is the Wess-Zumino model of a massive chiral multiplet (one propagating Majorana spinor, one scalar, one pseudo-scalar) with supercurrent

$$V_{\alpha\dot{\alpha}} = D_\alpha \phi \bar{D}_{\dot{\alpha}} \bar{\phi} + 2i \phi \overleftrightarrow{\partial}_{\alpha\dot{\alpha}} \bar{\phi} \tag{1.6}$$

where $\phi$ is the chiral superfield containing the fields of the model, $\bar{D}_{\dot{\alpha}} \phi = 0$. In the massive model, the field equations read

$$\bar{D}^2 \bar{\phi} = m \phi , \tag{1.7}$$

so the supercurrent is reducible and contains a chiral multiplet,

$$\bar{D}_{\dot{\alpha}} V^{\alpha\dot{\alpha}} = - \frac{m}{2} D^\alpha(\phi^2) . \tag{1.8}$$

An example in which the trace supermultiplet is linear is the massless model whose fields form a linear multiplet, where the transverse vector is the field strength for a gauge antisymmetric tensor. This model contains the same spin content as the Wess-Zumino model, which is based upon a chiral multiplet, but with the pseudoscalar replaced by the antisymmetric tensor. Due to the presence of the antisymmetric tensor, the model is not conformally invariant even though it is massless. The supercurrent is

$$V_{\alpha\dot{\alpha}} = D_\alpha G \bar{D}_{\dot{\alpha}} G , \tag{1.9}$$

where G is the real linear superfield containing the scalar, spinor and conserved vector field strength for the antisymmetric tensor, $D^2 G = \bar{D}^2 G = 0$.

The field equations read

$$\partial^{\alpha\dot{\beta}} \bar{D}_{\dot{\beta}} G = 0 , \qquad (1.10)$$

and in consequence the supercurrent is reducible and contains a linear multiplet,

$$\bar{D}_{\dot{\alpha}} V^{\alpha\dot{\alpha}} = \frac{1}{2} \bar{D}^2 D^{\alpha}(G^2) . \qquad (1.11)$$

The bilinear superfield $G^2$ is reducible, containing both a superspin 1/2 and a superspin 0 part, but the operator $\bar{D}^2 D^{\alpha}$ picks out only the linear, superspin 1/2 part.

Both types of non-conformal supercurrents also occur in theories with anomalies. This can be seen without regard to any particular regularization scheme in the BPHZ formalism (Clark et al. 1978). It can also be verified directly by performing quantum calculations using different regularization schemes. For example, the massless, interacting Wess-Zumino model

$$I = \int d^4x\, d^4\theta\, \bar{\phi}\, \phi + g \int d^4x\, d^2\theta\, \phi^3 \qquad (1.12)$$

can be quantized while maintaining manifest supersymmetry using Pauli-Villars regularization or using higher-derivative regularization. Pauli-Villars regularization adds a single massive ghost multiplet K and replaces the interaction term by $(\phi + K)^3$. This regularization breaks the conformal invariance and the associated special supersymmetry and also the axial U(1) invariance of (1.12). The result for the anomalies, given in Grisaru (1979), is of the form (1.3),

$$\bar{D}_{\dot{\alpha}} V^{\alpha\dot{\alpha}} = c\, D^{\alpha}(\phi\, \bar{D}^2\, \bar{\phi}) , \qquad (1.13)$$

where c is the anomaly coefficient.

On the other hand, if one uses higher derivative regularization, the action (1.12) is supplemented by the regular term $-1/m^2 \int d^4x\, d^4\theta\, \bar{\phi}\, \Box\, \phi$. The propagators for all the ordinary physical fields now have denominators $k^2(k^2 + m^2)$ ; in addition, the auxiliary fields of (1.12) now have ghost propagators $(k^2 + m^2)^{-1}$. If one separates the physical field propagators into partial fractions, the theory can be seen to effectively propagate one massless physical multiplet and two massive ghost multiplets. So, Pauli-Villars and higher derivative regularization differ in that the higher derivative scheme effectively introduces two Pauli-Villars regulator ghost multiplets. While this still breaks conformal invariance and the special supersymmetry, it leaves the axial U(1) invariance unbroken. This can easily be seen since the regulator term is invariant under the same phase rotations of $\phi$ and $\theta$ as the original action (1.12). Since the axial invariance is respected by this regularization, it cannot be broken by anomalies, so the anomaly equation must be of the form (1.4),

$$\bar{D}_{\dot{\alpha}}\, V^{\alpha\dot{\alpha}} = c\, \bar{D}^2\, D^{\alpha}\, (\phi\, \bar{\phi})\ . \tag{1.14}$$

Both (1.13) and (1.14) give the anomaly multiplet as the kinetic part of the Lagrangian multiplet, but in two forms corresponding to Lagrangian that differ by a total derivative. The kinetic terms can be written as in (1.12) or they may be written in terms of an integral over $d^2\theta$ ,

$$I^{kin} = \int d^4x\, d^2\theta\, (\phi\, \bar{D}^2\, \bar{\phi})\ , \tag{1.15}$$

corresponding to (1.4) and (1.3) respectively.

In the case of N = 1 supersymmetric Yang-Mills theory, there is only one possible structure for the anomaly multiplet due to the added constraint of gauge covariance. Although it is possible to write the action as an integral over all superspace, the resulting superspace Lagrangian is not gauge covariant, and so is unsuitable to use in the right-hand side of (1.4). The form (1.3) can be written covariantly, however, using the chiral Lagrangian for the theory,

$$\bar{D}_{\dot{\alpha}}\, V^{\alpha\dot{\alpha}} = d\, D^{\alpha}\, (\mathrm{tr}\, W^{\beta}\, W_{\beta}) \tag{1.16}$$

where $W_\beta$ is the field strength multiplet containing the spinor field, the Yang-Mills field strength and the D auxiliary field.

The different forms of non-conformal supercurrents correspond to different sets of auxiliary fields for supergravity, which must couple to the supercurrent. Thus, the non-conformal supercurrent containing a chiral multiplet, (1.3), couples to the minimal representation of N = 1 supergravity (Stelle & West 1978; Ferrara & van Nieuwenhuizen 1978). The non-conformal supercurrent (1.4) incorporating a linear multiplet couples to the new minimal representation (Sohnius & West 1981 b). In this case the constrained vector in the linear trace supermultiplet must couple to an axial gauge field. It is also possible to have both linear and chiral multiplets in the supercurrent, as for example in the free Wess-Zumino model with general coefficients for the scalar and pseudoscalar improvement terms. In this case the coupling must be made to the non-minimal supergravity representation (Breitenlohner 1977 a,b; de Wit & Grisaru 1978).

## 2 N = 2 SUPERSYMMETRY

In N = 2 supersymmetry, the analysis of the supercurrent and its non-conformal structure must take into account the internal symmetry properties of the multiplets as well. Irreducible representations are now classified by two quantum numbers, superspin and superisospin, the latter referring to the representations assigned under the internal SU(2) which is an automorphism of the N = 2 supersymmetry algebra. For the classification of extended supersymmetry multiplets, see Ferrara (1980) ; Ferrara et al. (1981) ; Taylor (1981) and Rittenberg & Sokatchev (1981).

The conformal supercurrent in N = 2 supersymmetry and one of the possible trace supermultiplets have been discussed by Sohnius (1979). The conformal supercurrent is the superspin 1, superisospin 0 multiplet,

Table 4
N=2 SS=1 SIS=0

| Spin | SL(2, C) | SU(2) | dim | identification |
|---|---|---|---|---|
| 2 | (1, 1) | 1 | 4 | $T_{\mu\nu}$ : $\partial^\mu T_{\mu\nu} = 0$ |
| 3/2 | (1/2, 1) + (1, 1/2) | 2 | 7/2 | $J_{\mu\alpha}{}^i$ : $\partial^\mu J_{\mu\alpha}{}^i = 0$ |
| 1 | (1/2, 1/2) | 3 | 3 | $J_\mu{}^{ij}$ : $J_\mu{}^{ij} = 0$ |
| | (1/2, 1/2) | 1 | 3 | $J_\mu{}^5$ : $\partial^\mu J_\mu{}^{(5)} = 0$ |
| | (0, 1) + (1, 0) | 1 | 3 | $A_{\mu\nu}$ |
| 1/2 | (1/2, 0) + (0, 1/2) | 2 | 5/2 | $\lambda_\alpha{}^i$ |
| 0 | (0, 0) | 1 | 2 | C |

Superfield : $V(x, \theta^i, \bar\theta_j)$ real, with $D^{\alpha i} D_\alpha{}^j V \equiv D^{(ij)} V = \bar D^{(ij)} V = 0$ .
The index-i runs from 1 to 2.

An example of a theory with a supercurrent of the above form is the "scalar hypermultiplet" (Fayet 1976), which is described in superspace (Sohnius 1978 b) by a Lorentz scalar, SU(2) isodoublet superfield $\phi_i(x, \theta, \bar\theta)$, satisfying the field equations

$$D_\alpha{}^i \, \phi_j = \frac{1}{2} \delta^i{}_j \, D_\alpha{}^k \, \phi_k$$
$$\bar{D}_{\dot{\alpha} i} \, \phi_j = \frac{1}{2} \varepsilon_{ij} \, \bar{D}_{\dot{\alpha}}{}^k \, \phi_k \, . \qquad (2.1)$$

The field content of the theory is a scalar isodoublet and a Dirac spinor. In order to formulate the theory off-shell it is necessary to include a dependence of $\phi_i$ upon a central charge coordinate $x^5$ as well, keeping (2.1) as constraints which no longer imply the field equations but instead allow only one more field, an auxiliary scalar isodoublet. In the following, we shall not need to consider the central charge coordinate because all the fields must satisfy their field equations in order for the supercurrent multiplet to satisfy the various trace and conservation properties given in Table 4. The supercurrent for the scalar hypermultiplet is

$$V = \bar{\phi}^i \, \phi_i \qquad (2.2)$$

which satisfies the differential constraints required for irreducibility.

Another example of an N = 2 superconformal theory is the N = 2 supersymmetric Yang-Mills theory, whose field strength multiplet is a chiral scalar superfield W satisfying the field equations (Grimm et al. 1978)

$$D^{ij} \, W = 0 \qquad ; \qquad \bar{D}_{\dot{\alpha} i} \, W = 0 \, . \qquad (2.3)$$

The supercurrent for this theory is

$$V = \bar{W} \, W \, . \qquad (2.4)$$

A detailed general account of the supercurrents in extended supersymmetric theories is given in Howe et al. (1981 b). This article also presents methods for extracting the component content of such extended superfields, keeping track of Lorentz and internal symmetry representations by Young tableau techniques.

Non-conformal N = 2 theories have a reducible supercurrent, as in N = 1. The representations available to act as trace supermultiplets are limited by the requirement that they contain no new unwanted high spin currents. The irreducible representations of N = 2 supersymmetry contain more spins than those in N = 1 theories. For example, the basic representation is the superspin 0, superisospin 0 representation (or "N = 2 linear multiplet"),

Table 5
N=2 SS=0 SIS=0

| Spin | SL(2, C) | SU(2) | dim | identification |
|---|---|---|---|---|
| 1 | (1/2, 1/2) | 1 | 4 | $V^{\mu}\ :\ \partial^{\mu} V_{\mu} = 0$ |
| 1/2 | (1/2, 0) + (0, 1/2) | 2 | 7/2 | $\sigma^{\mu} J_{\mu}$ |
| 0 | (0, 0) | 1 + 1 | 4 | $T^{\mu}{}_{\mu}\ ,\ \partial^{\mu} J_{\mu}{}^{5}$ |
|  | (0, 0) | 3 | 3 | $X^{ij}$ |

Superfield : $L^{(ij)} = \varepsilon^{ir}\,\varepsilon^{js}\,\bar{L}_{rs}$ with $D_{\alpha}{}^{(i}\,L^{jk)} = \bar{D}_{\dot{\alpha}}{}^{(i}\,L^{jk)} = 0.$

This is the representation that occurs as the trace supermultiplet for the scalar hypermultiplet theory with a mass. In this case, the supersymmetry algebra must include a central charge transformation, and the new conserved vector component is the current for the central charge transformations. The reducibility equation for the supercurrent is

$$D^{ij}\,V = \bar{D}^{ij}\,V = L^{ij} \tag{2.5}$$

Note that the axial current $J_{\mu}{}^{5}$ is not conserved in this case.

Other types of trace supermultiplets can appear in N = 2 supersymmetry. Higher superspin representations would introduce higher spin components satisfying conservation conditions, and we wouldn't have an interpretation for these within the context of N = 2 supersymmetry, but there still remains the possibility of different superisospins. As an example, consider the theory whose fields fall into the "N = 2 linear multiplet" shown in Table 5. In this case, the transverse vector must be the field strength for an antisymmetric tensor gauge field. The physical fields of the theory include a spinor isodoublet and a scalar isotriplet in addition to the antisymmetric tensor. This is thus a dual form of the scalar hypermultiplet theory, just as the N = 1 linear multiplet theory was a dual form of the N = 1 chiral multiplet theory. If we denote the linear multiplet of covariant fields by $G_{ij}$ , then the action for this theory is

$$I = \int d^4x\, d^2\theta^{ij}\, d^2\,\bar{\theta}^{k\ell}\,[G_{(ij}\,G_{k\ell)}]\ . \tag{2.6}$$

The Lagrangian in this action is integrated over a submanifold of N = 2 superspace. The action is invariant because of the constraints on $G_{ij}$, given in Table 5, which imply the following structure for $L_{ijk\ell} = G_{(ij}G_{k\ell)}$,

Table 6
N=2 SS=0 SIS=$\underset{\sim}{3}$

| Spin | SL(2, C) | SU(2) | dim | identification |
|---|---|---|---|---|
| 1+0 | (1/2, 1/2) | $\underset{\sim}{3}$ | 3 | $H_\mu^{ij}$ : $\partial^\mu H_\mu^{ij} \neq 0$ |
| 1/2 | (0, 1/2) + (1/2, 0) | $\underset{\sim}{2}$ | 7/2 | $\psi_\alpha{}^i$ |
| | (0, 1/2) + (1/2, 0) | $\underset{\sim}{4}$ | 5/2 | $\lambda_\alpha{}^{(ijk)}$ |
| 0 | (0, 0) | $\underset{\sim}{1}$ | 4 | T |
| | (0, 0) | $\underset{\sim}{3} + \underset{\sim}{3}$ | 3 | $N^{ij}$ , $\bar{N}_{ij}$ |
| | (0, 0) | $\underset{\sim}{5}$ | 2 | $L_{ijk\ell}$ (pseudo-real) |

Superfield : $L_{(ijk\ell)}(x, \theta, \bar{\theta}) = \varepsilon_{ii'}\, \varepsilon_{jj'}\, \varepsilon_{kk'}\, \varepsilon_{\ell\ell'}\, \bar{L}^{i'j'k'\ell'}$ with

$D_{\alpha(r}\, L_{ijk\ell)} = \bar{D}_{\dot{\alpha}(r}\, L_{ijk\ell)} = 0$ .

The invariance of the action (2.6) is proved subject to the constraints on $L_{ijk\ell}$ given in Table 6 in Sohnius et al. (1981) and in Howe et al. (1981 a). The latter article gives a general formalism for constructing supersymmetric invariants by integrating over various submanifolds of superspace, provided the Lagrangians satisfy appropriate constraints, such as the one given above

The supercurrent for the N = 2 linear multiplet is the only dimension 2 scalar superfield that can be formed,

$$V = G^{ij}\, G_{ij} \tag{2.7}$$

Although the fields of the theory are massless, there is no conformal invariance due to the presence of the antisymmetric tensor. Just as for the N = 1 linear multiplet, as shown in (1.11), the trace supermultiplet turns out to be the Lagrangian multiplet : upon use of the field equations

$$D^{ij}\, G_{ij} = \bar{D}^{ij}\, G_{ij} = 0 \tag{2.8}$$

and the constraints on $G_{ij}$ from Table 5, we obtain

$$D^{ij}\, V = -\frac{4}{5}\, D_{k\ell}\, L^{ijk\ell} . \tag{2.9}$$

Inspection of Table 6 shows that there is only one singlet of dimension 4 in the multiplet $L^{ijk\ell}$, and this must be the trace of $T_{\mu\nu}$. As in the case of the N = 1 linear multiplet, the axial current $J_\mu{}^5$ remains conserved even in a non-conformal theory. It can be verified directly that the action (2.6) is invariant under the full U(2) internal symmetry.

There are other forms of non-conformal supercurrents in N = 2 supersymmetry as well. In the examples given, we have constructed theories that possessed the full SU(2) internal symmetry. An example of a theory that has only SO(2) internal symmetry is the N = 2 non-linear sigma model in four dimensions of Curtwright & Freedman (1980). The structure of the supercurrent for this model is not known, but it must reflect the fact that two of the three currents of SU(2) must not be conserved.

Every possible supercurrent structure requires its own corresponding supergravity representation. To date, only the auxiliary fields corresponding to the supercurrent (2.5) have been worked out (Fradkin & Vasiliev 1979 a,b; de Wit et al. 1979, 1980). In order to build auxiliary field sets for N = 2 supergravity, it is necessary to add yet another multiplet to eliminate the gauge field character of the SU(2) triplet of vectors that couple to the SU(2) conserved currents in the supercurrent. However, the existence of the alternative trace supermultiplet (2.9) indicates the existence of another representation. What needs to be verified is the existence of an invariant supergravity action using this representation.

The anomalies in N = 2 theories might in principle avail themselves of the forms (2.5) or (2.9), but no examples are known of (2.9). There are no renormalizable models involving the scalar hypermultiplet, although this is the basis for the non-linear sigma model mentioned above. The smallest renormalizable theory is N = 2 supersymmetric Yang-Mills theory. Just as in N = 1 we have the added constraint of gauge invariance for this theory, which requires us to write the Lagrangian in terms of the field strength multiplet, W. Then the only available form for the anomaly equation is (2.5),

$$D^{ij}\,V \;=\; \bar{D}^{ij}\,V \;=\; d'[D^{ij}(W^2) + \bar{D}^{ij}(W^2)] \;, \qquad (2.10)$$

where the right-hand side satisfies the constraints required of an N = 2 linear multiplet (Table 5) due to $D_\alpha{}^{(i}\,D_\beta{}^j\,D_\gamma{}^{k)} \equiv 0$ and the chirality of W, $\bar{D}_{\dot\alpha}\,W = 0$. As in the N = 1 vector theory, this form of the anomaly equation gives a trace to $T_{\mu\nu}$ and a divergence to $J_\mu{}^5$, these being the two dimension four singlets shown in Table 5.

## 3 THE SUPERCURRENT IN N = 4 SUPERSYMMETRY

In N = 4 supersymmetry, the larger size and range of spins in the irreducible representations gives rise to a new feature : the irreducible representation corresponding to the conformal supercurrent is itself the smallest irreducible representation of the supersymmetry, of superspin 0 and a super-SU(4) singlet. So there are no smaller representations to take as trace supermultiplets. The conformal supercurrent multiplet has been given in the abstract by Taylor (1981) and concretely in terms of bilinears of the fields of N = 4 supersymmetric Maxwell theory by Bergshoeff et al. (1981). The structure of the multiplet is as follows :

Table 7
N=4, SU(4) SS = 0 Super SU(4) = 1

| Spin | | SL(2, C) | SU(4) | dim | identification |
|---|---|---|---|---|---|
| 2 | | (1, 1) | $\underset{\sim}{1}$ | 4 | $T_{\mu\nu} : \partial^{\mu} T_{\mu\nu} = 0$ |
| 3/2 | | (1, 1/2) + (1/2, 1) | $\underset{\sim}{4} + \underset{\sim}{4}^{*}$ | 7/2 | $J_{\mu\alpha}{}^{i} : \partial^{\mu} J_{\mu\alpha}{}^{i} = 0$ |
| 1 | { | (1/2, 1/2) | $\underset{\sim}{15}$ | 3 | $J_{\mu}{}^{i}{}_{j} : \partial^{\mu} J_{\mu}{}^{i}{}_{j} = 0$ |
| | | (0, 1) + (1, 0) | $\underset{\sim}{6} + \underset{\sim}{6}$ | 3 | $A_{\mu\nu}^{[ij]}$ |
| 1/2 | { | (1/2, 0) + (0, 1/2) | $\underset{\sim}{4} + \underset{\sim}{4}^{*}$ | 7/2 | $\chi_{\alpha}{}^{i}$ |
| | | (1/2, 0) + (0, 1/2) | $\underset{\sim}{20}^{*} + \underset{\sim}{20}$ | 5/2 | $\lambda_{\alpha}{}^{ij,k}$ |
| 0 | { | (0, 0) | $\underset{\sim}{1} + \underset{\sim}{1}$ | 4 | I, P |
| | | (0, 0) | $\underset{\sim}{10} + \underset{\sim}{10}^{*}$ | 3 | $H^{(ij)}$ |
| | | (0, 0) | $\underset{\sim}{20}'$ | 2 | $V^{ij,k\ell}$ |

Superfield : $V^{ij,k\ell}(x, \theta, \bar{\theta})$ in $\underset{\sim}{20}'$ = , $\bar{V}_{ij,k\ell} = \varepsilon_{ijrs}\,\varepsilon_{k\ell mn}\,V^{rs,mn}$

with the constraints $D_{\alpha}{}^{r}\, V^{ij,k\ell} \sim$ = $\underset{\sim}{20}^{*}$ only ie, = 0

$\bar{D}_{\dot{\alpha}r}\, V^{ij,k\ell} \sim$ = $\underset{\sim}{20}$ only

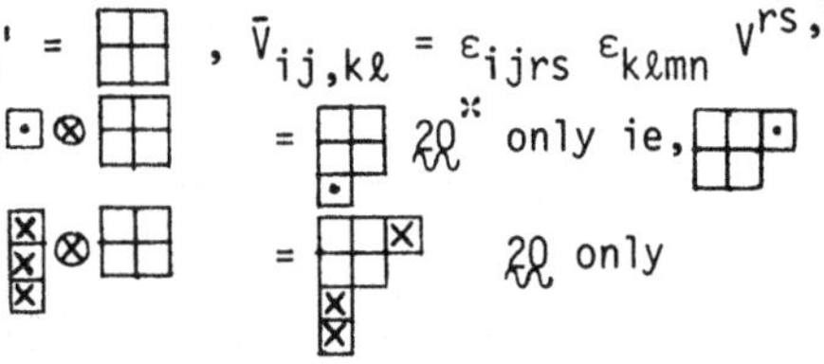

The N = 4 supersymmetric Yang-Mills theory has a supercurrent with the above structure. The theory has an SU(4) $\underset{\sim}{6}$ of scalar fields, a $\underset{\sim}{4}$ of Weyl spinors and a $\underset{\sim}{4}^{*}$ of their complex conjugates plus an SU(4) singlet vector, with all fields in the adjoint representation of some gauge group. The field strength superfield for the theory (Sohnius 1978 a) is a scalar carrying the $\underset{\sim}{6}$ representation of SU(4) : $\phi^{[ij]} = \frac{1}{2}\varepsilon^{ijk\ell}\,\bar{\phi}_{k\ell}$ , satisfying the field equations

$$\begin{aligned} D_{\alpha}^{(k}\,\phi^{i)j} &= 0 \\ \bar{D}_{\dot{\alpha}k}\,\phi^{ij} &= \tfrac{1}{3}(\delta^{i}{}_{k}\,\bar{D}_{\dot{\alpha}\ell}\,\phi^{\ell j} - \delta^{j}{}_{k}\,\bar{D}_{\dot{\alpha}\ell}\,\phi^{\ell i}). \end{aligned} \tag{3.1}$$

It is not known how to weaken the set of constraints (3.1) to avoid their implying field equations while still maintaining an invariant action. Subject to the constraints (3.1), however, the following action is invariant

$$I = \int d^4x\, d^2\theta_{ik}\, d^2\theta_{j\ell}\, [\phi^{ij}\,\phi^{k\ell} - \tfrac{1}{12}\,\varepsilon^{ijk\ell}\,\phi^{pq}\,\bar{\phi}_{pq}]\ . \tag{3.2}$$

The invariance is proved in Howe et al. (1981 a).
The classical supercurrent for this theory is

$$V^{ij,k\ell} = \phi^{ij}\,\phi^{k\ell} - \tfrac{1}{12}\,\varepsilon^{ijk\ell}\,\phi^{pq}\,\bar{\phi}_{pq}\ , \tag{3.3}$$

which satisfies the constraints required in Table 7 by virtue of the constraints-field equations (3.1). Note that the supercurrent multiplet (3.3) is exactly the same as the Lagrangian multiplet in (3.2). In Table 7, there are two scalar singlets of dimension four. One of these is the x-space Lagrangian tr $F_{\mu\nu}\,F^{\mu\nu} + \ldots$, while the other is the integrand for the Pontryagin index tr $F_{\mu\nu}\,{}^{*}F^{\mu\nu} + \ldots$ .

Since there is no irreducible representation of lower superspin to serve as a trace supermultiplet, there is no way to write an equation like (1.3), (1.4), (2.5) or (2.9) for a non-conformal supercurrent. But something else can happen. The multiplet shown in Table 7 contains a $\underset{\sim}{20}'$ of dimension 2 scalars. If one relaxes the assuption of SU(4) internal symmetry to some subgroup of SU(4), the $\underset{\sim}{20}'$ breaks up into several irreducible representations of the subgroup. When one of these is a singlet, it can disappear and reappear as the trace of $T_{\mu\nu}$ . This is similar to what happened with the N = 2 representation shown in Table 6, which is not simply a multiplication of $\underset{\sim}{3}$ into all the SU(2) representations shown in Table 5, because the vector triplet is not transverse, showing that three scalars have been absorbed.

This type of rearrangement within a multiplet can give a non-conformal supercurrent which is still irreducible, but with less than SU(4) internal symmetry. This phenomenon was found by Howe & Lindstrom (1981), who constructed the supercurrent for the dual form of the N = 4 Maxwell theory in which one of the scalars is replaced by an antisymmetric tensor (Sohnius et al. 1980). In this case, the remaining internal symmetry is SP(4), and the 20' breaks up into a 14 plus a 5 plus a 1. The singlet and the 5 disappear from the multiplet, while $T_{\mu\nu}$ acquires a trace and 5 of the 15 $J_{\mu}{}^{i}{}_{j}$ acquire divergences, leaving just 10 conserved internal symmetry currents for the generators of SP(4). At the same time, one of the six antisymmetric tensors is replaced by two transverse vectors, one of which becomes the current for the central charge that exists in the SP(4) theory, while the other is identically conserved.

Another rearrangement of the multiplet given in Table 7 preserves only SO(4) internal symmetry. In this case, the 20' breaks up into a 10 and a 9 and a 1. The 9 and the 1 disappear, while 9 SU(4) currents acquire divergences and $T_{\mu\nu}$ acquires a trace :

Table 8
N=4, SO(4) SS=0 Super SO(4)=1

| Spin | SL(2, C) | SO(4) | dim | identification |
|---|---|---|---|---|
| 2 + 0 | (1, 1) + (0, 0) | 1 | 4 | $T_{\mu\nu}\,;\,\partial^{\mu}T_{\mu\nu}=0,\ T^{\mu}{}_{\mu}\neq 0$ |
| $\frac{3}{2}+\frac{1}{2}$ | (1,1/2)+(0,1/2) + cc. | 4 | 7/2 | $J_{\mu\alpha}{}^{i};\partial^{\mu}J_{\mu\alpha}{}^{i}=0,\ \sigma^{\mu}_{\alpha\dot{\alpha}}J_{\mu}{}^{\dot{\alpha}i}\neq 0$ |
| 1 + 0 | (1/2, 1/2) | 9 | 3 | $J_{\mu}{}^{(ij)};\partial^{\mu}J_{\mu}{}^{(ij)}\neq 0$ |
| 1 | (1/2, 1/2) | 6 = 3 + 3 | 3 | $J_{\mu}{}^{[ij]};\partial^{\mu}J_{\mu}{}^{[ij]}=0$ |
| | (1, 0) + (0, 1) | 6 + 6 | 3 | $A_{\mu\nu}{}^{[ij]}$ |
| 1/2 | (0, 1/2) + (1/2, 0) | 4 | 7/2 | $\chi_{\alpha}{}^{i}$ |
| | (0, 1/2) + (1/2, 0) | 16 | 5/2 | $\lambda_{\alpha}{}^{ij,k}$ |
| 0 | (0, 0) | 1 + 1 | 4 | I.P. |
| | (0, 0) | 1+1+9+9 | 3 | $H,\ H^{*},\ H^{(ij)},\ H^{*(ij)}$ |
| | (0, 0) | 10 | 2 | $C^{ij,k\ell}$ |

Superfield : $C^{ij,k\ell}$ in the 10, real, with the constraint that $D_{\alpha}{}^{r}\,C^{ij,k\ell}$ and $\bar{D}_{\dot{\alpha}r}\,C^{ij,k\ell}$ contain only the 16.

The spinor fields have been rearranged as well in this multiplet ; the $\underset{\sim}{20}^{*}$ + $\underset{\sim}{20}$ of spinors at dimension 5/2 have broken up into $\underset{\sim}{16}$ + $\underset{\sim}{16}$ + $\underset{\sim}{4}$ + $\underset{\sim}{4}$, and the $\underset{\sim}{4}$ 's have disappeared while the supersymmetry currents have acquired σ-traces.

Can the multiplet shown in Table 8 describe a theory with anomalies ? The question is relevant to an argument that has been given, directed toward showing that the N = 4 Yang-Mills theory is finite to all orders (Ferrara & Zumino 1978; Sohnius & West 1981 a). As we have seen, in lower N supersymmetric gauge theories, the trace anomaly is always accompanied by an anomaly in a U(1) axial current. Since the six scalar fields of the N = 4 gauge theory are in a real representation of SU(4), this symmetry cannot be extended to U(4). Correspondingly, the conformal supercurrent shown in Table 7 contains only SU(4) and not U(4) currents. Therefore, to be consistent with the lower supersymmetry results, the axial current that acquires an anomaly must come from one of the 9 axial currents within SU(4). So SU(4) must be broken if there is a trace anomaly, ie. if the theory needs a coupling constant renormalization. This is corroborated by the impossibility of giving a trace to $T_{\mu\nu}$ while keeping SU(4) symmetry in the multiplet of Table 7.

If there were a trace anomaly, what could SU(4) be broken to ? The most likely residual symmetry would seem to be SO(4), which preserves the 6 non-chiral currents of SU(4) and gives up the 9 chiral ones. At a purely abstract level, there does not appear to be any way to exclude such a possibility from a consideration of available supermultiplets. The multiplet shown in Table 8 could correspond to a non-conformal supercurrent under N = 4 supersymmetry, with SU(4) internal symmetry broken to SO(4).

It is necessary to enquire more closely into the structure of the quantum supercurrent if one is to have a clearer understanding of the finiteness of the N = 4 supersymmetric Yang-Mills theory. In our earlier discussions of anomalies in N = 1 and N = 2 supersymmetric theories, we always found that the trace supermultiplet of anomalies was the Lagrangian multiplet, for the conformal trace anomaly itself must contain the Yang-Mills Lagrangian. In every case, the trace anomaly was accompanied by a U(1) chiral anomaly whose explicit form was the appropriately supersymmetrized integrand for the Pontryagin index, tr $F_{\mu\nu}^{*} F^{\mu\nu}$ + ... . This term is obviously a singlet under whatever internal symmetry group might be assumed to persist. Such a term is also contained as a singlet in the classical N = 4 Yang-Mills theory's Lagrangian multiplet, Table 7. The multiplet of Table 8 can also describe the Lagrangian multiplet for this theory, with the Lagrangian differing by a total derivative and with the classical stress tensor differing by improvement terms for 9 scalar fields, in comparison to the multiplet of Table 7. Nonetheless, the singlet scalar of dimension 4 that corresponds to tr $F_{\mu\nu} {}^{*}F^{\mu\nu}$ + ... remains in the multiplet's structure, although it too will be modified by total derivatives. Moreover, it is clear that this singlet will remain in any possible rearrangement of the multiplet of Table 7, for it is this component that corresponds to the imaginary part of the Clifford vacuum of the multiplet, whose spin and internal symmetry representation define the multiplet. Consequently, any supermultiplet of anomaly equations must give some singlet axial vector current an anomaly at the same time as it gives a trace to $T_{\mu\nu}$. The existence of a singlet axial vector current is also necessary for consistency with the anomaly equations under N = 1 and N = 2 supersymmetry.

Neither SO(4) nor SP(4) allow for a singlet spin one current. Under the former, the 15 vectors fall into a $\underset{\sim}{6}$ plus a $\underset{\sim}{9}$, while under the latter they fall into a $\underset{\sim}{10}$ plus a $\underset{\sim}{5}$. Thus, the multiplet of Table 8 cannot correspond to an anomalous supercurrent after all, independently of the details of how one might write a multiplet of anomaly equations. The largest internal symmetry that yields a singlet axial vector is SU(3) x U(1). If it were not for the four supersymmetries, SU(3) x U(1) could indeed be the residual symmetry for an N = 4 theory with anomalies. At this point the discussion is most clearly carried out by leaving the supercurrent and following more closely the details of the quantum calculations in the N = 4 Yang-Mills theory.

## 4 THE FINITENESS OF N = 4 SUPERSYMMETRIC YANG-MILLS THEORY

The calculation of quantum amplitudes in supersymmetric theories is vastly simplified by the use of two theoretical tools : superfield Feynman rules and the background field method. Superfield Feynman rules have made the calculation of the 3-loop β-function tractable by hand (Grisaru et al. 1980, 1981; Caswell & Zanon 1981). They may also be combined with use of the background field method to guarantee gauge invariance of the quantum corrections (Grisaru et al. 1979). The background field method can also be used to arbitrary loop order ('t Hooft 1975; DeWitt 1980; Boulware 1980; Abbott 1981). In order to enjoy the advantages of the N = 1 supersymmetric and gauge invariant formalism, a regularization scheme that respects supersymmetry and gauge invariance is needed. While its full consistency is still a subject of debate, the technique of regularization by dimensional reduction (Siegel 1979) has given correct results in the practical calculations carried out so far. We shall assume that its use will continue to guarantee supersymmetry at higher loop orders.

Using the above techniques together with the reasonable assumption that the extended supersymmetries which are not manifest in the N = 1 superfield formalism nonetheless govern the structure of the infinite counterterms, we can see the underlying reason for the vanishing of the β-function in the N = 4 supersymmetric Yang-Mills theory. The reason is the set of simultaneous constraints imposed upon the renormalization constants by gauge invariance, extended supersymmetry and the non-renormalization theorem for chiral superspace integrals that holds for N = 1 superspace Feynman rule calculations.

As may be seen most clearly in the superspace Feynman rules of Grisaru et al. (1979), all quantum corrections to the effective action of a supersymmetric theory can be written as a single $\int d^4\theta$ integral over the full N = 1 superspace. This result follows from the possibility of writing every vertex, for chiral or non-chiral superfields, as an integral over $d^4\theta$, and from an accounting of all the fermionic delta functions $(\theta_i - \theta_j)^4$ that occur on each superfield propagator. In consequence, terms in the original Lagrangian that cannot be written as full superspace integrals, in terms of the superfields used in the Feynman rules, must remain unrenormalized. This means particularly the mass and self-interaction terms for chiral scalar superfields, $\int d^2\theta\ \phi^n$, and is the reason for the non-renormalization of such terms that was found in the Wess-Zumino model (Wess & Zumino 1974; Iliopoulos & Zumino 1974).

If we knew a full set of auxiliary fields for the N = 4 theory, and a corresponding N = 4 supersymmetric gauge fixing procedure, then the calculation of quantum corrections would automatically respect N = 4 supersymmetry. Undoubtedly, in that case there would be a non-renormalization theorem that would bear upon the N = 4 Yang-Mills theory. Unfortunately, no one has found the full set of auxiliary fields for this theory, and we are reduced to quantizing it in terms of N = 1 superfields.

We assume that the extended N = 4 supersymmetry of the theory remains valid for the fully renormalized theory. Since we don't know the extended auxiliary fields, the three extended supersymmetry transformations are nonlinear and contain the gauge coupling constant. Thus, we must take into account the possibility that the gauge coupling constant may be renormalized, necessitating a renormalization of this constant also in the three extended supersymmetry transformations. But this and the wave function renormalizations of the fields are the only changes that we shall consider to be possible in the form of the extended supersymmetry transformations which leave the renormalized action invariant. Moreover, the wavefunction renormalizations of fields in the same extended supersymmetry multiplet must be the same.

In summary, we assume the constraints upon the various coupling constant and wavefunction renormalizations that would follow automatically if one were to have a full N = 4 invariant formalism, upon subsequent elimination of all but the N = 1 auxiliary fields. In principle, these constraints should be derivable from the Ward identities for the extended supersymmetries without knowledge of the extended auxiliary field structure. However, the complexity of the Ward identities for these non-linear transformations has so far impeded such a derivation.

In terms of N = 1 superfields, the classical action for the N = 4 Yang-Mills theory is written using one general scalar and three chiral superfields (Fayet 1979), all in the adjoint representation of some gauge group :

$$I = \mathrm{tr}\left[\frac{1}{64g^2}\int d^4x\, d^2\theta\, W^\alpha W_\alpha + \int d^4x\, d^4\theta\, e^{-gV}\,\bar\phi^i\, e^{gV}\,\phi_i + \left\{\frac{ig}{3!}\int d^4x\, d^2\theta\, \varepsilon^{ijk}\,\phi_i\,[\phi_j, \phi_k] + \mathrm{h.c.}\right\}\right] \qquad (4.1)$$

where $W_\alpha = \bar{D}^2(e^{-gV} D_\alpha e^{gV})$. The first term is real as it is written, because its imaginary part is the integral of a total divergence. The action (4.1) is invariant under linearly realized rigid SU(3) transformations on the indices of the chiral superfields, plus a U(1) "R invariance" that combines phase rotations on $\phi_i$ with phase rotations of $\theta_\alpha$ , plus the manifest N = 1 supersymmetry and local gauge invariance. The SU(3) x U(1) and local gauge transformations are given by

$$\begin{aligned} &\delta_\Omega\, \phi_i = \Omega_i{}^j\, \phi_j \quad , \qquad \delta_\Omega\, V = 0 \\ &\delta_r\, \phi_i = ir\,\phi_i \quad , \quad \delta_r V = 0 \quad , \quad \delta_r\, \theta_\alpha = -\tfrac{3}{2} ir\,\theta_\alpha \\ &\delta_\Lambda\, e^{gV} = i(\bar\Lambda\, e^{gV} - e^{gV}\,\Lambda) \quad , \quad \delta_\Lambda\, \phi_i = i[\Lambda, \phi_i] \; . \end{aligned} \qquad (4.2)$$

In addition, there are three non-linearly realized supersymmetry transformations and the rigid transformations of SU(4)/SU(3) x U(1). These may be written together in the fashion of Grisaru et al. (1981),

$$\begin{aligned} \delta_\chi\, e^{gV} &= ig[\chi_i\, \bar{\phi}^i\, e^{gV} - e^{gV} \bar{\chi}^i\, \phi_i] \\ \delta_\chi\, \phi_i &= \frac{i}{8g}\, W^\alpha\, D_\alpha\, \chi_i - \frac{1}{4}\, \varepsilon_{ijk}\, \bar{D}^2\, [\bar{\chi}^j\, e^{-gV}\, \bar{\phi}^k\, e^{gV}] \end{aligned} \tag{4.3}$$

where $\chi_i(\theta)$ is an x-independent chiral superfield

$$\begin{aligned} \chi_i(\theta) &= Z_i + \theta^\alpha\, \varepsilon_{\alpha i} + \theta^2 \omega_i \quad , \\ \bar{D}_{\dot\alpha}\, \chi_i &= 0 \quad . \end{aligned} \tag{4.4}$$

The $Z_i$ are off-shell central charge transformations, which reduce to gauge transformations on-shell with parameter $g^{-1}\, \bar{Z}^i\, \phi_i$ , while the $\varepsilon_{\alpha i}$ are the parameters of the three extended supersymmetry transformations and the $\omega_i$ those of the SU(4)/SU(3) x U(1) transformations. Subject to the field equations, the full SU(4) invariance of the theory becomes linearly realized.

In quantizing the theory described by (4.1), one must go through the Faddeev-Popov procedure of gauge fixing and addition of a ghost action. We shall not need to be involved with all of this machinery here, for we are concerned with the renormalization of the fields and coupling constant appearing in (4.1). If we use the background field method as given in Grisaru et al. (1979), we do not need to consider renormalization of fields that occur only in closed loops such as the ghosts and the "quantum" parts of the physical fields. The counterterms will be functionals only of the background fields, and will only contribute to a renormalization of the fields and coupling constant appearing in (4.1). In addition, in order to avoid infrared divergences, the Fermi-Feynman gauge is chosen, adjusting the gauge fixing constant $\alpha$ in the gauge fixing term

$$I_{g.fix.} = -\frac{1}{16\alpha}\, \mathrm{tr} \int d^4x\, d^4\theta\, (D^2\, V)(\bar{D}^2\, V) \tag{4.5}$$

(background covariant D)

so as to exclude $1/(k^2)^2$ terms in the vector propagator. Since the transverse parts of the vector propagator receive quantum corrections while the longitudinal part does not, the constant $\alpha$ must be readjusted order by order.

The consequence of our assumption that the four supersymmetries are preserved in the quantum corrections is that the fully renormalized

Lagrangian must retain the form of (4.1), although renormalization of the fields and coupling constant is a priori allowed. However, there can be only one overall wavefunction renormalization, and one coupling constant renormalization, common to the coupling constants appearing in the three terms of (4.1), keeping them equal. Thus,

$$\begin{aligned} g_0 &= Z_g\, g \\ V_0 &= Z^{1/2}\, V \quad , \quad \phi_{0i} = Z^{1/2}\, \phi_i \quad . \end{aligned} \tag{4.6}$$

In addition, it will be necessary to renormalize the gauge parameter, $\alpha_0 = Z_\alpha\, \alpha$ , but this will not affect any physical quantities.

The use of the background field method guarantees that counterterms will be manifestly gauge invariant with respect to the background gauge. This means that the kinetic term for V in (4.1) can only be renormalized by an overall divergent constant. Since this term is non-linear in (g V), this requires that

$$Z_g\, Z^{1/2} = 1 \quad , \tag{4.7}$$

thus allowing only one independent renormalization constant.

We get a final constraint on the renormalization constants from the presence of the third term in (4.1), the chiral multiplet self-interaction. Since this term can only be written as an integral over $d^2\theta$, it will not be renormalized due to the non-renormalization theorem discussed above. This imposes the final constraint on the renormalization constants

$$Z_g\, Z^{3/2} = 1 \quad . \tag{4.8}$$

The constraints (4.7) and (4.8) allow only the solution

$$Z_g = Z = 1 \quad , \tag{4.9}$$

thus showing that the β-function for the theory is zero since the coupling constant g is not renormalized. Thus, the theory remains conformally invariant and SU(4) invariant even when quantized.

The discussion presented above relies on the preservation of the four supersymmetry transformations, guaranteeing the existence of only one wavefunction renormalization and one coupling constant renormalization. This may also be obtained if one assumes the preservation of SO(4) internal symmetry if the N = 1 auxiliary fields are eliminated. This confirms the analysis of the possible structures for the quantum supercurrent given in section three.

The approach to understanding the cancellation of ultraviolet infinities in the N = 4 supersymmetric Yang-Mills theory presented in this article should be applicable to other supersymmetric theories. In particular, the vanishing of the one-loop β-function in N $\geq$ 5 gauged extended supergravities should be understood along these lines, with the possibility of extending the result to all orders.

This article is based in part upon a lecture given at the 1981 Nuffield Quantum Gravity Workshop. The work on the structure of the quantum supercurrent was done together with Paul Howe and Paul Townsend. The discussion in Section 4 is based on a remark in Grisaru et al. (1980). Conversations with Marc Grisaru and Paul Townsend on the subject of this section are also gratefully acknowledged.

REFERENCES

Abbott, L.F. (1981). The background field method beyond one loop. Nucl. Phys. B185, 189-203.
Bergshoeff, E., de Roo, M. & de Wit, B. (1981). Extended conformal supergravity. Nucl. Phys. B182, 173-204.
Boulware, D. (1980). Gauge dependence of the effective action. University of Washington preprint RLO-1388-322 (1980).
Breitenlohner, P. (1977a). A geometric interpretation of local supersymmetry. Phys. Lett. 67B, 49-51.
Breitenlohner, P. (1977b). Some invariant Lagrangians for local supersymmetry. Nucl. Phys. B124, 500-510.
Caswell, W.E. & Zanon, D. (1981). Zero three-loop beta function in the N = 4 supersymmetric Yang-Mills theory. Nucl. Phys. B182, 125-143.
Clark, T.E., Piguet, O. & Sibold, K. (1978). Supercurrents, renormalization and anomalies. Nucl. Phys. B143, 445-484.
Curtwright, T.L. & Freedman, D.Z. (1980). Nonlinear σ-models with extended supersymmetry in four dimensions. Phys. Lett. 90B, 71-74.
de Wit, B. & Grisaru, M.T. (1978). Auxiliary fields and ultraviolet divergences in supergravity. Nucl. Phys. B139, 531-544.
de Wit, B. & van Holten, J.W. (1979). Multiplets of linearized SO(2) supergravity. Nucl. Phys. B155, 530-542.
de Wit, B., van Holten, J.W. & van Proyen, A. (1980). Transformation rules of N = 2 supergravity multiplets. Nucl. Phys. B167, 186-204.
De Witt, B.S. (1980). A gauge invariant effective action, in Quantum Gravity II, eds. Isham, C., Penrose, R. & Sciama, D. (Oxford, to be published).
Fayet, P. (1976). Hypersymmetry. Nucl. Phys. B113, 135-155.
Fayet, P. (1979). Spontaneous generation of massive multiplets and central charges in extended supersymmetry theories. Nucl. Phys. B149, 137-169.

Ferrara, S. (1980). Aspects of supergravity theories. CERN preprint TH 2957 (1980).
Ferrara, S., Savoy, C.A. & Zumino, B. (1981). General massive multiplets in extended supersymmetry. Phys. Lett. 100B, 393-398.
Ferrara, S. & van Nieuwenhuizen, P. (1978). The auxiliary fields of supergravity, Phys. Lett. 74B, 333-335.
Ferrara, S. & Zumino, B. (1975). Transformation properties of the supercurrent. Nucl. Phys. B87, 207-220.
Ferrara, S. & Zumino, B. (1978). Unpublished.
Fradkin, E.S. & Vasiliev, M.A. (1979a). Lett. Nuovo Cimento 25, 79.
Fradkin, E.S. & Vasiliev, M.A. (1979b). Minimal set of auxiliary fields in SO(2)-extended supergravity. Phys. Lett. 85B, 47-56.
Grimm, R., Sohnius, M.F. & Wess, J. (1978). Extended supersymmetry for gauge theories. Nucl. Phys. B133, 275-284.
Grisaru, M.T. (1979). Anomalies in supersymmetric theories, in Recent Developments in Gravitation, eds. Lévy, M. & Deser, S. (Plenum, New York).
Grisaru, M.T., Rocek, M. & Siegel, W. (1980). Zero value for the three-loop β-function in N=4 supersymmetric Yang-Mills theory. Phys. Rev. Lett. 45, 1063-1066.
Grisaru, M.T., Rocek, M. & Siegel, W. (1981). Superloops 3, beta 0. Nucl. Phys. B183, 141-156.
Grisaru, M.T., Siegel, W. & Rocek, M. (1979). Improved methods for supergraphs. Nucl. Phys. B159, 429-450.
Howe, P. & Lindstrom, U. (1981). Higher order invariants in extended supergravity. Nucl. Phys. B181, 487-501.
Howe, P.S., Stelle, K.S. & Townsend, P.K. (1981a). Superactions. Nucl. Phys. B191, 445-464.
Howe, P.S., Stelle, K.S. & Townsend, P.K. (1981b). Supercurrents. Nucl. Phys. B192, 332-352.
Iliopoulos, J. & Zumino, B. (1974). Broken supergauge symmetry and renormalization. Nucl. Phys. B76, 310-332.
Rittenberg, V. & Sokatchev, E. (1981). Decomposition of extended superfields into irreducible representations of supersymmetry. Bonn University preprint BONN-HE-81-5 (May, 1981).
Siegel, W. (1979). Supersymmetric dimensional regularization via dimensional reduction. Harvard University preprint HUTP-79/A006 (March, 1979).
Sohnius, M.F. (1978a). Bianchi-identities for supersymmetric gauge theories. Nucl. Phys. B136, 461-474.
Sohnius, M.F. (1978b). Supersymmetry and central charges. Nucl. Phys. B138, 109-121.
Sohnius, M.F. (1979). The multiplet of currents for N = 2 extended supersymmetry. Phys. Lett. 81B, 8-10.
Sohnius, M.F., Stelle, K.S. & West, P.C. (1980). Off-mass-shell formulation of extended supersymmetric gauge theories. Phys. Lett. 92B, 123-12
Sohnius, M.F., Stelle, K.S. & West, P.C. (1981). Representations of extended supersymmetry, in Superspace and Supergravity, eds. Hawking, S.W. & Rocek, M. (Cambridge University Press, 1981).
Sohnius, M.F. & West, P.C. (1981a). Conformal invariance in N = 4 supersymmetric Yang-Mills theory. Phys. Lett. 100B, 245-250.
Sohnius, M.F. & West, P.C. (1981b). An alternative minimal off-shell version of N = 1 supergravity. Phys. Lett. 105B, 353-357.
Stelle, K.S. & West, P.C. (1978). Minimal auxiliary fields for supergravity. Phys. Lett. 74B, 331-332.
't Hooft, G. (1975). In Acta Universitas Wratislaviensis n° 38, 12th Winter School of Theoretical Physics in Karpacz; Functional and Probabilistic Methods in Quantum Field Theory, Vol. I.

Tarasov, O.V., Vladimirov, A.A. & Zharkov, A. Yu. (1980a). The Gell-Mann-Low function of QCD in the three-loop approximation. Phys. Lett. 93B, 429-432.
Tarasov, O.V. & Vladimirov, A.A. (1980b). Three-loop calculations in non-Abelian gauge theories. Dubna preprint E2-80-483 (1980).
Taylor, J.G. (1981). Extended superfields in linearized supersymmetry and supergravity, in Superspace and Supergravity, eds. Hawking, S.W. & Rocek, M. (Cambridge University Press, 1981).
Wess, J. & Zumino, B. (1974). A Lagrangian model invariant under supergauge transformations. Phys. Lett. 49B, 52-54.

# DUALITY ROTATIONS

B. Zumino
Lawrence Berkeley Laboratory and Department of Physics,
University of California, Berkeley, CA 94720, USA
On leave from CERN, Geneva, Switzerland

## INTRODUCTION

The invariance of Maxwell's equations under "duality rotations" has been known for a long time. These are rotations of the electric and magnetic fields into each other or, in relativistic notation, rotations of the electromagnetic field strength $F_{\mu\nu}$ into its dual

$$\tilde{F}_{\mu\nu} = \frac{1}{2} \varepsilon_{\mu\nu\lambda\sigma} F^{\lambda\sigma} \tag{1.1}$$

This invariance can be easily extended to the case when the electromagnetic field is in interaction with the gravitational field, which does not transform under duality (Misner and Wheeler 1957). Minimal electromagnetic couplings violate duality invariance and it is also easy to see that the Yang-Mills equations do not admit an invariance of this type (Deser and Teitelboim 1976).

Non-minimal couplings of the magnetic moment type can be duality invariant and, in some cases, this invariance generalizes to a non-abelian group. This happens in extended supergravity theories without gauging of the SO(N) symmetry (Ferrara, Scherk and Zumino 1977). The assumption that the theory is invariant under duality rotations can be used to simplify the construction of the correct supersymmetric Lagrangian (Cremmer and Scherk 1977, Cremmer et al. 1977). For $N = 4$ supergravity the U(4) duality extends to a larger $SU(4) \times SU(1, 1)$ non-compact duality invariance (Cremmer et al. 1978) and a similar situation occurs for $N > 4$; in particular for $N = 8$ the theory is invariant under a non-compact $E_7$ duality (Cremmer and Julia 1979). A non compact duality invariance arises when there are scalar fields in the theory, which can transform non-linearly.

Irrespective of supersymmetry, it is interesting to understand the special properties of theories admitting duality rotations. As we shall see, the Lagrangian of such a theory is not invariant under the transformations, nor does it change by a total derivative, but it transforms in a particular way which implies that the system of the equations of motion is invariant and that observables, such as the energy momentum tensor and therefore the total energy and momentum, are invariant. In this lecture I describe the main results of a recent paper on the properties of theories admitting duality rotations written in collaboration with M. K. Gaillard (1981).

As an example, consider the Lagrangian (Ferrara et al. 1977)

$$L = -\frac{1}{4} F_{\mu\nu}F^{\mu\nu} - \frac{i}{2} \bar{\psi}\gamma^{\mu}\overleftrightarrow{\partial}_{\mu}\psi + \frac{a}{2} F_{\mu\nu}\bar{\psi}\sigma^{\mu\nu}\psi + b(\bar{\psi}\sigma_{\mu\nu}\psi)(\bar{\psi}\sigma^{\mu\nu}\psi), \quad (1.2)$$

where $F_{\mu\nu}$ is the curl of a vector potential and $\psi$ is a massless Dirac spinor. (Our gamma matrices are real, $(\gamma_5)^2 = -1$, $\sigma_{\mu\nu}\gamma_5 = \tilde{\sigma}_{\mu\nu}$. The ordinary space-time derivative is denoted by $\partial_\mu$.) The equations of motion for $F_{\mu\nu}$ are

$$\partial_\mu(F^{\mu\nu} - a\bar{\psi}\sigma^{\mu\nu}\psi) = 0 \quad (1.3)$$

together with the Bianchi identities

$$\partial_\mu \tilde{F}^{\mu\nu} = 0. \quad (1.4)$$

Clearly (1.3) and (1.4) transform into each other by the duality rotation

$$\delta F_{\mu\nu} = \lambda\tilde{F}_{\mu\nu} - \lambda a\bar{\psi}\sigma_{\mu\nu}\gamma_5\psi \quad (1.5)$$

$$\delta\psi = -\frac{\lambda}{2}\gamma_5\psi \quad (1.6)$$

where $\lambda$ is an infinitesimal real parameter. In order for the theory to be invariant under (1.5) and (1.6), one must check that the equation of motion for $\psi$ is also invariant. It is easy to verify by explicit calculation that this is true if the coupling constant b is related to the magnetic moment coupling a by

$$b = \frac{1}{8}a^2. \quad (1.7)$$

Duality invariance gives relations among couplings. It is not an invariance of the Lagrangian, and it is also easy to see that the Lagrangian does not simply change by a divergence (except if one uses the equations of motion, but for the vector fields only). One can also verify quite easily (and it follows from the general argument below) that the energy momentum tensor (both the canonical and the symmetric) is invariant. Since the total energy and the equations of motion are invariant, it follows that the S-matrix is invariant under the transformation which operates on the "in" and "out" fields as in (1.5) (1.6) but with the coupling constant a set equal to zero. This is, of course, a purely formal statement, which ignores difficulties in the definition of the S-matrix, due to the vanishing masses and the non-renormalizability of the theory.

The duality transformations (1.5) (1.6) should not be confused with the chiral transformations

$$\delta F_{\mu\nu} = 0 \tag{1.8}$$

$$\delta\psi = \alpha\gamma_5\psi^* \tag{1.9}$$

($\alpha$ infinitesimal constant parameter) under which the Lagrangian is actually invariant.

## DUALITY ROTATIONS

Consider a Lagrangian which is a function of n real field strength $F^a_{\mu\nu}$ and of some other fields $\chi^i$ and their derivatives $\chi^i_\mu = \partial_\mu\chi^i$

$$L = L(F^a, \chi^i, \chi^i_\mu). \tag{2.1}$$

Since

$$F^a_{\mu\nu} = \partial_\mu A^a_\nu - \partial_\mu A^a_\mu, \tag{2.2}$$

we have the Bianchi identities

$$\partial^\mu \tilde{F}^a_{\mu\nu} = 0. \tag{2.3}$$

On the other hand, if we define

$$\tilde{G}^a_{\mu\nu} = \frac{1}{2}\varepsilon_{\mu\nu\lambda\sigma}G^{a\lambda\sigma} \equiv 2\frac{\partial L}{\partial F^{a\mu\nu}}, \tag{2.4}$$

we have the equations of motion

$$\partial^\mu \tilde{G}^a_{\mu\nu} = 0. \tag{2.5}$$

We consider an infinitesimal transformation of the form

$$\delta\begin{pmatrix}F\\G\end{pmatrix} = \begin{pmatrix}A & B\\C & D\end{pmatrix}\begin{pmatrix}F\\G\end{pmatrix}, \tag{2.6}$$

$$\delta\chi^i = \xi^i(\chi), \tag{2.7}$$

where A, B, C, D are real $n\times n$ constant infinitesimal matrices and $\xi^i(\chi)$ functions of the fields $\chi^i$ (but not of their derivatives), and ask under what circumstances the system of the equations of motion (2.3) (2.5) as well as the equations of motion for the fields $\chi^i$ are invariant. The analysis of Gaillard and myself (1981) shows that this is true if the matrices satisfy

$$A^T = -D, \; B^T = B, \; C^T = C, \tag{2.8}$$

where the superscript T denotes the transposed matrix, and the Lagrangian changes under (2.6) and (2.7) as

$$\delta L = \frac{1}{4}(FC\tilde{F} + GB\tilde{G}). \tag{2.9}$$

One can also see that this is essentially the most general possibility. The relations (2.8) show that (2.6) is an infinitesimal transformation of the real non compact symplectic group Sp(2n, R) which has U(n) as maximal compact subgroup. Clearly, particular theories may be only invariant under subgroups of the above. This is the case, for instance, for N = 8 supergravity, where n = 28 (number of vector fields) but where the additional requirement of supersymmetry not only determines the particle spectrum but also restricts Sp(56, R) and U(28) to their subgroups $E_7$ and SU(8).

Observe that the equations of motion (2.5) imply the existence of vector potentials such that

$$G^a_{\mu\nu} = \partial_\mu B_\nu - \partial_\nu B_\mu. \tag{2.10}$$

Using (2.2) and (2.10) one can write the right hand side of (2.9) as a divergence

$$\delta L = \frac{1}{2}\partial_\mu(A_\nu C\tilde{F}^{\mu\nu} + B_\nu B\tilde{G}^{\mu\nu}), \tag{2.11}$$

but this is true only in virtue of the equations of motion for $F^a_{\mu\nu}$. Now the variation of the Lagrangian induced by a variation of the fields $F^a$ only is, by (2.6),

$$\delta_F L = \delta F^a \frac{\partial L}{\partial F^a} = \frac{1}{2}(FA^T + GB)\tilde{G} \tag{2.12}$$

which, by using again the equations for $F^a_{\mu\nu}$, can be written as

$$\delta_F L = \partial_\mu(A_\nu A^T\tilde{G}^{\mu\nu} + B_\nu B\tilde{G}^{\mu\nu}). \tag{2.13}$$

Therefore, using (2.8),

$$\begin{aligned}\delta_\chi L = (\delta - \delta_F)L &= \partial_\mu(\frac{1}{2}A_\nu C\tilde{F}^{\mu\nu} - \frac{1}{2}B_\nu B\tilde{G}^{\mu\nu} + A_\nu D\tilde{G}^{\mu\nu}). \\ &= \frac{1}{2}\partial_\mu(A_\nu C\tilde{F}^{\mu\nu} - B_\nu B\tilde{G}^{\mu\nu} + A_\nu D\tilde{G}^{\mu\nu} - B_\nu A\tilde{F}^{\mu\nu}) \\ &= -\frac{1}{2}\partial_\mu \hat{J}^\mu.\end{aligned} \tag{2.14}$$

where

$$\hat{J}^\mu \equiv -A_\nu C\tilde{F}^{\mu\nu} + B_\nu B\tilde{G}^{\mu\nu} - A_\nu D\tilde{G}^{\mu\nu} + B_\nu A\tilde{F}^{\mu\nu}. \tag{2.15}$$

So far we have used the equations of motion for $F^a_{\mu\nu}$. Now, by the standard argument due to Emmy Noether, we know that the equations of motion for $\chi^i$ imply

$$\partial_\mu\left(\delta\chi^i \frac{\partial L}{\partial \chi^i_\mu}\right) = \delta_\chi L \tag{2.16}$$

Therefore, using all equations of motion we see that the current

$$J^\mu = \xi^i \frac{\partial L}{\partial \chi^i_\mu} + \hat{J}^\mu \tag{2.17}$$

is conserved

$$\partial_\mu J^\mu = 0. \tag{2.18}$$

The current (2.17) is not invariant under the gauge transformations

$$A^a_\mu \to A^a_\mu + \partial_\mu \alpha^a \tag{2.19}$$

$$B^a_\mu \to B^a_\mu + \partial_\mu \beta^a$$

which leave invariant (2.2) and (2.10). Instead, it changes as

$$J_\mu \to J_\mu - \frac{1}{2}\partial_\nu(\alpha C\tilde{F}^{\mu\nu} - \beta B\tilde{G}^{\mu\nu} + \alpha D\tilde{G}^{\mu\nu} - \beta A\tilde{F}^{\mu\nu}) \tag{2.20}$$

The corresponding integrated charge $\int J^0 d^3x$ is gauge invariant. One can see that it is actually the generator of the duality transformations, by using a Coulomb-like gauge and developing the appropriate canonical formalism.

The fact that the duality currents are not gauge invariant and therefore, as operators, are not true Lorentz vectors, shows that one cannot apply here the usual arguments (Coleman and Witten 1980, Weinberg and Witten 1980) according to which massless spin one states carrying the associated charge cannot exist. This is relevant for some recent attempts to connect supergravity with particle phenomenology (Ellis et al. 1980) in which one postulates that such massless spin one bound states arise dynamically.

Although the Lagrangian is not invariant under the transformations (2.6) (2.7), the derivative of the Lagrangian with respect to an invariant parameter is invariant. Assume that L depends upon an invariant parameter $\lambda$. If $\xi^i(\chi)$ is independent of $\lambda$, we differentiate (2.9) with respect to $\lambda$ and obtain

$$\frac{\partial \delta L}{\partial \lambda} = \frac{1}{2}\tilde{G}B\frac{\partial G}{\partial \lambda} = \frac{\partial L}{\partial F}B\frac{\partial G}{\partial \lambda} \quad . \tag{2.21}$$

On the other hand, since

$$\delta L = (\xi^i \frac{\partial}{\partial \chi^i} + \chi^j_\mu \frac{\partial \xi^i}{\partial \chi^j}\frac{\partial}{\partial \chi^i_\mu} + (FA^T + GB^T)\frac{\partial}{\partial F})L, \tag{2.22}$$

it follows that

$$\frac{\partial \delta L}{\partial \lambda} = \delta\frac{\partial L}{\partial \lambda} + \frac{\partial G}{\partial \lambda}B^T\frac{\partial L}{\partial F} \; . \tag{2.23}$$

Comparing (2.21) and (2.23), we find

$$\delta\frac{\partial L}{\partial \lambda} = 0. \tag{2.24}$$

The parameter $\lambda$ could be a coupling constant. For instance, in the example of the introduction, differentiate L given by (1.2) with respect to a, with the condition (1.7),

$$\frac{\partial L}{\partial a} = \frac{1}{2}F_{\mu\nu}\bar{\psi}\sigma^{\mu\nu}\psi + \frac{1}{4}a(\bar{\psi}\sigma_{\mu\nu}\psi)(\bar{\psi}\sigma^{\mu\nu}\psi)\,. \tag{2.25}$$

It is easy to check that this expression is invariant under (1.5) (1.6). The result (2.24) provides a way of checking that a theory admits duality rotations or of constructing the Lagrangian for such a theory, by switching on couplings in an invariant way. The case when the $\xi^i$ depend on $\lambda$ is a little more delicate (see Gaillard and Zumino 1981).

If $\lambda$ represents an external gravitational field, (2.23) implies that the energy momentum tensor, which is the variational derivative of the Lagrangian with respect to the graviational field, is invariant under duality rotations.

A Lagrangian satisfying (2.9) can be constructed by observing that, from (2.6) (2.8),

$$\frac{1}{4}\delta(F\tilde{G}) = \frac{1}{4}(FC\tilde{F} + GB\tilde{G})\,. \tag{2.26}$$

Therefore

$$L = \frac{1}{4}F\tilde{G} + L_{inv}\,, \tag{2.27}$$

where $L_{inv}$ is actually invariant under (2.6) (2.7). For instance, one can easily check that (1.2), with (1.7), is of this form, with

$$\tilde{G}_{\mu\nu} = 2\frac{\partial L}{\partial F^{\mu\nu}} = -F_{\mu\nu} + a\bar{\psi}\sigma_{\mu\nu}\psi \tag{2.28}$$

and

$$L_{inv} = -\frac{i}{2}\bar{\psi}\gamma^\mu\overleftrightarrow{\partial}_\mu\psi + \frac{a}{4}F_{\mu\nu}\bar{\psi}\sigma^{\mu\nu}\psi + \frac{a^2}{8}(\bar{\psi}\sigma_{\mu\nu}\psi)(\bar{\psi}\sigma^{\mu\nu}\psi)\,. \tag{2.29}$$

In general (2.27) can be used to construct Lagrangians, as described in

Gaillard and Zumino (1981). What we need is the expression for G as a function of F and $\chi$. The algebra is considerably simplified if one introduces the operator j which changes an antisymmetric tensor into its dual (see Misner and Wheeler 1967, Cremmer and Julia 1979)

$$jT_{\mu\nu} = \tilde{T}_{\mu\nu}$$
$$(j)^2 = -1. \qquad (2.30)$$

For many purposes this operator can be used in much the same way as the usual imaginary unit i. Let us assume that G is linear in F and write

$$G = jKF + X \qquad (2.31)$$

where the matrix $K(\chi)$ and $X(\chi)$ are functions of the fields $\chi^i$ and may contain j. Now (2.6) and (2.7) imply that

$$\delta K = -jC - jKBK + DK - KA, \qquad (2.32)$$
$$\delta X = DX - jKBX. \qquad (2.33)$$

Introducing two antisymmetric Lorentz tensors $(H_{\mu\nu}(\chi), I_{\mu\nu}(\chi))$ which transform under (2.6) (2.7) like $(F_{\mu\nu}, G_{\mu\nu})$, the quantity

$$X = -jI - KH \qquad (2.34)$$

satisfies (2.33). The Lagrangian (2.27) becomes

$$L = -\frac{1}{4} FKF + \frac{1}{4} jFX + L_{inv}$$
$$= -\frac{1}{4} FKF + \frac{1}{4} F(I - jKH) + L_{inv}. \qquad (2.35)$$

We cannot assume that $L_{inv}$ depends only upon the fields $\chi^i$ because (2.35), using (2.4), must reproduce (2.31), Therefore $L_{inv}$ must contain the invariant, linear in F,

$$\frac{1}{4}(FI - GH) = \frac{1}{4} F(I - jKH) + \frac{1}{4} jH(I - jKH). \qquad (2.36)$$

The result is finally

$$L = -\frac{1}{4} FKF + \frac{1}{2} F(I - jKH) + \frac{1}{4} jH(I - jKH) + L_{inv}(\chi), \qquad (2.37)$$

where the last term depends only on the fields $\chi^i$ and their derivatives.

From the transformation property (2.32) and from (2.8) we see that the matrix K can be taken to be symmetric. As we shall see in the next section, it can be taken to be a function of scalar fields only and it has the form

$$K = 1 + \ldots \qquad (2.38)$$

where the dots represent terms which vanish with the scalar fields, so that the first term in the right hand side of (2.37) contains the kinetic term for the vector fields. When there are no scalars and the duality rotations are restricted to the compact subgroup, the matrix K is just equal to the unit matrix, as in the simple example described in the introduction.

SCALAR FIELDS

Scalar fields valued in the quotient (coset) space Sp(2n, R)/U(n) can be described by a group element of Sp(2n, R) represented by the matrix

$$g = \begin{pmatrix} \phi_o & \phi_1^* \\ \phi_1 & \phi_o^* \end{pmatrix}, \tag{3.1}$$

where $\phi_o$ and $\phi_1$ are complex n × n matrices satisfying

$$\phi_o^\dagger\phi_o - \phi_1^\dagger\phi_1 = 1, \tag{3.2}$$

$$\phi_o^T\phi_1 = \phi_1^T\phi_o. \tag{3.3}$$

One can insure that the scalars are in the quotient space by requiring the theory to be invariant under the gauge transformation

$$g(x) \to g(x)[k(x)]^{-1}, \tag{3.4}$$

where k(x) is an element of U(n) represented by the matrix

$$k = \begin{pmatrix} U & o \\ o & U^* \end{pmatrix}, \tag{3.5}$$

$$U^\dagger U = 1. \tag{3.6}$$

Alternatively, one can parameterize the quotient space by using the n × n matrix (symmetric in virtue of (3.3))

$$Z = \phi_1\phi_o^{-1}\ , \ Z^T = Z \tag{3.7}$$

which is invariant under the gauge transformation (3.3).

The effect of Sp(2n, R) on the quotient space is described by the rigid transformation

$$g(x) \to g_o g(x), \tag{3.8}$$

where $g_o$ belongs to Sp(2n, R). Therefore we require the Lagrangian to be invariant under (3.8) also. It is sometimes convenient to use (3.4) to go to a special gauge, in other words to choose a representative for the equivalence class. In order to reestablish the special gauge, (3.8) must then be accompanied by a suitable transformation of the type (3.4). This gives rise to a non-linear realization of Sp(2n, R). Here we shall work in an arbitrary gauge and require separate invariance under (3.4) and (3.8).

In order to construct the invariant Lagrangian (see Gaillard and Zumino 1981 and references therein) we note that $g^{-1}\partial_\mu g$ belongs to the Lie algebra of Sp(2n, R) and can be split into a part $Q_\mu$ which is in the Lie algebra of U(n) and a part $P_\mu$ perpendicular to it

$$g^{-1}\partial_\mu g = Q_\mu + P_\mu . \tag{3.9}$$

Under (3.8) this expression is invariant, while under (3.4) it transforms as

$$g^{-1}\partial_\mu g \to k(g^{-1}\partial_\mu g - k^{-1}\partial_\mu k)k^{-1}, \tag{3.10}$$

so that

$$Q_\mu \to kQ_\mu k^{-1} - \partial_\mu k\, k^{-1}, \tag{3.11}$$

$$P_\mu \to kP_\mu k^{-1}. \tag{3.12}$$

Consequently the Lagrangian

$$L = -\frac{1}{2}\,\mathrm{Tr}P_\mu^2 \tag{3.13}$$

is invariant and contains the kinetic term for the scalar fields. This formula can be made more explicit by observing that, from (3.2)(3.3),

$$g^{-1} = \begin{pmatrix} \phi_o^\dagger & -\phi_1^\dagger \\ -\phi_1^T & \phi_o^T \end{pmatrix}. \tag{3.14}$$

One then finds

$$Q_\mu = \begin{pmatrix} \phi_o^\dagger\partial_\mu\phi_o - \phi_1^\dagger\partial_\mu\phi_1 & 0 \\ 0 & -\phi_1^T\partial_\mu\phi_1^* + \phi_o^T\partial_\mu\phi_o^* \end{pmatrix}, \tag{3.15}$$

$$P_\mu = \begin{pmatrix} 0 & \phi_o^\dagger\partial_\mu\phi_1^* - \phi_1^\dagger\partial_\mu\phi_o^* \\ -\phi_1^T\partial_\mu\phi_o + \phi_o^T\partial_\mu\phi_1 & 0 \end{pmatrix}. \tag{3.16}$$

Using (3.16) in (3.13) one finds, with a little algebra,

$$L = -\,\mathrm{tr}\{\partial_\mu Z(1 - Z^\dagger Z)^{-1}\partial_\mu Z^\dagger(1 - ZZ^\dagger)^{-1}\} . \tag{3.17}$$

The infinitesimal form of (3.8) is

$$\delta g = \begin{pmatrix} T & V^* \\ V & T^* \end{pmatrix} g , \tag{3.18}$$

where

$$T = -\,T^\dagger = M - iN$$

$$V = V^T = R - iS \tag{3.19}$$

and the real matrices M, N, R, S are related to those of equation (2.6) by

$$A = M + R,\ B = S + N,\ C = S - N,\ D = M - R. \tag{3.20}$$

This corresponds to using the complex basis $F \pm iG$

$$\delta\begin{pmatrix} F + iG \\ F - iG \end{pmatrix} = \begin{pmatrix} T & V^* \\ V & T^* \end{pmatrix} \begin{pmatrix} F + iG \\ F - iG \end{pmatrix}. \tag{3.21}$$

The transformation on Z induced by (3.18) is easily worked out to be

$$\delta Z = R - ZRZ - iS - iZSZ + [M, Z] + i\,\{N, Z\}\ . \tag{3.22}$$

Note that the matrix $Z^{-1}$ transforms exactly like $Z^* = Z^\dagger$. From the definition (3.7) and from (3.2) it follows that

$$1 - Z^\dagger Z = (\phi_o \phi_o^\dagger)^{-1}\ . \tag{3.23}$$

Since the right hand side is a positive matrix, this means that the eigenvalues of $Z^\dagger Z$ are smaller than one. Note that, if one introduces

$$K = \frac{1 - Z^*}{1 + Z^*} \tag{3.24}$$

one finds, from (3.22),

$$\delta K = [M, K] - \{R, K\} - iK(S + N)K - i(S - N). \tag{3.25}$$

This is the same as (2.32), if one uses (3.20) and replaces $i \rightarrow j$. Therefore (3.24), with the replacement $i \rightarrow j$, gives the matrix K which is to be used in the Lagrangian (2.37).

It is not difficult to introduce other fields besides the scalars. Let the field $\psi$ be invariant under (3.8) and let it transform as

$$\psi(x) \rightarrow k(x)\psi(x) \tag{3.26}$$

under (3.4). From (3.11) we see that

$$D_\mu \psi = \partial_\mu \psi + Q_\mu \psi \tag{3.27}$$

is a covariant derivative (here k and $Q_\mu$ are matrices in the appropriate representation). Using (3.27) one can construct invariant Lagrangians for fields other than the scalars, e.g. spinors or Rarita - Schwinger fields.

ACKNOWLEDGEMENT

This work was supported by the Director, Office of Energy Research, Office of High Energy and Nuclear Physics, Division of High Energy Physics of the U.S. Department of Energy under Contract W-7405-ENG-48.

REFERENCES

Coleman, S. and Witten, E. (1980). Phys. Rev. Lett., 45, 100.

Cremmer, E. and Julia, B. (1979). Nucl. Phys., B159, 141.

Cremmer, E. and Scherk, J. (1977). Nucl. Phys., B127, 259.

Cremmer, E., Scherk, J. and Ferrara, S. (1977). Phys. Lett., 68B, 234.

Cremmer, E., Scherk, J. and Ferrara, S. (1978). Phys. Lett., 74B, 61.

Deser, S. and Teitelboim, C. (1976). Phys. Rev., D13, 1592.

Ellis, J., Gaillard, M. K. and Zumino, B. (1980). Phys. Lett, 94B, 143.

Ferrara, S., Scherk, J. and Zumino, B. (1977). Nucl. Phys., B121, 393.

Gaillard, M. K. and Zumino, B. (1981). Nucl. Phys., to be published.

Misner, C. and Wheeler, J. A. (1957). Annals of Phys., 2, 525.

Weinberg, S. and Witten, E. (1980). Phys. Lett., 96B, 59.

# PART III

# COSMOLOGY AND THE EARLY UNIVERSE

# ENERGY, STABILITY AND COSMOLOGICAL CONSTANT

S. Deser
Department of Physics, Brandeis University
Waltham, Massachusetts 02254

Abstract. The definition of energy and its use in studying stability in general relativity are extended to the case when there is a nonvanishing cosmological constant $\Lambda$. Existence of energy is first demonstrated for any model (with arbitrary $\Lambda$). It is defined with respect to sets of solutions tending asymptotically to any background space possessing timelike Killing symmetry, and is both conserved and of flux integral form. When $\Lambda<0$, the energy is shown to be positive, and hence stability established, for all systems tending to anti-De Sitter space. Supergravity methods are used here, upon defining spinorial charges as flux integrals which obey the global graded algebra. For $\Lambda>0$, small excitations about De Sitter space are stable inside the event horizon. Outside excitations can contribute negatively due to the Killing vector's flip at the horizon. This is a universal phenomenon associated with the possibility of Hawking radiation. Apart from this effect, the $\Lambda>0$ theory appears to be stable, also at the semi-classical level.

## 1. INTRODUCTION

The observed smallness of the cosmological constant $\Lambda$, or equivalently of the vacuum energy density of the Universe is one of the major problems in current physics. It is highly unnatural from the point of view of particle physics since it demands extreme fine tuning of parameters. A possible way to exclude the cosmological constant would be to show that it leads to some fundamental classical or semiclassical instabilities in the Einstein theory. This was one motivation of the present work; related to it was the challenge of extending to the $\Lambda\neq0$ case the energy concepts so important when $\Lambda=0$ and more generally of understanding when conserved quantities may be defined in general relativity. The results have been obtained in collaboration with L. F. Abbott and details may be found in a forthcoming joint paper (Nucl. Phys. B).

Stability of a bounded matter system in flat space is usually established by showing it to have positive energy with respect to a lowest, vacuum, state. For gravity, (with $\Lambda=0$) energy of any asymptotically flat solution is also perfectly definable with respect to flat space as vacuum. It turns out that this energy is always positive and that the theory (also in the presence of positive energy matter) is stable; quite general and rigorous results have been obtained in recent years (Brill & Deser 1968; Deser & Teitelboim 1977; Grisaru 1978; Schoen & Yau 1979b,c; Witten 1981). On the other hand, when $\Lambda \neq 0$, flat space is no longer an acceptable background (since it does not solve the Einstein equations), but must be replaced as vacuum by the "flattest", maximally symmetric solutions of the cosmological equations, namely De Sitter O(4,1) or anti-De Sitter O(3,2) space according to whether $\Lambda > 0$ or $\Lambda < 0$, respectively. At this point, a number of problems arise for the stability programme. First, can any reasonable physical substitute for energy be defined? Having lost asymptotic Poincare invariance, one is left with the asymptotic De Sitter or anti-De Sitter algebra at infinity, for which $P_\mu^2$ is no longer a Casimir operator, being replaced by the five-dimensional "rotations" $J_{ab} = -J_{ba}$ ($a,b = 0,\ldots,4$). One must therefore show that $J_{04}$, which becomes $P_0$ upon contraction ($\Lambda \to 0$), is acceptable and that this quantity is really definable as a flux integral at infinity so as to satisfy the asymptotic global algebra. This is accomplished in Section 2, through a general analysis of how to obtain and interpret conserved quantities with respect to a background which possesses symmetries. We will see that, associated to every such symmetry, there is a generator which is conserved, background-covariant, and of flux integral form. The preferred one, which we call the Killing energy, is that which is connected with a timelike Killing vector or symmetry. Fortunately, for both signs of $\Lambda$, the maximally symmetric backgrounds have timelike symmetries, associated with $J_{04}$. The other nine generators are not timelike (nor are $P_i$, $J_{\mu\nu}$ when $\Lambda = 0$). A brief review of the properties of De Sitter spaces in Section 3 will show that for $\Lambda > 0$, the necessary presence of an event horizon, where the timelike Killing vector becomes null and then spacelike requires a more careful analysis of stability. Within the horizon, however, and in all space for $\Lambda < 0$ (where there is no horizon) it will be seen in Section 4 that small oscillations about vacuum have positive energy. The possiblity of negative energy for excitations outside the horizon (for $\Lambda > 0$) is a reflection, in Hamiltonian form, of the generic features which lead to Hawking radiation (Gibbons & Hawking 1977). In Section 5, we shall discuss semi-classical stability, against quantum mechanical tunnelling, for $\Lambda > 0$, which would be violated in the presence of Euclidean "bounce" solutions (Coleman 1977; Coleman & Callan 1977; Perry 1981; Gross, Perry & Yaffe 1981; Witten

1981b); no evidence for these is found. Section 6 is devoted to a demonstration of stability for all asymptotically anti-De Sitter solutions when $\Lambda<0$ by showing that the full energy is positive in that case. Here one uses methods of supergravity, parallel to those which were used to establish (Deser & Teitelboim 1977; Grisaru 1978) positivity of the energy for $\Lambda=0$. To accomplish this it is first necessary to define spinorial charges which are conserved and also have flux integral form and show the corresponding existence of appropriate Killing spinors, whose presence is implicit but rather trivial when $\Lambda=0$.

We conclude that models with cosmological constant are quite similar, in their energy definition and consequent stability properties, to the usual $\Lambda=0$ case. [This is not to say that they do not have other, unrelated, idiosyncrasies (Hawking & Ellis 1973).] Thus, although we have failed to exclude $\Lambda\neq0$ on stability grounds, a unification has been achieved in our understanding both of these models and of the general Hamiltonian mechanism underlying the modifications in stability when event horizons are present.

## 2. CONSERVED QUANTITIES

Consider the physical system defined by the Einstein equations

$$G_{\mu\nu} + \Lambda g_{\mu\nu} = 0 \tag{2.1}$$

together with a background metric $\bar{g}_{\mu\nu}$ which satisfies (2.1); we decompose the full metric $g_{\mu\nu}$ according to

$$g_{\mu\nu} = \bar{g}_{\mu\nu} + h_{\mu\nu} \tag{2.2}$$

where $h_{\mu\nu}$ is not necessarily small, but does obey the boundary condition that it vanishes asymptotically at some appropriate speed. We will construct conserved quantities from $(\bar{g}_{\mu\nu}, h_{\mu\nu})$ corresponding to the symmetries of the background. Although we are primarily concerned with background De Sitter or anti-De Sitter spaces, the method is completely general, and leads to flux integral expressions for these generators, which are constructed from the gravitational stress-tensor and the appropriate Killing vectors. Our conventions are that $R_{\mu\nu} \equiv R^{\alpha}{}_{\mu\alpha\nu} \sim + \partial_{\alpha}\Gamma^{\alpha}{}_{\mu\nu}$, signature (+++−) and all operations such as covariant differentiation ($\bar{D}_{\mu}$) or index moving are with respect to $\bar{g}_{\mu\nu}$. We define the symmetric stress tensor $T^{\mu\nu}$ to be all terms of second and higher order in $h_{\mu\nu}$ when the decomposition (2.2) is inserted in (2.1):

$$G_L^{\mu\nu}(\bar{g},h) + \Lambda h_{\mu\nu} = T^{\mu\nu} = T^{\nu\mu} \quad . \tag{2.3}$$

The subscript L refers to terms linear in $h_{\mu\nu}$. Using the fact that $G_{\mu\nu}(\bar{g}) + \Lambda\bar{g}_{\mu\nu} = 0$ and that the left side of (2.3) therefore obeys the (exact) linearized Bianchi identity $\bar{D}_\mu(G^{\mu\nu} + \Lambda h^{\mu\nu}) = 0$, the field equations imply covariant conservation of $T^{\mu\nu}$:

$$\bar{D}_\mu T^{\mu\nu} = 0 \ . \tag{2.4}$$

To turn covariant into ordinary conservation, we have to define a conserved (background) contravariant vector density $J^\mu$ from the tensor $T^{\mu\nu}$ since then, and only then, is $\bar{D}_\mu J^\mu \equiv \partial_\mu J^\mu$. When $\bar{g}_{\mu\nu}$ has a symmetry, there exists a Killing vector $\bar{\xi}_\mu$, obeying

$$\bar{D}_\mu\bar{\xi}_\nu + \bar{D}_\nu\bar{\xi}_\mu = 0 \ . \tag{2.5}$$

Consequently,

$$\bar{D}_\mu(T^{\mu\nu}\bar{\xi}_\nu) = (\bar{D}_\mu T^{\mu\nu})\bar{\xi}_\nu + 1/2\ T^{\mu\nu}(\bar{D}_\mu\bar{\xi}_\nu + \bar{D}_\nu\bar{\xi}_\mu) = 0 \tag{2.6}$$

and $\sqrt{-\bar{g}}T^{\mu\nu}\bar{\xi}_\nu$ is the desired contravariant tensor density, for which true conservation holds:

$$\bar{D}_\mu(\sqrt{-\bar{g}}T^{\mu\nu}\bar{\xi}_\nu) \equiv \partial_\mu(\sqrt{-\bar{g}}T^{\mu\nu}\bar{\xi}_\nu) = 0 \ . \tag{2.7}$$

So to every Killing vector, there is associated a conserved generator

$$E(\bar{\xi}) = 1/8\pi G \int d^3x\sqrt{-\bar{g}}T^{0\nu}\bar{\xi}_\nu \tag{2.8}$$

In particular, if $\bar{\xi}_\nu$ is timelike, this is the Killing energy.

Despite the fact that $\Lambda\neq 0$, we now show that $E(\bar{\xi})$ can be written in flux integral form, just as for $\Lambda=0$. From (2.3) it follows that

$$T^{\mu\nu} \equiv \bar{D}_\alpha\bar{D}_\beta K^{\mu\alpha\nu\beta} + 1/2(\bar{R}^\mu{}_{\alpha\beta}{}^\nu H^{\alpha\beta} - \Lambda H^{\mu\nu}) \ . \tag{2.9}$$

Here the superpotential K is defined by

$$2K^{\mu\alpha\nu\beta} \equiv \bar{g}^{\mu\beta}H^{\nu\alpha} + \bar{g}^{\nu\alpha}H^{\mu\beta} - \bar{g}^{\mu\nu}H^{\alpha\beta} - \bar{g}^{\alpha\beta}H^{\mu\nu},$$

with (2.10)

$$H^{\mu\nu} \equiv h^{\mu\nu} - 1/2\ \bar{g}^{\mu\nu}\ h^{\alpha}{}_{\alpha}\ .$$

It has the algebraic symmetries of the Riemann tensor:

$$K^{\mu\alpha\nu\beta} = -K^{\alpha\mu\nu\beta} = K^{\alpha\mu\beta\nu} = K^{\nu\beta\mu\alpha}\ . \tag{2.11}$$

Symmetry and conservation of $T^{\mu\nu}$ in (2.9) can easily be checked using the background field equations and its derivative consequences,

$$\bar{D}_\beta\bar{R}_{\mu\nu\alpha}{}^{\beta} \equiv \bar{D}_\nu\bar{R}_{\alpha\mu} - \bar{D}_\mu\bar{R}_{\alpha\nu} = 0\ , \quad \bar{D}_\beta\bar{R}^{\mu\nu} = 0$$

but without any assumptions on the full background Riemann tensor.

Next, we form $T^{\mu\nu}\bar{\xi}_\nu$, and recast it into the expression

$$\sqrt{-\bar{g}}\ T^{\mu\nu}\bar{\xi}_\nu = \sqrt{-\bar{g}}\ \bar{D}_\alpha\left[(\bar{D}_\beta K^{\mu\alpha\nu\beta})\bar{\xi}_\nu - K^{\mu\beta\nu\alpha}\bar{D}_\beta\bar{\xi}_\nu\right]$$

(2.12)

$$\equiv \bar{D}_\alpha F^{\mu\alpha}\ .$$

It may be verified that all additional terms "miraculously" vanish upon use of the Killing identity $\bar{D}_\beta\bar{D}_\alpha\bar{\xi}_\nu + R^{\lambda}{}_{\beta\alpha}{}^{\nu}\bar{\xi}_\nu \equiv 0$. Furthermore the quantity $F^{\mu\alpha}$ is (almost obviously) an antisymmetric tensor density, and therefore its divergence is an ordinary one, $\bar{D}_\alpha\ F^{\mu\alpha} \equiv \partial_\alpha\ F^{\mu\alpha}$. Hence the desired result:

$$8\pi G E(\bar{\xi}) = \int d^3x\sqrt{-\bar{g}}\ T^{o\nu}\bar{\xi}_\nu = \oint dS_i F^{oi}\ . \tag{2.13}$$

Note also that the Killing generator is mainfestly background-covariant, i.e. independent of the coordinate choice used for the background metric. As a check, when $\Lambda$=0 and the background is chosen to be flat, introduction of Cartesian coordinates simplifies (2.13) to be the usual expression for the Poincare generators. In particular, when $\bar{\xi}_\mu = (1,\vec{0})$ we obtain the standard energy formula

$$16\pi G\ E = \oint dS_i(h_{ji,j} - h_{jj,i}) \tag{2.14}$$

which correctly reproduces the mass of any asymptotically Schwarzschild metric. That the ten Poincare generators obey the (global) Poincare algebra is an immediate consequence of the commutation properties of the Killing vectors.

When $\Lambda\neq0$, with $\bar{g}_{\mu\nu}$ a De Sitter or anti-De Sitter metric, we would obtain the 10 Killing generators corresponding to the background De Sitter or anti-De Sitter symmetries, and they automatically satisfy the appropriate global algebra. In particular, we get the timelike $E(\bar{\xi})$ expression by using the appropriate Killing vectors of Section 3. [A check here is to verify that the equivalent of the Schwarzschild solution for $\Lambda\neq0$ has energy m. For $\Lambda>0$, there are corrections because one must stay within the unavoidable event horizon (rather than go to infinity) in calculating the flux integral; apart from these, the correct result E=m emerges.] We also mention that this whole procedure could also have been carried out in first order form (used in Section 4) to yield $E(\bar{\xi})$ there as well.

We complete this section with a treatment of the graded algebra which can be introduced when $\Lambda<0$ (but not $\Lambda>0$!), in terms of spinorial charges Q. The resulting local supersymmetry is that of supergravity with a cosmological term (Townsend 1977) and a spin 3/2 "mass" term (Deser & Zumino 1977). This will be used in Section 6 to show that the supergravity energy operator is positive and from this establish stability for classical gravity with $\Lambda<0$. First, however, we must show that the spinor charges can be written as surface integrals as in (2.13) so as to satisfy the graded global algebra at infinity. The spinorial charge density is

$$Q^{\mu} = \varepsilon^{\mu\alpha\beta\nu}\gamma_5\,\bar{\gamma}_{\alpha}\tilde{D}_{\beta}\psi_{\nu} \quad . \tag{2.15}$$

Its origin may be understood, just like that of $T^{\mu\nu}$, in terms of a decomposition of the full Rarita-Schwinger equation into a "linear" part and a remainder. Here $\psi_\nu$ is the spin 3/2 field, $\bar{\gamma}_\alpha$ are the background covariant $\gamma$ matrices with respect to the background vierbein and the modified covariant derivative on a spinor, $\tilde{D}_\beta$, defined by

$$\tilde{D}_{\beta} \equiv \bar{D}_{\beta} + \frac{1}{2}\, m\, \bar{\gamma}_{\beta} \quad , \qquad m^2 \equiv \frac{1}{3}\,|\Lambda| \tag{2.16}$$

has the basic property that $[\tilde{D}_\beta,\tilde{D}_\alpha]=0$ for a background anti-De Sitter space. The current $Q^\mu$ satisfies $\tilde{D}_\mu Q^\mu = 0$, and to convert this to an ordinary conservation law, we introduce Killing spinors obeying $\tilde{D}_\mu\alpha=0$ (consistent with $[\tilde{D}_\nu,\tilde{D}_\mu]\alpha=0$). The quantity $\bar{\alpha}Q^\mu$ is easily seen to take the form

$$\bar{\alpha}Q^\mu = \bar{D}_\beta(\bar{\alpha}\varepsilon^{\mu\alpha\beta\nu}\gamma_5\bar{\gamma}_\alpha\psi_\nu) \equiv \partial_\beta(\bar{\alpha}\varepsilon^{\mu\alpha\beta\nu}\gamma_5\bar{\gamma}_\alpha\psi_\nu) \ . \tag{2.17}$$

The last equality follows because the quantity in parentheses is an antisymmetric tensor density. But since $\bar{\alpha}Q^\mu$ is a contravariant vector, we see immediately that its ordinary divergence vanishes identically, $\partial_\mu(\bar{\alpha}Q^\mu)\equiv 0$, so that the spinor charge is both conserved and has the flux form

$$Q(\alpha) \equiv \int d^3x\bar{\alpha}Q^\circ \equiv \oint dS_j\bar{\alpha}\varepsilon^{\circ ijk}\gamma_5\bar{\gamma}_i\psi_k) \tag{2.18}$$

This is the required analogue of (2.13) for the bosonic generators. [The analogy actually goes further, in that only one "Coulomb" component of $\psi_k$ enters in (2.18), corresponding to the "Coulomb" part of the metric in (the appropriate generalization of) Eq. (2.14)].

Each of the four independent Killing spinors $\alpha_{\beta'}{}^{(\beta)}$, where $\beta'$ is the spinor index and $(\beta)$ is the label of each spinor, defines a fermionic charge $Q_{(\beta)}$. These then satisfy the anticommutation relation

$$\{Q(\beta),\bar{Q}(\beta')\} = \frac{1}{2}\left(\gamma^{(\mu)}\right)_{(\beta\beta')} J_{(\mu 4)} + \left(\sigma^{(\mu\nu)}\right)_{(\beta\beta')}J_{(\mu\nu)} \tag{2.19}$$

if we take the $\alpha$'s to commute (for convenience). Having defined the required conserved quantities, we turn next to the choice of Killing vectors.

## 3. DE SITTER SPACES

We give a brief review of the symmetries of "vacuum" spaces when $\Lambda\neq 0$. De Sitter space corresponds to a four-surface $z_\mu^2 + z_4^2 = 3/\Lambda$, $\Lambda>0$, in flat five-space. Among the rotations of the embedding space are the boosts mixing $z_o$ with $(z_i,z_4)$. For example $\bar{\xi}_a = (z_4,0,0,0,z_o)$ is a timelike Killing vector when $|z_4| > |z_o|$, but signals the existence of an event horizon at $|z_4| = |z_o|$, where stability must be discussed separately, since $E(\bar{\xi})$ no longer acts like an energy beyond it. Of course, an observer will only interact with events inside the horizon, which means that $E(\bar{\xi})$ tests stability to excitations visible to the observer. It is illuminating to apply these ideas to a simple model, namely a scalar field in De Sitter space. Representing the metric in the form

$$ds^2 = -dt^2 + f^2(t)(dx^2 + dy^2 + dz^2) \ , \quad f(t) \equiv \exp\sqrt{\tfrac{\Lambda}{3}}\, t \qquad (3.1)$$

with the "timelike" vector

$$\bar{\xi}^\mu = (-1, \sqrt{\tfrac{\Lambda}{3}}\, \vec{x}) \quad , \quad \bar{\xi}^2 = -1 + \tfrac{\Lambda}{3}\, |\vec{x}f|^2 \qquad (3.2)$$

we see that the horizon appears (at any given time) for distances such that $|\sqrt{\Lambda/3}\ \vec{x}f| = 1$. The action and energy momentum densities for this theory are

$$\begin{aligned} I &= \int d^4x f^3 [\tfrac{1}{2}\ \dot{\phi}^2 - \tfrac{1}{2} f^{-2}(\nabla\phi)^2 - V(\phi)] \\ -\mathfrak{T}^o{}_o &= \tfrac{1}{2}\ f^{-3}\pi^2 + \tfrac{1}{2}\ f(\nabla\phi)^2 + f^3 V(\phi) \qquad (3.3) \\ -\mathfrak{T}^o{}_i &= \ \pi\partial_i\phi \end{aligned}$$

where $\pi \equiv f^2\dot{\phi}$. The energy density is positive [if $V(\phi)$ is] but time-dependent. The conserved energy (which is of course not a flux integral here) is still $E(\bar{\xi}) = \int d^3x\ \mathfrak{T}^{0\mu}\bar{\xi}_\mu$. The integrand has the form

$$\mathfrak{T}^{o\mu}\bar{\xi}_\mu = \{\tfrac{1}{2}\ [f^{-3}\pi^2 + f(\nabla\phi)^2] - \sqrt{\tfrac{\Lambda}{3}}\vec{x}\cdot\pi\vec{\nabla}\phi\} + f^3 V(\phi) \ . \qquad (3.4)$$

It will be positive provided the bracketed quantity is. The triangle inequality

$$\tfrac{1}{2}(\vec{A}^2+\vec{B}^2) \geq (\sqrt{\tfrac{\Lambda}{3}} f|\vec{x}|)\vec{A}\cdot\vec{B} \quad , \ \vec{A} \equiv f^{-3/2}\pi\hat{x}, \ \vec{B} \equiv f^{1/2}\vec{\nabla}\phi \ , \ \sqrt{\tfrac{\Lambda}{3}} f|\vec{x}| < 1 \qquad (3.5)$$

makes it easy to see that the positivity condition corresponds to $\bar{\xi}^2<0$ in (3.2), i.e., to excitations within the horizon. This correlation between the event horizon and positivity is an expression in Hamiltonian form of Hawking radiation (Gibbons & Hawking 1977), and will be seen later to be universal.

Anti-De Sitter space is the covering space for the surface $z_\mu^2 + z_4^2 = 3/\Lambda$, $\Lambda<0$. Here there is a global timelike Killing vector corresponding to $(z_0,z_4)$ rotation. It is $\bar{\xi}_a=(z_4,0,0,0,-z_0)$, $\bar{\xi}^2 = (z_\mu^2 + z_4^2)<0$, since $z_4 = z_0 = 0$ is excluded. There is one peculiar feature of anti-De Sitter space which should be mentioned. Specification of initial data on a complete spacelike surface does not lead to a unique prediction of the future state of a system (including gravity itself). Radiation not specified by the initial conditions can propagate in from infinity at a later time. Unlike the usual case where initial boundary conditions exclude incoming radiation thereafter, one is "safe" here only within ever more restricted regions of space at later times. Therefore, although the initial energy is perfectly well defined by the initial data, one can only extend the integration volume to all space at a later time if further (timelike) boundary conditions at infinity are imposed. It is in this sense that our stability results are to be understood, although the proof that energy is positive holds formally on any complete initial surface with no incoming radiation.

## 4. SMALL EXCITATIONS

We now apply the Killing energy together with the Killing vectors defined in the last two sections to discuss small oscillations about De Sitter or anti-De Sitter backgrounds. For this purpose, a canonical approach (Arnowitt, Deser & Misner 1959, 1960, 1962) is most useful to discuss the $O(h^2)$ part of $E(\bar{\xi})$. Indeed $T^0{}_\nu$ was derived canonically to this order long ago by Nariai and Kimura (1962). We will skip all details here, only noting that the excitations can be parametrized by the $h_{ij}$ and their conjugate momenta $p^{ij}$, both being transverse traceless with respect to $\bar{g}$.

For $\Lambda>0$, the Hamiltonian density can be cast into the form

$$\begin{aligned} -\mathfrak{T}^0{}_0 &= \frac{1}{2}\,(f^{-3}(P^{ij})^2 + f(\nabla Q^{ij})^2) \\ -\mathfrak{T}^0{}_i &= P^{ik}\partial_i Q_{jk} \end{aligned} \tag{4.1}$$

exactly as for the scalar field of (3.2), in terms of a canonically transformed set (Q,P). Not surprisingly, all the other features of the scalar model follow as well: $\int d^3x \mathfrak{T}^{0\nu}\bar{\xi}_\nu$ is conserved and is positive within the event horizon $\bar{\xi}^2<0$. However outside, when $\bar{\xi}^2>0$, the energy is no longer positive, because the triangle inequality no longer applies, as with (3.5). One would expect all physical systems to behave in this way: the free part of the energy is always $\sim 1/2\int\{\pi^2+(\nabla\phi)^2\}$ while the momentum density is $\sim\pi\nabla\phi$. For physical matter, the non-linear parts of the energy are, like $V(\phi)$ in the scalar case, positive so the critical condition arises primarily at the free field level, where excitations beyond the horizon can give negative contributions. In particular, if the higher terms in $T^0{}_0(h)$ are effectively positive (as in the $\Lambda=0$ case), then the only De Sitter instability would be that due to the horizon.

For $\Lambda<0$, the small excitations are straightforwardly treated; this time only $T^0{}_0$ is required, since we can pick a coordinate system which is static and in which $\bar{\xi}_i=0$, and there is no horizon. The energy density is positive,

$$T^0{}_0 \sim [p_{ij}^2 + \frac{1}{4}(\nabla h_{ij})^2 + |\frac{\Lambda}{6}| h_{ij}^2] \ , \qquad (4.2)$$

and the system is stable. The masslike term in (4.2) is an artifact just like that of the spin 3/2 field (Deser & Zumino 1977) in cosmological supergravity; the gravitons have only two degrees of freedom since $h_{ij}$ is transverse-traceless.

5. SEMI-CLASSICAL STABILITY FOR $\Lambda > 0$

Having shown that $\Lambda > 0$ solutions are stable to small fluctuations about the vacuum, at least within the horizon, one may make the further test of semi-classical stability in this case, i.e., look for Euclidean "bounce" solutions whose presence would signal quantum tunnelling instability. Of course even better would be proof that the total energy is positive, which will be given for $\Lambda<0$ in the next section; lacking this for $\Lambda>0$, we make some comments on bounces there. In general, topological effects can give stability problems in gravity (Brill & Deser 1973) and semi-classical instability has been found in other gravitational contexts (Perry 1981; Gross, Perry & Yaffe 1981; Witten 1981b).

A bounce solution here would be a metric which is asymptotically De Sitter and solves the Euclideanized Einstein equations. For example, the Euclidean continuation of the Schwarzschild-De Sitter metric would be a candidate. However, it is impossible to remove both the Schwarzschild and horizon singularities of this metric by the usual periodicity trick (Gibbons & Perry 1978). In terms of the Hawking picture, De Sitter space contains radiation at a temperature fixed by the value of $\Lambda$. If a black hole could form in this space with an intrinsic temperature less than this, it would grow forever by accretion. However, for the Schwarzschild-De Sitter black hole, the black hole temperature is always larger (Gibbons & Hawking 1977) than that of the exterior and the space is stable against this catastrophe.

Although it is doubtful on general grounds that bounce solutions exist for $\Lambda > 0$, it would clearly be desirable to extend the proofs of their absence (Witten 1981; Schoen & Yau 1979a) for $\Lambda = 0$ to this domain as well.

6. STABILITY FOR $\Lambda < 0$

We now show the Killing energy is positive for all excitations about the anti-De Sitter vacuum which vanish at infinity, and thereby establish stability in the $\Lambda < 0$ sector. We have already noted in Section 2 that all the generators of the graded anti-De Sitter algebra in supergravity are expressed as flux integrals, with the result that they obey the global algebra relations, in particular that of Eq. (2.19):

$$\{Q_{(\beta)},\bar{Q}_{(\beta')}\} = \frac{1}{2}\,\gamma^{(\mu)}_{(\beta\beta')}J_{(\mu 4)} + \sigma^{(\mu\nu)}_{(\beta\beta')}J^{(\mu\nu)}$$

We emphasize that in this expression, all indices are labels of particular Killing vectors or spinors. The explicit relations between the two are quite analogous to those holding in the Poincaré case, and indeed can be essentially reduced to it because the $\tilde{D}_\mu$ commute; there exists (Gursey & Lee 1963) a transformation $\alpha = S\eta$ which reduces the equations $\tilde{D}_\mu\alpha = 0$ to $\partial_\mu\eta = 0$. A basis for the latter is given by, e.g., $\eta_\beta = \delta_\beta{}^{(\beta')}$. In any case, we may now simply treat the spinor "labels" $(\beta)$, $(\beta')$ in (2.19), which refer to the particular Killing spinor defining the corresponding charge $Q_{(\beta)}$, as normal flat space spinor indices. Multiplying (2.19) by the numerical matrix $\gamma^{(o)}_{\beta\beta'}$ and tracing gives the positivity relation for the operator $J_{(04)}$:

$$J_{(04)} = \sum_{\beta=1}^{4} Q_{(\beta)}Q_{(\beta)} \geq 0 \qquad (6.1)$$

since the $Q_{(\beta)}$ are real Majorana spinor operators. Now we just proceed as in the $\Lambda = 0$ case (Deser & Teitelboim 1977; Grisaru 1978), taking matrix elements of (6.1) with no on-shell fermions and go to the tree limit, $\hbar\to 0$. This implies that $E(\bar{\xi})$, which is just this limit of $J_{(04)}$, is positive for classical $\Lambda < 0$ gravity.

We also believe, although we have not carried out the details, that the recent purely classical proof of Witten (1981a) that energy is positive for $\Lambda = 0$ gravity can also be applied here. His proof, inspired by the supergravity argument, is based on considering solutions of the Dirac equation $\not{D}\varepsilon\equiv\gamma^i D_i\varepsilon=0$ in an external metric satisfying $G_{o\mu}=0$. From the relations

$$0 = \varepsilon^*\not{D}^2\varepsilon \equiv \varepsilon^*(D^2 + G_{o\mu}\gamma^o\gamma^\mu)\varepsilon = \varepsilon^*D^2\varepsilon \ ,$$

it follows upon integration that

$$\oint dS_i\,\varepsilon^* D^i\varepsilon = \int|\nabla\varepsilon|^2 d^3x \geq 0 \ . \qquad (6.2)$$

The surface integral is then separately shown to be proportional to E, which establishes positivity of the latter. The same reasoning should apply here with $D_i$ replaced by $\tilde{D}_i$, and the metric now satisfying $G_{0\mu}+\Lambda g_{0\mu}=0$ provided, as is likely, the surface integral is again proportional to E. Similarly, it would be of interest to generalize the classical geometrical proof of Schoen and Yau (1979b, 1979c) to the $\Lambda < 0$ case. It may even be possible to establish full non-linear stability in the $\Lambda > 0$ case for excitations lying within the horizon by analytic continuation from $\Lambda < 0$, using the static form of the O(4,1) metric which covers the interior region only.

## 7. CONCLUSIONS

Our first task was to establish a unified and physically appropriate definition of energy in general relativity independent of the value of $\Lambda$. This was possible, for every $\Lambda$ and for each set of solutions which tend asymptotically to a common background solution equipped with a timelike Killing vector. The associated energy is conserved, manifestly background gauge invariant and expressible as a flux integral at spatial infinity. We next studied stability by considering the positivity properties of the energy. For $\Lambda=0$ it was of course known that it is always positive for arbitrary asymptotically flat excitations. This property turned out to be shared by the $\Lambda<0$ models, as was demonstrated by a parallel study of the embedding supergravity theory. There, spinorial charges are also flux integrals and obey the anti-De Sitter graded global algebra, which easily implies positivity. For $\Lambda>0$, there is no grading possible and we studied several different stages. The first was that of small oscillations about the De Sitter background, whose energy was positive within the intrinsic event horizon. Those outside could exhibit a negative character associated with the transition of the Killing vector from timelike to spacelike at the horizon. This phenomenon was seen to be entirely independent of the system and is a simple consequence of the form of any free-field Hamiltonian; it marks the onset of Hawking radiation. Apart from this property, we found no evidence for instability at the semiclassical level, and indeed we believe that energy inside the horizon is positive for general excitations there. The resemblance between these properties of gravity models, irrespective of whether $\Lambda=0$, provides a unified picture, but also implies that $\Lambda\neq0$ theory cannot be excluded on stability grounds alone.

## ACKNOWLEDGEMENTS

This research was supported in part by NSF grant PHY7809644A02.

REFERENCES

R. Arnowitt, S. Deser and C. W. Misner (1959), Phys. Rev. 116, 1322; (1960) 117, 1595 and in (1962), Gravitation: an introduction to current research, ed. L. Witten (Wiley, New York).

D. Brill and S. Deser (1968), Ann. Phys. 50, 548.

D. Brill and S. Deser (1973), Comm. Math. Phys. 32, 291.

S. Coleman (1977), Phys. Rev. D15, 2929.

S. Coleman and C. G. Callan (1977), Phys. Rev. D16, 1762.

S. Deser and C. Teitelboim (1977), Phys. Rev. Lett. 39, 249.

S. Deser and B. Zumino (1977), Phys. Rev. Lett. 38, 1433.

G. W. Gibbons and S. W. Hawking (1977), Phys. Rev. D10, 2738.

G. W. Gibbons and M. J. Perry (1978), Proc. Roy. Soc. A358, 467.

M. Grisaru (1978), Phys. Lett. 73B, 207.

D. Gross, M. J. Perry and L. Yaffe (1981), Princeton University preprint.

F. Gursey and T. D. Lee (1963), Proc. Nat. Acad. Sci. 49, 179.

S. W. Hawking and G. F. R. Ellis (1973), The Large Scale Structure of Space-time, (Cambridge).

H. Nariai and T. Kimura (1962), Progr. Theor. Phys. 28, 529.

M. J. Perry in (1981) Superspace and Supergravity, eds. S. W. Hawking and M. Rocek (Cambridge).

P. Schoen and S. T. Yau (1979a), Phy. Rev. Lett. 42, 547; (1979b), Comm. Math. Phys. 65, 45; (1979c) Phys. Rev. Lett. 43, 1457.

P. K. Townsend (1977), Phys. Rev. D15, 2802.

E. Witten (1981a), Comm. Math. Phys. 80, 381; (1981b), Princeton University preprint.

# PHASE TRANSITIONS IN THE EARLY UNIVERSE

T.W.B. Kibble
Blackett Laboratory, Imperial College
Prince Consort Road, London, SW7 2BZ

The idea that in its early history the universe underwent phase transitions stems from two basic hypotheses - grand unification and the hot big bang - which I shall take for granted.

In a grand unified theory (GUT) there is a fundamental gauge theory with symmetry group G (for example SU(5) or SO(10)) which is broken spontaneously, first to SU(3)xSU(2)xU(1) at an energy scale of around $10^{15}$ GeV and then at about 100 GeV to $SU(3)_{colour} \times U(1)_{electromagnetism}$. (In more complicated models there may be more than two transitions.) In GUTS, quark-lepton transitions are possible, so that in principle the proton can decay, with a predicted lifetime of $10^{30}$ years or more. If this prediction is correct, we should know quite soon. Once baryon number conservation is discarded, the way is open to an explanation of the baryon-to-photon ratio of the universe, observationally $10^{-9\pm1}$. This qualitative explanation is one of the great successes of GUTS, but one I shall not discuss here.

In most theories of spontaneous symmetry breaking, the symmetry is restored at high enough temperature. If we accept the hot big bang we should thus expect that at very early times the universe was above all critical temperatures and exhibited the full symmetry of the group G. As it expanded and cooled more or less adiabatically, it must have passed successively through various phase transitions at which the symmetry was reduced. My concern in this talk will be with the various topological structures that can be generated at these transitions - domain walls (which can easily be ruled out), strings, which have been suggested as a mechanism of galaxy formation, and monopoles, concerning which the main problem is to avoid an overabundance.

THE INITIAL STATE

Let me start by discussing the state of the universe shortly after the Planck time, when its temperature is only a little below the Planck mass $m_P = G^{-\frac{1}{2}} = 1.2 \times 10^{19}$ GeV. Despite the fact that the causal horizon contains only a few particles, I shall assume that the universe is in thermal equilibrium. Why this should be so is at present a mystery - discussed elsewhere in this meeting. It is commonly supposed to be somehow attributable to quantum gravity, but as yet we have no proper quantum gravity, so this notion remains speculative.

At these temperatures all masses are essentially negligible, so we may treat our system as a relativistic gas. Moreover, because of asymptotic freedom, it is a weakly-interacting gas, with a coupling strength $\alpha = g^2/4\pi \simeq 0.025$. Thus it may be treated as approximately ideal. At first sight, the notion of an ideal gas with a density vastly exceeding nuclear densities must seem odd at best. However it is consistent. In such a gas the number density of each boson species (counting the two helicity states of massless bosons with spin as two species) is

$$n_s = \pi^{-2}\zeta(3)T^3$$

For fermions there is an extra factor of $\frac{3}{4}$. Thus there is approximately one particle of each species within a thermal volume, of radius $1/T$. Since all cross-sections are of order $\sigma \simeq \alpha^2/T^2$, the mean free path $\lambda$ is given by

$$1/\lambda = n\sigma \simeq N_* \alpha^2 T$$

where $N_*$ is the number of species. In the simplest GUT based on SU(5), $N_* = 160.75$, so

$$N_* \alpha^2 \simeq 1/15 \ll 1$$

and thus $\lambda$ is large compared to the thermal radius, and a fortiori to the interparticle spacing. This is at least a partial justification for the ideal gas approximation. Unfortunately it is also true that until T falls to about $10^{17}$ GeV, $\lambda$ is large compared to the expansion time, which makes it all the harder to understand how the universe came to be in thermal equilibrium.

SPONTANEOUSLY BROKEN GAUGE THEORIES.

Next let me recall some basic facts about spontaneously broken gauge theories. Consider a gauge theory with gauge group G and a Higgs field $\phi$ belonging to some representation of G, interacting with itself via a potential $U(\phi)$, a G-invariant polynomial of degree 4 (to ensure renormalizability). For purposes of illustration, let us take $G = SO(N)$ and consider a Higgs field $\phi$ belonging to the fundamental N-dimensional (vector) representation. Then we may take

$$U(\phi) = \frac{1}{8} h^2 (\phi.\phi - \eta^2)^2.$$

The coefficient of the quadratic term has been chosen negative so that the surface M of minima of U is not the point $\phi = 0$ but the (N-1)-sphere $\phi.\phi = \eta^2$. In general, if $\phi$ is a point on M, and H is the corresponding isotropy subgroup of G, $H = \{g\varepsilon G: g\phi = \phi\}$, then M may be identified with the quotient space G/H. In our example, the constant term in U has been chosen so that U vanishes when $\phi$ lies on M. This is necessary to ensure that the cosmological constant is nearly zero, but is fundamentally a rather arbitrary procedure.

The particular case N=2 of this model is familiar in superconductivity as the Landau-Ginsberg model. In its relativistic incarnation it is the Higgs model.

When we go to a finite temperature the role of the potential U is taken by the effective potential $V(\phi)$, which is simply the minimum free energy density in states with a given expectation value $\langle\phi\rangle$ of $\phi$. For weak coupling we may compute it by adding to U the 1-loop (order $\hbar$) corrections. At high temperatures the most important contributions are

$$V(\phi) = U(\phi) - N_*(\pi^2/90)T^4 + (1/24)M_*^{\,2}(\phi)T^2 + O(T),$$

where $N_*$ is the total number of distinct helicity states of "light" particles, i.e. those whose masses are small compared to T, and $M_*^2$, which is $\phi$-dependent, is the sum of squared masses of these states. (Fermion contributions appear in $N_*$ multiplied by 7/8 and in $M_*^2$ by $\frac{1}{2}$).

We can now see how symmetry restoration arises. As T varies the nature of the minimum surface M of V may change. For example in our model the $\phi^2$ term is

$$\tfrac{1}{2}(-\tfrac{1}{2}h^2\eta^2 + AT^2)\phi^2$$

where A is a positive constant (a linear combination of the Higgs coupling constant $h^2$ and the square of the gauge coupling $g^2$). Thus there is a critical temperature $T_c$ of order $\eta$. For $T > T_c$ the coefficient of $\phi^2$ is positive, so we are in a symmetric phase, with $\langle\phi\rangle = 0$. When T falls below $T_c$, $\phi$ acquires a nonzero expectation value, lying on the sphere

$$\langle\phi\rangle^2 = \eta^2 (1-T^2/T_c^{\,2}).$$

We are then in an ordered phase. In this simplified version of the theory the transition is second-order. In reality, for various reasons I shall come back to later, it would almost certainly be first order.

Note that corresponding to the two coupling constants there are two characteristic lengths (familiar in superconductivity) in this problem, the correlation length $\xi = m_H^{-1}$ and the penetration depth $\lambda = m_X^{-1}$, where $m_H$ and $m_X$ are the masses of the scalar Higgs particle and the vector (gauge) particle, respectively. At zero temperature their values are

$$\xi_o = 1/h\eta \quad \text{and} \quad \lambda_o = 1/g\eta \ .$$

## HISTORY OF THE UNIVERSE

If the ideas presented here are broadly correct, it is likely that in the course of its early evolution the universe underwent at least three phase transitions. The first is the grand unification transition at about $10^{15}$ GeV, which occurs when the age of the universe is about $10^{-37}$s, and at a red-shift Z of $10^{28}$. Before this, the universe is in a fully symmetric state, with the symmetry of the grand unified group, for example SU(5). After it, the symmetry is broken to SU(3) x SU(2) x U(1).

The next transition need not occur until about $10^{-11}$s after the big bang, when the temperature has fallen to a mere 100 GeV. This is the electroweak transition, at which the SU(2)xU(1) symmetry breaks to the U(1) of electromagnetism. In the simplest models nothing of interest happens in the vast "desert" between $10^{15}$ GeV and 100 GeV. However, it is perfectly possible to construct more complicated theories in which the desert flowers with multiple transitions.

Even without such complications there is probably a third transition in the QCD part of the system, associated with quark confinement. This is the transition from a quark and gluon soup to a gas of identifiable individual hadrons. As yet, this last transition is rather poorly understood, but presumably it occurs at a temperature of some hundreds of MeV when the universe is aged about a microsecond. I shall ignore the confinement transition because it does not seem to be associated with symmetry breaking and so does not exhibit the same structures as the earlier ones.

What happens when the universe passes through a phase transition? For the moment let us assume that, as the simple theory suggests, the transition is second-order. Above the critical temperature $T_c$, the Higgs field $\phi$ has vanishing expectation value. As $T$ falls below $T_c$, the order parameter $\langle\phi\rangle$ becomes nonzero, and if equilibrium is maintained takes on a value corresponding to some point on the surface M of minima of V - but which point?

The situation here is similar to that of a ferromagnet cooled through its Curie point. It will acquire a spontaneous magnetization, but the direction of the magnetization may be randomly chosen (from within some set of allowed directions). The choice of direction would be determined in practice by stray magnetic fields or random fluctuations in the initial state.

The universe has a similar choice to make of a point on M, but it need not make the same choice everywhere. Indeed, it is hard to see how there could be any correlation between the choices in widely separated regions, certainly when they are farther apart than the current horizon distance. One should note however that there may be a conflict between this argument and the assumption that thermal equilibrium has been achieved over much greater distances. If we nevertheless accept it, we find that topological singularities may appear. There may be incompatibilities between the choices made in different regions and thus obstructions to the extension of $\langle\phi\rangle$ throughout space while keeping its value on M.

The various possibilities can readily be illustrated by

special cases of the SO(N) model I discussed earlier.

DOMAIN WALLS

First consider the case N=1, which is not a gauge theory at all, but simply a model of a single real scalar field $\phi$. The potential V in this case has for small T two minima close to $\phi = \pm\eta$. By random choice the universe may select $\langle\phi\rangle = \eta$ in some regions and $\langle\phi\rangle = -\eta$ in others. Thus we expect a domain structure with domains separated by walls. Across a wall $\langle\phi\rangle$ must go from $\eta$ to $-\eta$, and thus pass through 0, which is a maximum of V. The height of the potential hump is $\Delta f \approx h^2\eta^4$ while the thickness of the wall is of the same order of magnitude as the correlation length $\xi = 1/h\eta$. Thus the mass of wall per unit area, which in this relativistic case is also the surface tension $\sigma$, is $\sigma \approx \xi_0 \Delta f \approx h\eta^3$. It is easy to check that this is impossibly large. As was pointed out by Zel'dovich, Kobzarev and Okun (1974), even when $\eta$ is set at 100 GeV, the value appropriate to the electroweak transition, a single domain wall stretched across the visible universe would easily dominate its mass and cause quite unacceptable deviations from isotropy in the microwave background. If we start with an initially random distribution of the two domains, one of them may often locally grow at the expense of the other, so that the typical domain size grows. But if there is no underlying asymmetry between the two we must always expect to find some regions of each type in the universe. No physical processes can generate correlations extending beyond the causal horizon, so the maximum domain size will be of that order or less. Hence we must always expect at least one domain wall across the visible universe.

This argument allows us to rule out any theory involving the purely spontaneous breaking of a discrete symmetry. This is a remarkable, and powerful, restriction. One possible way of avoiding it might be to arrange that the symmetry is never restored at high temperature (a possibility I shall discuss in another context later). However this does not really avoid the problem; it merely removes it from an event at a finite time to the initial state. In the initial state a random choice was made of one minimum of V rather than the other. Why was the same choice made everywhere?

STRINGS AND MONOPOLE

Next, let us examine the model with N=2, the Landau-Ginsburg model. Here, M is a circle of radius $\eta$. We may encounter a string singularity characterized by the fact that as we follow $\langle\phi\rangle$ around a loop enclosing the string, the angle defining the point on M changes by $2\pi$. As in the previous case, $\langle\phi\rangle$ must vanish somewhere within this loop. The string therefore has a mass per unit length, or tension

$$\mu \simeq \xi_o^2 \, \Delta f \simeq \eta^2 .$$

This is still in conventional terms very large. A grand unified string has a mass per unit length of about ten tons per fermi! Nevertheless, because of the lower dimensionality strings are less catastrophic. Note that

$$G\mu \approx (\eta/m_P)^2 \approx 10^{-8} .$$

A single string stretched across the visible universe to the horizon, at distance 2t (in the radiation-dominated era), has a mass $4\eta^2 t$. Let us compare this to the mass of the universe, namely $(4\pi/3)\rho 8t^3$, where $\rho = .03/Gt^2$. Clearly the ratio of the two is approximately

$$4G\mu \simeq 10^{-7} ,$$

a comfortably small number.

It is easy to extend the discussion to the next case, N=3, where M is a 2-sphere. In this case point-like monopole singularities may appear, as in the famous solution of 't Hooft (1974) and Polyakov (1974). Following the same rather crude argument as before one would find for the monopole mass the estimate $\xi_o^3 \Delta f \simeq \eta/h$. However this is at best a lower limit because it completely ignores the contribution of the gauge fields to the energy. A better estimate, roughly correct except when h is very small compared to g, is (Bogomol'nyi 1976)

$$m_M \simeq 4\pi\eta/g$$

which for a grand unified monopole is of order $10^{16}$ GeV. In general, the conditions for producing domain walls, strings or monopoles may be stated in terms of the homotopy groups $\pi_n(M)$ of the space M of degenerate vacua. For the three cases we require $\pi_0(M)$, $\pi_1(M)$ or $\pi_2(M)$ respectively to be nontrivial. Since M may be identified with the space G/H of cosets, these groups are related to the homotopy groups of G and H by the familiar exact sequence. In particular if G is a connected and simply connected Lie group, so that $\pi_0(G)$ and $\pi_1(G)$ as well as $\pi_2(G)$ are trivial, then

$$\pi_1(M) = \pi_1(G/H) = \pi_0(H)$$

and

$$\pi_2(M) = \pi_2(G/H) = \pi_1(H).$$

The condition for strings to appear is that H be disconnected. Note that to apply this analysis to the N=2 model I discussed earlier, one must take for G the universal covering group of SO(2) namely R. Then H=Z, which is obviously disconnected.

For monopoles the condition is that H be non-simply connected. In particular any U(1) factor in H will yield a factor Z in $\pi_2(M)$, though there may also be finite factors such as $Z_2$. If we start with a simple or semisimple group G and end as we must with an unbroken subgroup containing a U(1) factor - the U(1) of electromagnetism - then it is certain that at some phase transition monopoles must be produced. In the simplest GUT they are produced at the first transition. By complicating the model one can postpone their creation but not easily eliminate them altogether. This is the origin of the problem to which I now turn.

NUMBER DENSITY OF MONOPOLES

This problem has been discussed by many authors, beginning with Zel'dovich and Khlopov (1978) and Preskill (1979).

To estimate the initial density of monopoles, it is useful to begin by examining the length scale of random fluctuations in the Higgs field. This is the correlation length $\xi = 1/m_H$, which in theory becomes infinite at the phase transition. In practice, infinite-range correlations cannot arise because the universe goes through the transition at a finite rate. However the correlation length will be large at $T_c$ and will fall thereafter until it finally reaches the zero-temperature value $\xi_0$.

So long as T is not much below $T_c$, the height $\Delta f$ of the central peak in the effective potential V will be fairly small. If

$\xi^3 \Delta f \lesssim T$ then fluctuations back to $\langle\phi\rangle = 0$ will be quite common. The temperature at which this becomes an equality in the Ginzburg temperature $T_G$ (Ginzburg 1960). If the Higgs coupling is week, $T_G$ lies not far below $T_c$, in fact $T_c - T_G \approx h^2 T_c$. The monopoles effectively freeze out at this temperature.

The simplest estimate of the initial monopole density is

$$n_M \approx p\xi(T_G)^{-3}$$

where p is a numerical factor rather less than unity (say 0.1) and $\xi(T_G) \approx 1/h^2\eta$ . An important parameter is the ratio

$$r = n_M/T^3.$$

In the absence of annihilation this ratio will be effectively constant during the subsequent evolution. (It would be more accurate to use the monopole-to-entropy ratio, but the difference is not important when discrepancies of many orders of magnitude are in question.) With the simple estimate ,

$$r \approx h^6 p.$$

which might be expected to be of order $10^{-7}$ or so.

It has been shown by Preskill (1979) and Zel'dovich and Khlopov (1978) that subsequent annihilation processes cannot lower r below about $10^{-10}$ if it is initially above this value (and will have little effect if it is below). For heavy monopoles this would be a disaster. The limit obtained by requiring that the present density of monopoles should not exceed the upper limit on the density of the universe is $r < 10^{-24}$. One might argue that in the recent universe monopoles have accumulated preferentially in special places, centres of galaxies for example, where their annihilation was enhanced. Even so there is another stringent limit. Helium synthesis is very sensitive to the expansion rate and hence to the density of the universe then. This imposes a limit $r < 10^{-19}$.

ELIMINATION OF MONOPOLES

To escape this problem one must either find a way of greatly enhancing the annihilation rate of monopoles (for a note of caution about estimates of the rate see Blagojevic et al. 1981), or alternatively find a mechanism that inhibits their initial production. Several have been proposed.

One of the most straightforward is due to Bais and Rudaz 1980). They argue that the estimate of initial monopole density above is over-simplified, and that a better estimate is obtained by taking the thermal equilibrium density at the Ginzburg temperature, i.e.

$$r = n_M/T^3 \simeq \left[m_M(T_G)/T_G\right]^{3/2} \exp\left[-m_M(T_G)/T_G\right].$$

(They justify the thermal equilibrium assumption by an estimate of relaxation time). They then show that

$$m_M(T_G)/T_G \simeq Ch/g = C\, m_H/m_X$$

where C is a numerical constant subject to considerable uncertainty, mainly because of uncertainty in the condition defining the Ginzburg temperature. However the factors in C that can be calculated are quite large, of order 30. This is not enough by itself to ensure a small value of r, but if we also assume that $h/g = 2.5$ or more, then we find $r \approx 10^{-32}$, comfortably below the limit. This explanation requires an uncomfortably large value of the Higgs mass. Moreover it has recently been severely criticized by Linde (1980a), on the grounds that monopole -antimonopole pairs have much lower energy and should be produced with a distance scale $\xi(T_G)$, i.e. with no exponential suppression.

Linde (1980b) has himself proposed an interesting hypothesis of the alternative type. He suggests that higher-order corrections to the effective potential (which cannot be reliably calculated because of intractible infrared problems) yield a "magnetic mass" for the gauge fields of order $g^2T$. Its effect, at a low enough temperature, would be to confine the monopoles by joining monopoles and antimonopoles with relatively light strings. This would lead to fairly rapid annihilation. (In this context see also Bais and Langacker 1981).

Another suggestion due to Langacker and Pi (1980) exploits the observation by Weinberg (1974) that symmetry may in some cases be lower on the high temperature side of a phase transition. Materials that exhibit such behaviour are known - an example is Rochelle salt.

Langacker and Pi constructed a model with three Higgs doublets with the symmetry breaking scheme

$$SU(5) \rightarrow SU(3)xSU(2)xU(1) \rightarrow SU(3) \rightarrow SU(3)xU(1).$$

Monopoles are produced at the first transition but then become unstable and disappear at the second. Finally at the third transition when a U(1) factor reappears another set of monopoles may be created but these will be rather light and of no obvious astrophysical significance. This mechanism is effective in eliminating monopoles but may be criticised on the grounds of arbitrariness and complexity.

Several groups of authors have invoked first-order phase transitions to deal with the monopole problem. Let me therefore turn now to this topic.

## FIRST-ORDER PHASE TRANSITIONS

For several reasons, the phase transitions are likely in fact to be first order.

Radiative corrections contribute to the effective potential even at T=0 a term of the form

$$+(m^2/8\pi^2)^2 \ \ell n \ (m^2/\mu^2)$$

where $\mu$ is a renormalization scale. In special cases this term may be important - if there are no quadratic terms in the Higgs potential U (Coleman and Weinberg 1973) or if the Higgs coupling is very small, $h^2 \approx g^4$ (Linde 1976). Then V will have a local minimum at $\phi = 0$ separated from the global minimum by a barrier, so the transition will be first order.

Sometimes there are cubic terms in U which will have a similar effect - for example in the breaking of SU(5) by a Higgs field in the adjoint representation, where U contains a term proportional to $\mathrm{tr}\phi^3$. More generally, there are cubic terms in the temperature-dependent corrections, of the form $-m^3T/12\pi$.

In a first-order transition we may expect supercooling. The transition does not actually happen until some way below the critical temperature and typically occurs by formation of bubbles of the new phase which then expand until the whole space has been converted. Just below

$T_C$ the bubbles may be formed simply by thermal fluctuations. The energy of a bubble of radius $r$ is essentially

$$E(r) = (4\pi/3)\, r^3 \Delta f + 4\pi\, r^2 \sigma \ ,$$

where $\sigma$ is the surface tension of the wall separating the phases.

Thus if the radius once reaches the critical value $r_c = 2\sigma/\Delta f$ the bubble will continue to grow. The probability of a fluctuation creating such a bubble is proportional to $\exp(-E(r_c)/T_c)$. If the exponent $E/T$ is small the transition will be rapidly completed and we will have a weakly first-order transition almost indistinguishable from a second-order one.

If thermal fluctuations are not enough, we may turn to quantum tunnelling. The probability of bubble formation by tunnelling is of the form $Ae^{-B}$ where B is the Euclidean action for the tunnelling process (Callan and Coleman 1977, Coleman 1977).

Strongly first-order transitions with supercooling by many orders of magnitude have been suggested as a means of eliminating the problem of super-abundant monopoles (Guth and Tye 1980, Lazarides and Shafi 1980, Einhorn, Stein and Toussaint 1980, Einhorn and Sato 1981, Billoire and Tamvakis 1981, Kennedy, Lazarides and Shafi 1981). The essential point is that in that case the density will be dominated by the vacuum energy $\rho_{vac} = h^2\eta^4$, which plays a role like that of the cosmological constant. In particular, it is accompanied by a negative pressure $p_{vac} = -\rho_{vac}$, which leads to an exponential expansion of the universe, $R(t) = R(0)e^{kt}$ with $k = h\eta^2/m_P$. When the transition finally occurs it generates a lot of entropy which effectively dilutes the monopole density - by increasing the denominator of the monopole-to-entropy ratio. Of course any pre-existing baryon number is similarly diluted, so the released latent heat must reheat the universe at least to the Higgs-boson mass to ensure the generation of a baryon asymmetry.

There are however difficulties with this scenario. It has been pointed out by Sher in 1981 that taking account of the energy dependence of the running coupling constant much reduces the expected scale of supercooling. Another crucial observation has been made by Hut and Klinkhamer (1981). During the period of exponential expansion the universe behaves like a de Sitter universe. In particular there are event horizons. The distance to the event horizon is $k^{-1} = m_P/h\eta^2$. Note that mode mixing will occur for wavelengths exceeding this distance, i.e. for frequencies $\omega \lesssim k$. Once T falls below k, this mode mixing will induce the transition. For the simplest GUT this will happen at $T = 10^{11}$ GeV. Thus supercooling is limited to about four orders of magnitude - not nearly enough to reduce the monopole density to an acceptable level. Essentially the same conclusion has been reached by Horibe and Hosoya (1981) who argue that particle creation similar to that in a de Sitter universe will prevent the temperature falling below about $10^{11}$ GeV.

Finally it may be possible to escape the monopole problem in a model with several successive phase transitions. For example we might consider the breaking of SU(5) with an intermediate stage,

$$SU(5) \rightarrow SU(4) \times U(1) \rightarrow SU(3) \times SU(2) \times U(1).$$

Steinhardt (1980) has pointed out that the monopoles produced in the first transition may later become unstable to radial expansion. In effect the core of the monopole may make a transition to the third phase which then expands at the expense of the second. Thus the monopoles nucleate the second phase transition. The disappearance of the original monopoles does not automatically solve the problem, for new monopoles of the third phase can be created. However they may be lighter and therefore somewhat less troublesome.

An interesting detailed analysis of what happens to the monopoles of one phase at a second transition has been given by Bais (1981). Depending on the topological relationships between the groups involved they may become unstable, remain unaltered, decay to lighter monopoles of the new phase, or acquire flux tubes that pull them together and lead to rapid annihilation.

## GALAXY FORMATION

I now want to turn to my final topic, the possibility that strings might be used to explain galaxy formation.

Let me first recall briefly some ideas about how galaxies were formed. Everyone agrees that the basic mechanism is gravitational condensation. The minimum length scale above which such condensations will grow is the Jeans length

$$L_J \approx c_s/(G\rho)^{\frac{1}{2}} \approx c_s t$$

where $c_s$ is the sound velocity. Initially in the radiation-dominated era and when matter and radiation are still tightly coupled, $c_s$ has the value $1/\sqrt{3}$, so that $L_J$ is comparable with the horizon distance. At the decoupling time, when $T = 4000K$, the electrons and protons combine to form neutral hydrogen, and $c_s$ drops suddenly to the value appropriate to atomic hydrogen gas namely $(5T/3m_H)^{\frac{1}{2}}$.

The real question is, where do the initial density perturbations come from that later grow into galaxies? There are essentially two existing theories, based respectively on adiabatic and isothermal perturbations.

In the first, the density contrast $(\delta\rho/\rho)_{rad} = (4/3)\times(\delta\rho/\rho)_{matter}$ grows as $t$ so long as the wavelength exceeds the horizon distance. These perturbations are heavily damped by photon diffusion at the decoupling time on all scales smaller than the Silk mass, which is at least $10^{12}$ solar masses. The initial structures formed are thus larger than galaxies. The theory due principally to Zel'dovich (1970) suggests the formation of "pancakes" within which the galaxies subsequently condense. There are a number of serious problems with this scenario. It may be difficult to avoid excessive secondary isothermal perturbations generated at the decoupling time. The theory predicts an anisotropy in the microwave background close to the present observational upper limit. Most serious of all perhaps, there is no real explanation for the origin of the initial spectrum of density perturbations. One rather speculative suggestion has been made by Press (1979) who attributes it to an early phase transition. However his explanation requires that the Higgs field interact with gravity in a special, unusual way, and may create as many problems as it solves.

The second theory of galaxy formation is based on isothermal perturbations, for which $(\delta\rho/\rho)_{rad} = 0$. In this case $(\delta\rho/\rho)_{matter}$ is essentially constant before coming inside the horizon. These perturbations are not damped at decoupling and so can survive. On scales larger than the subsequent Jeans mass (about $10^5$ solar masses) they can start to grow like $t^{2/3}$. The difficulty with this theory is that baryon number generation destroys any initial isothermal perturbations, and it is difficult to see where they can originate.

Since neither theory is wholly satisfactory, it is natural to ask whether strings might provide an alternative explanation. This has been suggested by Zel'dovich (1980) and more recently in a different form by Vilenkin (1981b).

THE STRING THEORY

Let me start by discussing the evolution of strings. Initially, they are formed in a random tangle but for energetic reasons will tend to straighten and shorten. Strings may also cross and exchange partners, and closed loops may shrink to a point and disappear. These mechanisms lead to an increase in the typical length scale L. In the very early stages, the strings are heavily damped and will move slowly. One can show (Kibble 1976) that this process will continue until the damping time becomes comparable with the expansion time, when in fact

$$L \simeq t_d \simeq t \simeq m_P^2/h\eta^2 \simeq 10^{-32}\text{s}.$$

Thereafter there is little damping and the strings reach relativistic speeds.

The length scale L certainly cannot exceed the horizon distance t. If in fact $L \approx t$, then the total length of string per unit volume goes as $t^{-2}$, so the ratio of the average mass density of strings to the total mass density remains constant at

$$\rho_s/\rho \approx G\mu \approx 10^{-7}.$$

In this picture the strings must be losing energy. Vilenkin (1981b) argues that the dominant mechanism of energy loss comes from the formation of closed loops which then radiate gravitationally. If we start with a loop of radius $\ell$(but not of perfectly circular shape) it will oscillate with a typical frequency $\omega=1/\ell$ . Since the mass is $M \simeq 2\pi\ell\mu$, the rate of energy loss by gravitational radiation is

$$dM/dt \approx -GM^2\omega^2 \approx -(2\pi\mu)^2G .$$

Hence the lifetime of the string is of order $\ell/2\pi G\mu = 10^6\ell$. Thus the loops live a long time, so long in fact that their total mass far exceeds that of the infinite or very long strings: $\rho_{loops}/\rho \approx 10^{-3}$.

This figure is just about what one needs to have as a density contrast at the decoupling time to provide the starting point for gravitational condensation. Moreover if one looks at the range of scales available one finds that it readily spans the galactic masses.

There are however several problems with this attractive theory. Strings do not carry electric charge, but it is far from obvious that they do not radiate electromagnetically. Even if higher multipole moments are involved, this may still yield a more rapid energy loss mechanism, hence a shorter lifetime for loops and a lower density. Moreover in its random oscillations a loop may fold over on itself and break into two smaller loops whose lifetimes will be shorter. If this process happens frequently - say, every few oscillations - then again the lifetime of loops will be much reduced. If we have to accept a reduced lifetime, the only way of saving the scenario is the one proposed by Zel'dovich (1980) who suggests that the strings were formed at a transition even closer to the Planck time - say at $10^{17}$ GeV.

One must also ask whether it is possible to find a unified theory that will produce strings. Unlike monopoles they do not appear in most simple models. I do not have a fully realistic GUT to offer, but it is certainly possible to construct models that exhibit strings. Consider for example an SO(10) theory broken down to SO(6)xSO(4) by a Higgs field in the 54-dimensional symmetric tensor representation. Here the unbroken-symmetry subgroup H is really $S[O(6)\times O(4)]$ which consists of two disconnected pieces - the nontrivial piece contains for example rotations through $\pi$ in planes such as (67). In this model therefore $\pi_1(G/H) = \pi_0(H) = Z_2$, so we have "mod-2" strings. This model is undesirable for other reasons, especially that as a left-right symmetric model it is liable to generate domain walls when that symmetry is broken. However it does illustrate the possibilities.

One intriguing feature of strings is their unusual gravitational field (Vilenkin 1981a). The space around a string is flat, but slightly cone-shaped, as though a small wedge of angle $8\pi G\mu$ had been removed. Thus we might see double images of the same source around the two sides of the string - a possible explanation, it has been suggested, for double quasars.

CONCLUSIONS

If we accept the ideas of grand unification and the hot big bang we are led naturally to the concept of phase transitions in the early universe. Each of the three types of structures that could be formed at such a transition has interesting features. Domain walls can be excluded on the grounds of their unacceptably large gravitational effects, thereby casting doubt on any theory involving spontaneous symmetry breaking of a discrete symmetry. Monopoles are formed in almost all models. The problem then is to avoid having far too many of them. Finally strings can be produced in at least some models, and might possibly be used to explain galaxy formation. Whether it is possible to arrange this within a realistic GUT remains to be seen.

REFERENCES

Bais, F.A. (1981). The topology of monopoles crossing a phase boundary. Physics Letters 98B, 437-40.

Bais, F.A. and Langacker, P. (1981). Energy loss mechanisms and the annihilation of confined monopoles. CERN report, TH.3142-CERN.

Bais, F.A. and Rudaz, S. (1980). On the suppression of monopole production in the very early universe. CERN report, TH.2885-CERN.

Billoire, A. and Tamvakis, K. (1981). The Coleman-Weinberg mechanism in early cosmology. CERN report, TH-3019-CERN.

Blagojević, M., Meljanec, S., Picek, I and Senjanović, P. (1981). Radiation effects in monopole pair creation. Rudjer Boskovic Institute report IRB-TP-2-81.

Bogomol'nyi, E.B. (1976). The stability of classical solutions. Yad.Fiz. 24, 861-70 (Sov.J.Nuc.Phys. 24, 449-54).

Callan, C.G. and Coleman, S. (1977). Fate of the false vacuum II. First quantum corrections. Phys. Rev. D16, 1762-8.

Coleman, S. (1977). Fate of the false vacuum: semiclassical theory. Phys. Rev. D15, 2929-36.

Coleman, S. and Weinberg, E. (1973). Radiative corrections as the origin of spontaneous symmetry breaking. Phys. Rev. D7, 1888-1910.

Einhorn, M.B. and Sato, K. (1981). Monopole production in the very early universe in a first-order phase transition. Nucl. Phys. B180, 385-404.

Einhorn, M.B., Stein, D.L. and Toussaint, D. (1980). Are grand unified theories compatible with standard cosmology? Phys. Rev. D21, 3295-8.

Ginzburg, V.L. (1960). Some remarks on phase transitions of the second kind and the microscopic theory of ferroelectric materials, Fiz.Tverdogo Tela 2, 2031-43 (Sov.Phys.Solid State 2, 1824-34).

Guth, A.H. and Tye, S.H. (1980). Phase transitions and magnetic monopole production in the very early universe. Phys. Rev. Letters. 44, 631-5.

Horibe, M. and Hosoya, A. (1981). Particle creations by expanding universe and suppression of supercooling. Osaka University report OU-HET 41.

Hut, P. and Klinkhamer, F.R. (1981). Global space-time effects on first-order phase transitions from grand unifications. Phys. Letters 104B, 439-43.

Kennedy, A., Lazarides, G. and Shafi, Q. (1981). Decay of the false vacuum in the very early universe. Phys. Letters 99B, 38-44.

Kibble, T.W.B. (1976). Topology of cosmic domains and strings. J. Phys. A 9, 1387-98.

Kibble, T.W.B. (1980). Some implications of a cosmological phase transition. Physics Reports 67, 183-99.

Langacker, P. and Pi, S-Y. (1980). Magnetic monopoles in grand unified theories. Phys. Rev. Letters 45, 1-4.

Lazarides, G. and Shafi, Q. (1980). The fate of primordial magnetic monopoles. Physics Letters 94B, 149-152.

Linde, A.D. (1976). Dynamic reconstruction of symmetry and limits on the masses and coupling constants in the Higgs model. Zh. Eksp. Teor.Fiz. Pis'ma Red. 23, 73-6 (JETP Letters 23, 64-7).

Linde, A.D. (1980a). Confinement of monopoles at high temperatures: a solution of the primordial monopole problem. Lebedev Physical Institute report No:125 (BI-TP 80/20). Published in part in Linde (1980b).

Linde, A.D.(1980b). Confinement of monopoles at high temperatures. Physics Letters 96B, 293-6.

Polyakov, A.M. (1974). Particle spectrum in quantum field theory. Zh.Eksp.Teor.Fiz.Pis'ma Red. 20, 430-3 (JETP Letters 20, 194-5).

Preskill, J.P. (1979). Cosmological production of superheavy magnetic monopoles. Phys. Rev. Letters 43, 1365-8.

Press, W.H. (1979). Spontaneous production of the Zel'dovich spectrum of cosmological fluctuations. Physica Scripta 21,702-707.

Steinhardt, P.J. (1980). Monopole dissociation in the early universe. Harvard University preprint.

't Hooft, G. (1974). Magnetic monopoles in unified gauge theories. Nucl. Phys. B79, 276-284.

Vilenkin, A. (1981a). Gravitational field of vacuum domain walls and strings. Phys. Rev. D23, 852-7.

Vilenkin, A. (1981b). Cosmological density fluctuations produced by vacuum strings. Phys Rev. Letters, 46, 1169-76, 1496(E).

Weinberg, S. (1974). Gauge and global symmetries at high temperature. Phys. Rev. D9, 3357-78.

Zel'dovich, Ya. B. (1970). Gravitational instability: an approximate theory for large density perturbations. Astron. Astrophys. 5, 84-9.

Zel'dovich, Ya. B. (1980). Cosmological fluctuations produced near a singularity. Mon.Not.R. Astr.Soc. 192, 663-7.

Zel'dovich, Ya.B. and Khlopov, M.Y. (1978). On the concentration of relic magnetic monopoles in the universe. Physics Letters 79B, 239-241.

Zel'dovich, Ya. B., Kobzarev, I. Ya. and Okun, L.B. (1974). Cosmological consequences of a spontaneous breakdown of a discrete symmetry. Zh. Eksp. Teor.Fiz. 67, 3-11 (JETP 40,1-5).

# COMPLETE COSMOLOGICAL THEORIES

L.P.Grishchuk
Shternberg Astronomical Institute, Moscow 117234
Ya.B.Zeldovich
Institute of Applied Mathematics, Moscow 124047
U.S.S.R.

## 1. INTRODUCTION

Modern cosmology successfully describes the main features of the Universe but uses for that some specific initial data which do not yet have a reasonable explanation. There is no theory capable not only of describing the observable world but giving also an answer why and for what reasons the Universe has these or other properties. The most important unsolved issue is the nature of the cosmological singularity whose existence is predicted by the classical (not quantum) relativistic theory of gravity. For a thorough analysis of the singularity one needs a quantum theory of gravity which is yet to be constructed. Because of the lack of such a theory one has to start doing cosmology from classical stages by introducing different initial conditions and comparing their consequences with real observations.

In this talk an attempt is suggested to give a rational explanation of the origin of those initial conditions which lead to the so called standard cosmological model. Cosmological theories which pretend to describe the Universe from the "very beginning" (including the quantum-gravitational stage) and up to the present time we shall call complete. It is clear that there is not yet any finished theory of such a kind, but the very fact that there are some reasonable considerations and arguments (presented below) which permit one to make an outline of such a theory we consider as some progress. The scenario suggested below contains many separate ingredients known already in the literature; however the full picture as far as we know, has not yet been presented. Being conscious of the vague nature of some of the suggested conclusions, we consider this talk as a challenge to the highbrow experts on quantum gravity gathered here, as an attempt to call their attention to the problems which appear to be important for cosmology.

## 2. PRESENT STATE OF THE UNIVERSE

It is useful to recall some properties of the observed world which seem to reflect conditions in the very early Universe. These properties can be used as landmarks in the process of extrapolating backward in time. On the other hand, they should represent the outcome of a complete cosmological theory.

It is believed that the most important properties of the observable world are the following (see e.g. Zeldovich and Novikov 1975, Weinberg 1972, Peebles 1971):

1. Large scale homogeneity and isotropy

The distribution of galaxies in phase space is such that they do not show any noticable inhomogeneity or anisotropy after averaging over scales of the order of 100 Mpc. The most convincing manifestation of the large scale homogeneity and isotropy is the absence of angular variations in the microwave background radiation, $\Delta T/T \lesssim 10^{-4}$. The uniformity of the radiation emitted by the primordial plasma and coming to us from different directions is quite mysterious. According to the standard cosmological model elements of the primordial plasma sufficiently separated could not have been causally connected in the past even if one extrapolates their histories up to the cosmological singularity. This is known as the horizon problem.

2. Closeness of the mean density to the critical density

A variety of astronomical data indicate that the mean matter density, $\rho$ , is less than the critical one, $\rho_c$ , and $\rho \approx (10^{-1} \div 3 \cdot 10^{-2})\,\rho_c$ . Other data (for instance the possible existence of massive neutrinos) do not exclude the possibility that $\rho$ may in fact be of the same order of magnitude or even higher than $\rho_c$ . In any case, the ratio $\Omega = \rho/\rho_c$ seems to be close to unity. In terms of Newtonian mechanics one can say that the kinetic energy of expansion is almost equal, at the present time, to the gravitational potential energy. Such a balance between kinetic and potential energy is quite mysterious since it implies that, say, in the epoch of the nucleosynthesis the equality of kinetic and potential energy was satisfied with an incredible relative accuracy, of the order of $10^{-15}$. This fact is known as the flatness problem, though it would be more appropriate to call it the problem of the closeness of $\Omega$ to unity.

3. Existence of structure in the form of galaxies and their clusters

The Universe is obviously inhomogeneous at scales characteristic of galaxies and their clusters. According to contemporary views this structure was formed as a result of the growth of small initial perturbations. In order to produce the actual structure, the initial perturbations had to have a specific spectrum and a specific amplitude. Thus, a complete cosmological theory should contain an explanation of the overall properties of the Universe as well as origin of its structure.

4. Baryon asymmetry and specific entropy

The observable matter consists of baryons, and there is no indication of the presence in the Universe of any large amount of antibaryons. The ratio of the number density of photons, $n_\gamma$ , to the number density of baryons, $n_b$ is $n_\gamma / n_b \simeq 10^9$ . The large value of the specific entropy $s = 4n_\gamma / n_b$ motivates the use of words "hot Universe". Baryon asymmetry and the actual value of the specific entropy will probably find a satisfactory explanation within the scope of grand unified theories provided the application of the standard cosmological model is justified up to the energies of the order of $10^{17}$ Gev.

## 3. OUTLINE OF A COMPLETE SCENARIO

Let us present briefly a complete scenario, leaving the details and references for the other parts of the paper.

We assume that in the initial state there was

nothing except zero (vacuum) fluctuations of all physical fields including gravitational. Since the notions of space and time are classical, in the initial state there were no particles, no space, no time. Speaking about time, one can say, loosely, that there was a time when there was no time.

It is assumed that as a result of a fluctuation there appeared a classical 3-dimensional closed geometry. The finiteness of the 3-volume is a necessary condition for such a process. Since there were no real particles yet, the dynamical evolution of this geometry was governed by vacuum polarization effects caused by the external gravitational field. For the first time there appears a notion of classical space-time. It is natural to expect that all the characteristic parameters of the world were Planckian, i.e. classical space-time comes into being at the limit of applicability of classical general relativity.

During some interval of its evolution the Universe existed as a space-time very close to that of De-Sitter. It is well known that De-Sitter space-time is as symmetric as Minkowski space-time; it admits a 10-parameter group of motions. The line element of De-Sitter space-time is

$$ds^2 = c^2 dt^2 - a^2(t)\left[dr^2 + \sin^2 r\left(d\theta^2 + \sin^2\theta\, d\varphi^2\right)\right] \quad (3.1)$$

where

$$a(t) = r_0 \,\mathrm{ch}\, ct/r_0 \, , \quad r_0 = \mathrm{const}.$$

According to our scenario the moment of appearance of the Universe corresponds to $t=0$ (see Fig.)

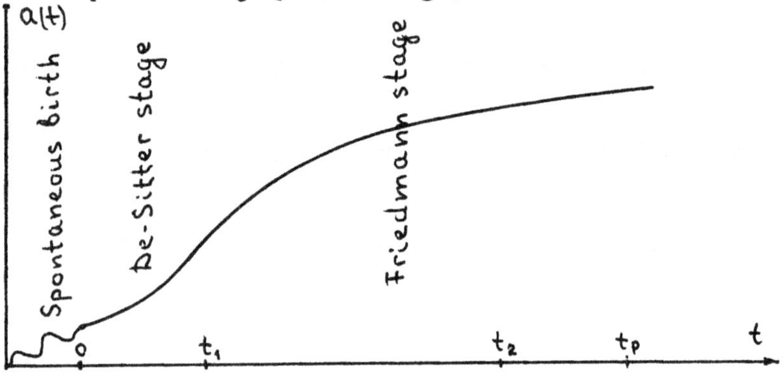


It is reasonable to suppose that the world appeared with

small deviations, $h_{\mu\nu}$ , from the metric tensor (3.1). Similarly to the scale factor $a(t)$ , perturbations $h_{\mu\nu}$ have also induced the vacuum polarization of all physical fields and were governed by it. In a closed space one can represent $h_{\mu\nu}$ as a discrete set of mode functions. As is usual in problems of this kind one can expect that after a certain time the lowest mode will be the most excited one.

We assume that by the time $t = t_1$ , during a short transition period, the expansion law of the Universe will change from De-Sitter to Friedmann regime, $a(t) \sim t^{1/2}$ , and the Universe will get filled with various particles having dominant equation of state $p_r = 1/3\, \varepsilon_r$ . Such a drastic change could happen for the following reasons. The lowest mode, after a sufficient increase, insures the transition of the scale factor everywhere in the space simultaneously. Apart from this, the perturbations $h_{\mu\nu}$ make possible the creation of different particles by the external gravitational field. As is known, particle creation is a quadratic effect in $h_{\mu\nu}$ while the vacuum polarization is already present in the linear approximation. Created particles fill the Universe with the matter. One can expect that, after $t = t_1$ , the Universe will be described by the standard radiation-dominated cosmological model, which contains small perturbations of matter, radiation and gravitational field.

All these processes occur long before the time when the baryon number asymmetry is generated and the possible phase transitions take place. Thus, early nucleosynthesis, transition to the matter dominated stage, growth of density perturbations, and galaxy formation could proceed according to the usual ideas about these events.

What are the achievements of the proposed scenario? First, it answers the question about the initial state of the Universe. According to this scenario, the Universe originates, in a certain sense, from nothing. Second, it includes a sufficiently long De-Sitter stage, which resolves the horizon problem and the $\Omega \approx 1$ problem. Third, it provides a natural explanation of the origin of density perturbations with spectrum and amplitude necessary for the formation of the observed large scale structure. These perturbations can grow from the initial fluctuations $h_{\mu\nu}$ appearing in the De-Sitter stage. Moreover, it appears to be sufficient to take $h_{\mu\nu}$ with the minimal possible amplitude - i.e. at the level of quantum fluctuations.

## 4. SPONTANEOUS BIRTH OF THE CLOSED WORLD

A widespread point of view is that the cosmological singularity can be avoided by quantum effects in an external gravitational field. Singularity free solutions of this kind do exist but they are unstable; they cannot be extrapolated back in time up to $t = -\infty$. We take another point of view. We assert that the cosmological singularity must be replaced by an essentially quantum-gravitational prcess which can be called spontaneous birth of the Universe.

A quantitative formulation of this process can be, probably, done by using Feynman's path integral approach. For quantum gravity it was developed by Wheeler (1962), Hawking (1978) and others.

The states of the gravitational field can be characterized by 3-dimensional geometries - the initial one, $g_1^{(3)}$, and the final one, $g_2^{(3)}$. The probability amplitude for passing from the initial state to the final state is given by the path integral

$$< g_2^{(3)} \mid g_1^{(3)} > = \int d[g^{(3)}] \exp[i\, I(g^{(3)})]$$

For simplicity we restrict ourselves to Friedmann-Robertson-Walker metrics

$$ds^2 = c^2 dt^2 - a^2(t)\left[dr^2 + f^2(r)(d\theta^2 + \sin^2\theta\, d\varphi^2)\right]$$

(Such a restriction brings to mind Penrose's hypothesis on the vanishing of the Weyl tensor at the initial singularity; Penrose, 1979). As the initial state we accept a 3-geometry with zero volume, i.e. $a_i = 0$. As the final state we accept a 3-geometry with finite volume, of the order of the cube of Planckian length, i.e. $a_f \approx c t_{Pl}$. (In the line element (1) this state corresponds to $t = 0$, $r_0 \approx c t_{Pl}$ ). Compactness of the 3-space is a necessary condition for such a transition. Indeed, the probability amplitude $A$ can be expressed as

$$A \sim \int d[g^{(3)}] \exp[-\int \mathcal{L}\, dv\, d\tau]$$

A nonvanishing value for $A$ can only be obtained if the integration over 3-volume gives a finite quantity. Thus, only a closed world could, probably, appear as the result of such a quantum-mechanical jump (Zeldovich, 1981).

An additional argument in favor of such a process is the fact that the birth of a closed world preserves the quantum numbers of the vacuum: zero value for the total energy, total electric charge, etc. (Tryon, 1973). According to present views the baryon charge is not strictly a conserved quantity. Therefore, the creation of the Universe with zero total baryon charge does not prevent the generation of baryon asymmetry during subsequent evolution.

In some papers the birth of an expanding universe is regarded as the result of quantum-mechanical tunnelling from an initial configuration which is considered as the ground state of the system. Specifically, the question has been studied whether quantum decay of Minkowski space-time or De-Sitter space-time with constant cosmological $\Lambda$ -term is possible (Brout et al, 1978; Atkatz and Pagels, 1981; Gott, 1981). In our opinion these suggestions have the following disadvantages. First, they do not avoid the question of the origin of the initial classical space-time. Second and more important, within the scope of standard general relativity, Minkowski and De-Sitter space-times possess minimal energy and, therefore, are stable classically as well as quantum-mechanically (Witten, 1981a; Abbott and Deser, 1981). Nevertheless, in the context of more complicated theories (like Kaluza-Klein theory, for example) the ground state ( $M^4 \times S^1$ ) may happen to be unstable with respect to quantum mechanical tunnelling (Witten, 1981b).

## 5. INTERMEDIATE DE-SITTER STAGE

According to the scenario the created world was governed by the vacuum polarization of all physical fields. It means that its dynamical evolution was described by a selfconsistent solution of the Einstein equations:

$$R_{\mu\nu} - \frac{1}{2} g_{\mu\nu} R = \frac{8\pi G}{c^4} \langle 0 | T_{\mu\nu} | 0 \rangle \tag{5.1}$$

The evaluation of the right-hand side of these equations and the search for selfconsistent solutions, in particular, De-Sitter solutions, were undertaken by Dowker and Critchley (1976); Davies (1977); Fischetti et al (1979); Davies et al (1977); Mamaev and Mostepanenko (1980); Starobinsky (1980). In his work Starobinsky considered selfconsistent solution as a means to avoid the cosmological singularity. He has also pointed out the possibility of transition of such a solution into Friedmann stage (see, also Gurovich and Starobinsky, 1979). In our scenario, the De-Sitter stage has not only an end but also a beginning.

In the papers mentioned above the expectation value $\langle 0 | T_{\mu\nu} | 0 \rangle$ was determined by the conformal anomaly of the full energy-momentum tensor. At the present time it is not clear if the conformal anomaly is an inevitable feature of quantum field theory in curved space-time (see, DeWitt, 1979; Christensen et al, 1980). However, the appearance of a De-Sitter stage for one reason or another seems justifiable.

It is well known (Hawking and Ellis, 1973) that the De-Sitter solution can be represented as a hyperboloid with arbitrary radius $r_0$, embedded into 5-dimensional pseudo-Euclidean space-time. The frame of reference (3.1) covers the whole hyperboloid, i.e. the whole De-Sitter space-time. Spatial sections $t = \text{const}$ are 3-dimensional spheres. The section $t=0$ is different from the others in that it has the minimal volume. However the points of this section are by no means distinguished from other points of the space-time. Because of the high symmetry of De-Sitter space-time it can be covered by an infinite number of other frames of reference where the metric will again have the form (3.1) but the spatial section of minimal volume will be different. In fact such a section can include an arbitrary point of the space-time and can have an arbitrarily oriented time-like normal vector in it.

In De-Sitter space-time one can also introduce frames of reference with flat or open (hyperbolic) space sections. These frames of reference do not cover the whole De-Sitter space-time.

One or another choice of time (i.e. choice of spatial sections) is very important since, by assumption, at certain moments changes will occur in the expansion law, and one particular moment of time, $t=0$ , marks the act of spontaneous creation of the world. Which spacelike sections should be chosen? According to views put forward here, the created world was closed. Therefore, the privileged spatial sections are closed spaces corresponding to $t=const$ in the presentation (3.1) of De-Sitter space-time. The 3-sphere $t=0$ is physically distinguished by the very event of creation. It seems natural to assume that the transition to the Friedmann stage occurs at a hypersurface $t=t_1$ which is "parallel" to the hypersurface $t=0$ .

If the selfconsistent De-Sitter solution was governed by the conformal anomaly than it follows from the Einstein equations (5.1) that $r_0 = c/H$ , where $H^{-1}$ is proportional to $t_{pl}$ and depends on contributions of all fields. It is important to note that if the duration of the De-Sitter stage was 70-100 characteristic periods $H^{-1}$ than, as will be shown below in detail, the horizon and $\Omega \approx 1$ problems could be completely eliminated.

## 6. TRANSITION TO THE FRIEDMANN STAGE

Ignoring, at the present time, the possible causes by which the De-Sitter stage is replaced by a Friedmann one, let us consider the joining of these solutions at $t=t_1$ (see Fig.). During the De-Sitter stage vacuum polarization effects give an effective energy density $\varepsilon_v$ and pressure $p_v$ where $\varepsilon_v = \frac{3c^2H^2}{8\pi G}$ and $p_v = -\varepsilon_v$ . During the Friedmann radiation-dominated stage the equation of state is $p_r = 1/3\, \varepsilon_r$ and the energy density behaves as $\varepsilon_r \sim a^{-4}$ . After transition, at $t=t_2$ , to matter-dominated stage the density of matter goes like $\rho_m \sim a^{-3}$. Let us denote the present values of density and scale factor by $\rho_p$ , $a_p$ .

The $\Omega \approx 1$ problem is solved as follows: $\Omega_p$ satisfies the relation

$$\Omega_p - 1 = \frac{k}{\frac{8\pi G}{3c^2}\rho_p a_p^2 - k}$$

where $k=+1$ for closed world. In order to have $\Omega_p \approx 1$ it is necessary that $\frac{8\pi G}{3c^2}\rho_p a_p^2$ be of the order of unity. By using the junction condition and known parameters it can be shown that this implies $e^{Ht_1} \approx 10^{30}$. It follows that $\Omega_p$ will be of the order of unity if $Ht_1 \gtrsim 70$.

The horizon problem is eliminated by the fact that the particle horizon increases practically up to the future event horizon already during the time interval from $t=0$ to $t=t_1$ (i.e. during the De-Sitter stage). Therefore, all the particles now accessible for observations could have been in causal contact long ago. It should be mentioned that the usefulness of the De-Sitter solution for eliminating the horizon and $\Omega \approx 1$ problems was first considered by Guth (1981) though in his case De-Sitter solution was realized in a different cosmological epoch and for different reasons.

## 7. SMALL PERTURBATIONS AND GALAXY FORMATION

Similarly to the background De-Sitter solution small variations to it also satisfy equations (5.1). It is convenient to introduce the variable $\eta$ by the relation $d\eta = c\,dt/a(t)$. Then, $a(\eta) = r_0/\cos\eta$. If the function $f(\eta)$ represents a perturbation with the proper wave-number $n$ then the typical equation for it is

$$f'' + f\left[n^2 - \frac{2+C^2}{\cos^2\eta}\right] = 0$$

where $C^2$ is a constant depending on the actual parameters of the conformal anomaly. Equations of this form occur often in the theory of amplification of classical waves and quantum particle creation in external gravitational fields. On the basis of past experience it can be expected that perturbations with low eigenvalues will guarantee the departure of the scale factor from that of the De-Sitter scale factor, while the (density) perturbations with large $n$ (which correspond to the size of the present day clusters of galaxies) will increase up to the necessary amplitude. This important issue is not yet worked out though some encouraging suggestions have been proposed in recent papers (Mukhanov and Chibisov, 1981; Kompaneetz et al, 1981; Starobinsky, 1981).

## 8. LOCALIZED CREATION OF AN OPEN WORLD

The instability of the background De-Sitter solution with respect to the lowest mode, $a(t) \to a(t) + \delta a(t)$, implies that the transition to the radiation-dominated stage will occur simultaneously in the whole space. In this case the spatial sections corresponding to the comoving frame of reference during the Friedmann stage will be closed. However, it can not be excluded that an instability in the background solution will develop locally, starting from a small region. The reason for such an instability could be related to the fact that in the polarized vacuum there is $\varepsilon_v > 0$, $p_v < 0$ and zero entropy, $S = 0$. It can happen that for this medium it will be energetically more favorable to form an expanding bubble (Coleman, 1977) inside of which there will be $\varepsilon_r > 0$, $p_r > 0$, $S > 0$ after certain time. The walls of the bubble expand almost with the velocity of light and "cut out" a region of the background space-time restricted by a light cone. It is natural to think that, inside this region, the transition to the radiation-dominated stage will take place at so-some moment of time $\tau = \tau_1 = const$, where $\tau$ is measured along the world lines which emanate from the event $\tau = 0$ where the localized perturbation first originated. But the spatial sections $\tau = const$ are hyperbolic, open. In terms of $\tau$ - time a piece of De-Sitter space-time inside the light cone can be described by

$$ds^2 = c^2 d\tau^2 - r_0^2 \, sh^2 \, c\tau/r_0 \left[d\chi^2 + sh^2\chi (d\theta^2 + \sin^2\theta \, d\varphi^2)\right].$$

By assumption, the perturbation has originated in the point $\tau = 0$, $\chi = 0$. If the transition happens at $\tau = \tau_1$ then the spatial sections with uniform density, pressure and temperature will be open. It means that in this case we would find ourselves living in an open universe. Again, in order to have $\Omega_p \approx 1$, i.e. in order that the Universe not be "too open", it is necessary to have a transition to the radiation-dominated stage not earlier than at $\tau=\tau_1$, where $e^{H\tau_1} \approx 10^{30}$

Does a possibility of such a localized perturbation mean that the idea of the spontaneous birth of the world is superfluous? Can one maintain the conception of a complete De-Sitter space-time which exists from $t=-\infty$ and which is a background for many localized perturbations transforming later in many open "universes"? (Such an idea is considered by Gott (1981)). It seems to us that the idea of the spontaneous birth of a closed world is necessary, i.e., one can not regard De-Sitter stage as existing from $t=-\infty$. This conclusion is connected with the observation that the origin of localized perturbations should be characterized by a finite probability per unit of time per unit of volume. Because of the full symmetry of De-Sitter space-time this probability should be the same at all world points. But in this case the total probability is infinite since the integration should include all previous moments of time, up to $t=-\infty$. One can avoid this difficulty only by assuming that classical stage of the complete De-Sitter solution starts not from $t=-\infty$ but from some finite $t=0$.

## 9. GRAVITONS AND GRAVITINI IN THE EARLY UNIVERSE

Apparently the only way to check hypothesis on physical conditions in the very early Universe will be the search for a primordial gravitational-wave background. Gravitational waves and gravitons (in contrast to other known massless fields and particles) have the remarkable ability to be amplified and to be created by conformally flat gravitational field (Grishchuk, 1974). As a consequence of this process a nonthermal background of gravitons should exist now. Predictions on their spectrum and intensity depend on specific properties of the gravitational field in the very early Universe (Grishchuk, 1977; Starobinsky, 1979). These conclusions about gravitons follow from standard general relativity; however, they also hold in supergravity theories (Grishchuk and Popova, 1979). Contemporary experimental power is not sufficient for detection of such a gravitational-wave background; however the situation can, hopefully, improve in the near future.

We are indebted to B.S. DeWitt for a discussion.

REFERENCES

Abbott, L.F. & Deser, S. (1981). Preprint TH.3136-CERN

Atkatz, D. & Pagels, H. (1981). Preprint Report Number RU 81/B/2.

Brout, R., Englert, F. & Gunzing, E. (1978). Ann. Phys. 115, 78.

Christensen, S.M., Duff, M.J., Gibbons, G.W. & Rocek, M. (1980). Phys. Rev. Lett. 45, 161.

Coleman, S. (1977). Phys. Rev. D15, 2929.

Davies, P.C.W. (1977). Phys. Lett. 68B, 402.

Davies, P.C.W., Fulling, S.A., Christensen, S.M. & Bunch, T.S. (1977). Ann. Phys. 109, 108.

DeWitt, B.S. (1979). In "General Relativity", Eds. S.W.Hawking & W.Israel. Cambridge University Press.

Dowker, J.S. & Critchley, R. (1976). Phys. Rev. D13, 3224.

Fischetti, M.V., Hartle, J.B. & Hu, B.L. (1979). Phys. Rev. D20, 1757.

Gott, J.R. (1981). Preprint. Princeton.

Grishchuk, L.P. (1974). Zh.E.T.F. 67, 825.

Grishchuk, L.P. (1977). Ann. N.Y. Acad. Sci. 302, 439.

Grishchuk, L.P. & Popova, A.D. (1979). Zh.E.T.F. 77, 1665.

Gurovich, V.Ts. & Starobinsky, A.A. (1979). Zh.E.T.F. 77, 1699.

Guth, A.H. (1981). Phys. Rev. D23, 347.

Hawking, S.W. (1978). Nucl. Phys. B138, 349.

Hawking, S.W. & Ellis, G. (1973). The large scale structure of space-time. Cambridge University Press.

Kompaneetz, D.A., Lukash, V.H. & Novikov, I.D. (1981). Preprint. Pr.-652.

Mamaev, S.G. & Mostepanenko, V.M. (1980) Zh.E.T.F. 78, 15.

Mukhanov, V.F. & Chibisov, G.V. (1981). Pis'ma Zh.E.T.F. 33, 549.

Peebles, P. (1971). Physical Cosmology. Princeton University Press.

Penrose, R. (1979). In "General Relativity". Eds. S.W.Hawking & W.Israel. Cambridge University Press.

Starobinsky, A.A. (1979).Pis'ma Zh.E.T.F. 30, 719.
Starobinsky, A.A. (1980). Phys. Lett. 91B, 99.
Starobinsky, A.A. (1981). Pis'ma Zh.E.T.F. 34, 460.
Tryon, E.P. (1973). Nature 246, 396.
Weinberg, S. (1972). Gravitation and Cosmology. Wiley, N.Y.
Wheeler, J.A. (1962). Geometrodynamics. Academic Press, N.Y.
Witten, E. (1981)[a]. Comm. Math. Phys. 80, 381.
Witten, E. (1981)[b]. Preprint. Princeton.
Zeldovich, Ya.B. (1981). Pis'ma Zh.E.T.F. 7, 579.
Zeldovich, Ya.B. & Novikov, I.D. (1975). Structure and Evolution of the Universe (In Russian) Moscow.

# THE COSMOLOGICAL CONSTANT AND THE WEAK ANTHROPIC PRINCIPLE

S.W. Hawking
Department of Applied Mathematics and Theoretical Physics,
Silver Street, Cambridge, CB3 9EW

## INTRODUCTION

Observations of distant galaxies and of the microwave background radiation indicate that the universe is described by a Friedman-Robertson-Walker model to a high degree of accuracy. In such a model the Einstein equations give

$$\frac{3\ddot{R}}{R} = -\frac{4\pi}{m_p^2}\left(\mu + 3p\right) + \Lambda$$

where R is the scale factor, μ and p are the energy density and pressure of the universe, Λ is the cosmological constant and $m_p$ is the Planck mass in units in which $\hbar = c = 1$ . In principle the deceleration parameter $q_0 = -(R\ddot{R}/\dot{R}^2)$ may be determined from the shape of the magnitude-redshift curve for galaxies, but possible evolutionary changes in the brightness of the galaxies mean that we can place only an upper limit on $|q_0|$ of about 2. The universe is dominated by non-relativistic matter at the present time so the effective value of the pressure p is small compared to the energy density μ. Measurements of the density are again a bit uncertain but the matter in the galaxies and clusters of galaxies seem to correspond to a density of about $10^{-30}$gm/cc or about 1/10 of the critical density $3m_p^2\dot{R}^2/(8\pi R^2)$ There might be other forms of matter which have not been observed but it would be very difficult to accommodate more than about ten times the observed amount. Thus one can place an upper limit of about $10^{-32}$eV on Λ or about $10^{-60}$ on $|\Lambda/m_p^2|^{1/2}$ .

The effective cosmological constant is thus observed to be zero with an accuracy better than for any other quantity in physics. For example, the observational upper limit on the mass of the photon given by space-craft measurements of the earth's magnetic field is only about $10^{-16}$ eV or $m_\gamma/m_e < 10^{-22}$ . Even so, we do not believe that the photon mass is so small merely by accident or by the fine tuning of some adjustable parameter. Rather we invoke gauge invariance to make the photon mass exactly zero. By contrast, even if the bare cosmological constant were zero, there does not seem any similar reason why the effective induced cosmological constant should be zero. Indeed one would expect it to be very large for the following reason:

1) There will be an induced cosmological constant from the diagrams

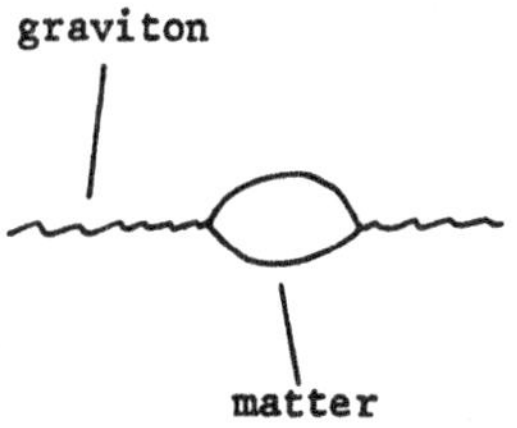


$$\sim \frac{\mu^4}{m_p^2}$$

where $\mu$ is the cut-off. The natural cut-off would be the Planck mass. This would give $\Lambda \sim m_p^2$ which would have to be balanced very accurately by a similar negative bare cosmological constant.

2) If there are Higgs scalar fields which break the grand unified or electro weak symmetries, there will be a contribution to $\Lambda$ of $8\pi V(\phi_c)/m_p^2$ where V is the effective potential and $\phi_c$ is the expectation value of the scalar field. It would require very fine tuning of V i.e.

$$|V(\phi_c)/\phi_c^4| < 10^{-48}$$

in order not to produce a cosmological constant above the observed upper limit.

3) It is believed that at high temperatures QCD behaves like a Coulomb gas but that at about T ~ 200MeV there is a Bose condensation which causes confinement. One might expect that this Bose condensation would give rise to a vacuum energy density of about $(200\text{ MeV})^4$ or a $\Lambda^{\frac{1}{2}}$ of about $10^{-12}$eV. This would have to be cancelled by a bare cosmological term of an accuracy of better than 1 part in $10^{20}$ in order not to exceed the observed upper limit.

4) In a supersymmetric theory the numbers of bosons and fermions are equal, and their infinite contributions to the cosmological constant cancel each other. However, we know from observation that supersymmetry must be broken at energies lower than $10^3$GeV. This would give rise to an induced value of $(m_p{}^2\Lambda)^{1/4}$ of the same order. One could possibly cancel this positive $\Lambda$ with a negative $\Lambda$ arising from gauging the O(N) symmetry in extended supergravity (Ferrara et al. 1979 ), but this would require very fine tuning.

One way of explaining why the observed cosmological constant is so small would be to appeal to the Anthropic Principle, that the theory which describes the universe must be such as to allow the development of intelligent life: if it did not, there would not be anyone to observe the theory. In the case of the cosmological constant one could argue that a value slightly more negative than the observed limit would have caused the universe to recollapse before life could have developed while a more positive value would have caused the universe to expand too rapidly to allow the formation of galaxies.

Many people including myself, are rather unhappy about the strong version of the Anthropic Principle which requires that the theory which describes the universe should be specially chosen to allow for our existence. On the other hand, most people would accept the Weak Anthropic Principle which states that intelligent life can exist only in certain regions of a given universe with given physical laws. This weak version has been used to explain the numerical coincidence, first noted by Dirac, that the ratio of the present Hubble radius of the universe to the classical radius of the electron, and the ratio of the electrical to gravitational forces between two electrons are both about $10^{40}$. In the conventional theory the Hubble radius of the universe is a function of time whereas the other quantities referred to are supposed to be constant. Dirac (1938) thought that it was too much of a coincidence that these two large dimensionless numbers should be of the same order just at the present time. He therefore invented a theory with a variable gravitational constant in which they were equal at all times. However, Dicke, (1961) and Carter (1974) pointed out that we do not live at an arbitrary time in the history of the universe: life could not evolve until after a first generation of massive stars in which the heavier elements were produced and life is not likely to survive after all the stars have burnt out. The life times of stars are determined by the fundamental constants such as the proton mass in just such a way that Dirac's coincidence is bound to hold at the time in history of the universe when life is around to observe it.

I would like to suggest an explanation in the same spirit for the very small apparent value of the cosmological constant. It is based on the idea that the path integral for the universe has to be evaluated over compact Euclidean (ie positive definite) metrics with all values of the Euclidean 4-volume, V. I shall argue that the path integral is dominated by metrics which may be highly curved and very complicated topologically on scales of the Planck length or less but which appear smooth and nearly flat when viewed on larger scales. On these scales they would appear to have a cosmological constant of order $V^{-\frac{1}{2}}$, whatever the value of the bare or induced cosmological constant. Only those universes with a large Euclidean volume would contain life to observe that the apparent cosmological constant was small.

The Anthropic argument would indicate that the apparent cosmological constant must be small, of the order of the observational upper limit, but it could not show that it should be exactly zero because life could have developed in a universe whose 4-volume V was of the order of $H^{-4}$, where $H = \dot{R}/R$ is the present rate of expansion of the universe. However it is reasonable to suppose that N(V), increases sharply with V where N(V) is the density of states of the gravitational field with Euclidean 4-volume (This will be defined more precisely in the next section). In this case one might be able to argue that universes with large values of V were more numerous than those with small values and were therefore, in some sense, more probable. The most probable value of the apparent cosmological constant would then be exactly zero.

The apparent cosmological constant would be small or zero in the Euclidean regime. However, this does not necessarily imply it would be small or zero everywhere when one analytically continues back to real physical spacetime. In a K = +1 Friedman universe the Euclidean section intersects the physical space-time at the moment of maximum expansion. Thus one would expect the apparent cosmological constant to be small at late stages in the expansion of the universe. However, it might be large in the early stages if there were, for example, a phase transition above the Grand Unified energy.

## THE VOLUME ENSEMBLE

I shall adopt the Euclidean approach (Hawking 1979a & b ) in which the path integral is evaluated over all positive definite metrics, g, on space-time manifolds of all topologies. If the space-time manifold were not compact, one would have to include a boundary term in the action (Gibbons & Hawking 1977 ). We know what this term is for asymptotically flat spaces but it is fairly clear that the universe is not asymptotically flat. I shall therefore consider only compact space-time manifolds. This is not to say that space-time is actually compact but it can be viewed as a normalization condition like periodic boundary conditions. In the class of the spaces with very high topology that I shall be considering, the action of a compact and non-compact manifold will differ only by a relatively small amount.

Let $N(V_o)dV_o$ denote the number of states of the gravitational field with Euclidean 4-volumes between $V_o$ and $V_o + dV_o$. One can calculate $N(V_o)$ by inserting a $\delta(V-V_o)$ in the path integral:

$$N(V_o) = \frac{1}{2\pi i}\int_{-i\infty}^{i\infty} d\left(\frac{m_p^2\Lambda}{8\pi}\right)\int d[g]\,d[\phi]\,\exp\left\{-I[g,\phi] - \frac{m_p^2\Lambda}{8\pi}(V-V_o)\right\}$$

where $I[g,\phi]$ is the Euclidean action of the positive definite metric, g, and the matter fields, $\phi$. The term $\exp\left\{-\frac{m_p^2\Lambda V}{8\pi}\right\}$ can be absorbed into the Euclidean action where it acts as a cosmological constant. One can define the volume partition function as

$$Z[\Lambda] = \int d[g]\,d[\phi]\,\exp\left\{-I[g,\phi,\Lambda]\right\}.$$

Formally, this is the Laplace transform of N(V).

$$Z[\Lambda] = \int_0^\infty N(V)\exp\left\{-\frac{m_p^2\Lambda V}{8\pi}\right\}dV.$$

One expects that Log(N(V)) should be proportional to $V^{\frac{1}{2}}$ for large V, so this should converge. The density of states is then given by the inverse Laplace transform.

$$N(V_o) = \frac{1}{2\pi i}\int_{-i\infty}^{i\infty} d\left(\frac{m_p^2\Lambda}{8\pi}\right) Z[\Lambda]\exp\left\{\frac{m_p^2\Lambda V_o}{8\pi}\right\}.$$

The integral in $\Lambda$ should be taken to the right of the essential singularity in $Z[\Lambda]$ at $\Lambda = 0$.

One might think that the volume of a solution of the Einstein equations would be of order $|\Lambda|^{-2}$. However, the examples of the Einstein-Kahler metrics (Hawking 1978 ) show that there can be solutions with very large volumes and very high Euler numbers, $\chi$. These spaces seem to appear nearly flat on length scales, L, such that $(V_o)^{\frac{1}{4}} \gg L \gg (V_o|\chi)^{\frac{1}{4}} \sim \Lambda^{-\frac{1}{2}}$. Their action is of the form

$$I[g,\Lambda] = -\frac{m_p^2}{8\pi\Lambda}\cdot a + b$$

where 'a' and 'b' depend on the solution but not on $\Lambda$ (at the tree level). The 'a' term arises from the usual Einstein Lagrangian. $a \geqslant 0$, $a \sim c\chi$ for large $\chi$, where c ~ O(1). The 'b' term is non zero, if there are quadratic curvature terms in the action, $b \geqslant 0$ for stability, and $b \propto \chi$ for large $\chi$.

One can show how these solutions contribute to $Z[\Lambda]$ by consider the Borel transform (Zim-Justin 1981 ).

$$Z[\Lambda] = \int_0^{\infty} B(x) \exp\left\{-\frac{m_p^2 x}{8\pi\Lambda}\right\} d\left(\frac{m_p^2 x}{8\pi\Lambda}\right)$$

The Borel transform $B(x)$ will have singularities at the values of $x$ equal to the quantity $(-a)$ in the classical solutions. Thus $Z[\Lambda]$ is Borel summable because all the singularities of $B(x)$ lie on the negative $x$-axis apart from a possible $\delta$-function contribution to B from the singularity at $x = 0$ corresponding to the K-3 solution.

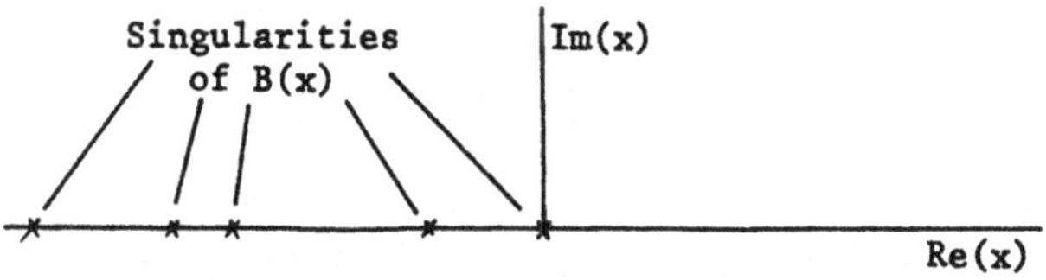


If $B(x)$ is suitably behaved for large x, this implies that $Z[\Lambda]$ is non-singular for $\mathrm{Re}\Lambda > 0$, and that $N(V) = 0$ for $V < 0$. One can express $N(V)$ in terms of $B(x)$ by (Zim-Justin 1981).

$$N(V) = (m_p^2/8\pi)^2 \int_0^{\infty} B(x)\, J_0\left(\frac{m_p^2 x^{1/2} V_0^{1/2}}{4\pi}\right) dx$$

If there were a bare or induced cosmological constant, $\Lambda_0$, already present in the action, $I[g, \phi\ \Lambda]$ it could be absorbed into a shift in the dummy parameter $\Lambda$ and would give a factor $\exp\{-m_p^2 \Lambda_0 V_0/8\pi\}$ in $N(V_0)$. If $\Lambda_0$ were negative, this would make $N(V)$ increase even faster with V, so the argument that the most probable universe would have infinite V and zero apparent cosmological constant would still hold. If $\Lambda_0$ were positive, it might still be that $N(V)$ increased with V. Even if it did not, it would still be true that life would exist only in those states for which V was large. Thus one could appeal to the Weak Anthropic Principle to argue that the apparent cosmological constant should be very small.

## SUPERGRAVITY

The action for $N = 1$ supergravity can be written as an integral of the curvature superfield R over chiral superspace (Siegel & Gates 1979 )

$$I = \frac{m_p^2}{8\pi}\int d^4x\, d^2\theta\, R \det\tilde{E}$$

where $\tilde{E}$ is the restriction of the achtbein to the chiral subspace. One can add a term

$$(m_p^3 e/8\pi)\int d^4x\, d^2\theta \det\tilde{E}$$

proportional to the volume of chiral superspace, where $e$ is a dimensionless coupling constant. This gives a theory with a cosmological constant $\Lambda = -e^2 m_p^2$.

One can now consider the number of states $N(V_o)dV_o$ with chiral super-volumes between $V_o$ and $V_o + dV_o$.

$$\mathcal{N}(V_o) = \frac{1}{2\pi i}\int_{-i\infty}^{i\infty} d\left(\frac{m_p^3 e}{8\pi}\right) \exp\left\{\frac{m_p^3 e V_o}{8\pi}\right\} Z[e]$$

where

$$Z[e] = \int d[U] \exp\{-I[U,e]\}$$

In a similar manner one can define the Borel transform by

$$Z[e] = \int_0^\infty B(x) \exp\left\{\frac{x}{8\pi e^2}\right\} d\left(\frac{x}{8\pi e}\right).$$

The argument goes through much as in the ordinary gravity case. There is however a difference in that one might expect the fermionic fluctuations to cancel the bosonic ones and so make $Z[0] = 1$. In this case

$$1 = \int_0^\infty \mathcal{N}(V)\, dV.$$

This would suggest that $N(V) \to 0$ as $V \to \infty$. On dimensionless grounds one might guess

$$\log\{\mathcal{N}(V)\} \propto -m_p^2 V^{2/3}.$$

We do not yet have a superspace formulation of the higher N supergravities, but if the N=8 theory can be derived by dimensional reduction from N=1 supergravity in 11-dimensions, the volume of majorana superspace

$$\int d^{11}x\, d^{16}\theta \det \tilde{E}$$

has the right dimensions of $[x]^3$ to be added to the action as a cosmological term. In fact all that one needs is that it should be possible to write the action for the gauged O(N) theory as

$$I[e] = I[0] + \left(\frac{m_p^3 e}{8\pi}\right) . J$$

where J is some invariant integral over superspace. One can then consider $N(J_o)dJ_o$ the number of states between $J_o$ and $J_o + dJ_o$.

## THE SMALL-SCALE STRUCTURE OF SPACE-TIME

In order to define what it means to say that the space-time has a certain structure on a length scale, L, suppose (for the moment) that space-time is a smooth compact manifold with positive definite metric. Then one can define a distance function, d(x,y), and can cover the manifold by a finite collection, C(L), of balls of radius L. One can regard C(L) as a simplical complex with a topology given by Céch Cohomology. Then one can define the Euler number, $\chi(L)$, as the minimum $\chi$ of these complexes. One is interested in $\chi(L)/V$, the density of Euler number on the scale L.

### Higher Derivative Theories

If the action contains terms quadratic in the curvature, these will dominate over the Einstein and cosmological terms at short distances. If the coefficients of the higher derivative terms were constant, the Euler number density would be scale invariant, ie. proportional to $L^{-4}$. However it seems that one loop effects will cause these coefficients to increase at small L (Fradkin & Tseytlin 1981 ). This will damp out topological fluctuations below some scale, $L_o$, and the space-time manifold will be smooth below this scale.

### Supergravity

If the effective action is just the classical action, one would expect the Euler number density to go up two powers of $L^{-1}$ faster than in the scale invariant case, ie. to be of order $L^{-6}$. In this case space-time would be a fractal and not a smooth manifold. However there could be terms of order $(\text{curvature})^4$ and higher in the effective action even if the theory is finite to all orders. These might provide a cut-off to topological fluctuations below some length-scale, $L_o$.

One can ask how particles would propagate through such a foamy space-time. On general grounds and from some particular examples it seems as if a topological fluctuation of scale, $\rho$, will scatter a particle of spin, s, and momentum, $k \ll m_p$ by an amount (Hawking et al. 1980 ) $\rho^2(\rho k)^{2s}$

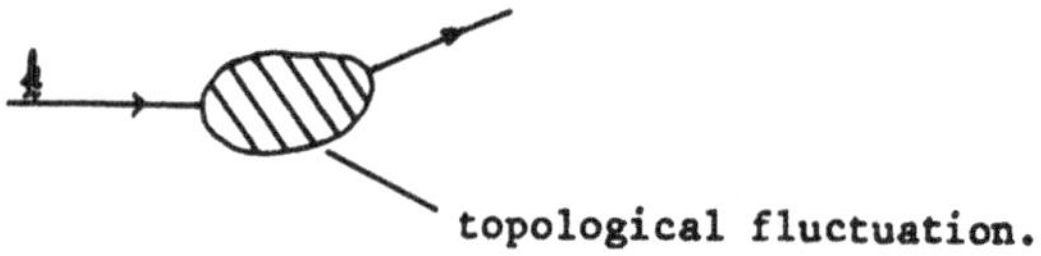

If the spectrum of topological fluctuations is $\sim (m_p L)^{-\gamma} m_p^4$ for $m_p^{-1} > L > L_0$ then scalar particles will acquire an effective mass of order $[(m_p L_0)^{-\gamma} - 1] m_p$ but fermions will be protected by gauge invariance from acquiring a mass.

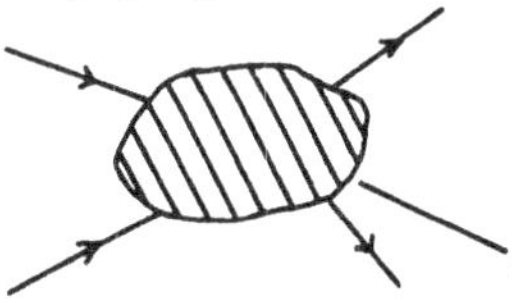

topological fluctuation.

The simultaneous scattering of two particles by a topological fluctuation will give rise to an effective 4-point vertex of the order of

$$m_p^{-4s} \left[ (m_p L_0)^{4+4s-\gamma} - 1 \right]$$

If $\gamma = 4$, there will be very little effect on particles of spin $s > \frac{1}{2}$, even if $L_0 = 0$, i.e. even if there is no cut-off to the fluctuations. If $\gamma = 6$, the vertex will be small for $s \geq 1$ but will be logarithmically divergent as $L_0 \to 0$ for $s = \frac{1}{2}$.

It follows from these results that if there is a lower limit to the scale of topological fluctuations, then scalar particles will acquire a very large mass, but particles of spin $s \geq \frac{1}{2}$. will propagate almost as if in flat space. Any observed scalar particles would have to be bound states. The same conclusions would hold even is there were no lower cut-off, $L_0$, to the scale of topological fluctuations, provided that $\gamma < 6$. However, if $\gamma = 6$, then the observed spin-$\frac{1}{2}$ particles would also have to be bound states. In this case, one might believe, like Ellis, Gaillard and Zumino (1980) that all observed particles, except possibly the graviton, were bound states or collective excitations of the foam. Such excitations would form representations of the Poincaré group corresponding to the almost flat structure of space-time on scales larger than the Planck length. They would thus appear as particles of spin, S=0, $\frac{1}{2}$ and 1. Higher spin excitations would not obey renormalizable effective theories and so might not be observable at low energies.

Acknowledgements

I would like to thank Murray Gell-Man for arousing my interest in the problem of the cosmological constant and Bernard Whiting for help in preparing and delivering this talk.

## REFERENCES


Ferrara et al. (1979) Nucl.Phys. B147, 105-131.

Dirac, P.A.M. (1938) Proc.Roy.Soc.London, A165, 199-208.

Dicke, R.H. (1961) Nature 192, 440-441.

Carter, B. (1974) "Large number coincidences and the Anthropic Principle in Cosmology". In Confrontation of Cosmological Theories with Observational Data, ed. M.S. Longair, p.291-298, Reidel, Dordrecht.

Hawking, S.W. (1979a) In General Relativity : An Einstein Centenary Survey ed. S.W. Hawking & W. Israel, Cambridge University Press.

Hawking, S.W. (1979b) In Recent Developments in Gravitation, Plenum Press.

Gibbons & Hawking (1977) Phys.Rev. D15, 109-167.

Hawking, S.W. (1978) Nucl.Phys. B144, 349.

Zim-Justin, J. (1981) Phys. Reps. 70, 109-167

Siegel & Gates (1979) Nucl.Phys. B147, 77-104.

Fradkin & Tseytlin (1981) Higher Derivative Quantum Gravity : One-loop Counterterms & Asymptotic Freedom, Lebedev Institute preprint.

Hawking et al. (1980) Nucl.Phys. B170, 283-306.

Ellis, Gaillard & Zumino (1980) Phys.Letts. 94B, 343-348.